Drosophila suzukii Management

Flávio Roberto Mello Garcia
Editors

Drosophila suzukii Management

Springer

Editor
Flávio Roberto Mello Garcia
Instituto de Biologia
Universidade Federal de Pelotas
Pelotas, Brazil

ISBN 978-3-030-62691-4 ISBN 978-3-030-62692-1 (eBook)
https://doi.org/10.1007/978-3-030-62692-1

This Springer imprint is published by the registered company Springer Nature Switzerland AG
The registered company address is: Gewerbestrasse 11, 6330 Cham, Switzerland

Acknowledgments

I would like to thank the Brazilian National Council for Scientific and Technological Development (CNPq) research productivity grant and the International Atomic Energy Agency (IAEA) for the project funding entitled "Evaluating the use of Sterile Insect and pupal parasitoids to managing *Drosophila suzukii* in greenhouse", which I am the coordinator. I would like to thank the 35 contributors who made this book a reality.

Contents

About the Editor

Flávio Roberto Mello Garcia is professor at Universidade Federal de Pelotas (UFPel), Brazil. He did master's degree and a PhD in Zoology (Entomology) at Pontifical Catholic University of Rio Grande do Sul (PUCRS), Brazil. He did post-doctoral training at the University of Florida (UF) and trained at the US Department of Agriculture (USDA). He is a CNPq ()Brazilian National Council for Scientific and Technological Development researcher. He is an author of more than 200 articles and chapters and 9 books. Project Coordinator "Evaluation of the use of Sterile Insects and pupae parasitoids for the management of *Drosophila suzukii* in greenhouse" approved and encouraged by the International Atomic Energy Agency (IAEA). Advise postdoctoral, doctorate, masters, and scientific initiation. He coordinated more than 34 research projects in Brazil, Africa, Uruguay, and the United States.

Contributors

Jocelyn A. Berry Ministry for Primary Industries, Wellington, New Zealand

Kostas Bourtzis Insect Pest Control Laboratory, Joint FAO/IAEA Programme of Nuclear Techniques in Food and Agriculture, Vienna, Austria

María J. Buonocore-Biancheri LIEMEN, División Control Biológico de Plagas, PROIMI Biotecnología, CCT NOA Sur-CONICET, Avda. Belgrano y Pje. Caseros, T4001MVB San Miguel de Tucumán, Tucumán, Argentina

Hannah Burrack North Carolina State University, Raleigh, NC, USA

Carlos Cáceres Insect Pest Control Laboratory, Joint FAO/IAEA Programme of Nuclear Techniques in Food and Agriculture, Vienna, Austria

Dong H. Cha USDA-ARS, U.S. Pacific Basin Agricultural Research Center, Hilo, HI, USA

Kent M. Daane University of California Berkeley, Berkeley, CA, USA

Peter A. Follett USDA-ARS, U.S. Pacific Basin Agricultural Research Center, Hilo, HI, USA

Claudia F. Funes INTA Estación Experimental Agropecuaria Famaillá, Tucumán, Argentina

Flávio Roberto Mello Garcia Departamento de Ecologia, Zoologia e Genética, Instituto de Biologia, Universidade Federal de Pelotas, Pelotas, RS, Brazil

David Gonthier Entomology Department, S-225 Agriculture Science Center, University of Kentucky, Lexington, KY, USA

Alberto Grassi Center for Technology Transfer, Fondazione Edmund Mach, San Michele all'Adige (TN), Italy

Kim A. Hoelmer USDA ARS Beneficial Insects Introduction Research Unit, Newark, DE, USA

Syeda A. Jaffri Department of Insect Biotechnology in Plant Protection, Justus-Liebig-University Giessen, Institute for Insect Biotechnology, Giessen, Germany

Daniel S. Kirschbaum INTA Estación Experimental Agropecuaria Famaillá, Tucumán, Argentina

Universidad Nacional de Tucumán, Facultad de Agronomía y Zootecnia, Tucumán, Argentina

Ryan Kuesel Entomology Department, S-225 Agriculture Science Center, University of Kentucky, Lexington, KY, USA

Byung-Ho Lee Institute of Life Science, Gyeongsang National University, Jinju, Republic of Korea

Jana C. Lee USDA ARS Horticultural Crops Research Unit, Corvallis, OR, USA

Oscar E. Liburd University of Florida, Gainesville, FL, USA

Greg Loeb Cornell University, Geneva, NY, USA

Katerina Nikolouli Department of Forest and Soil Sciences, Boku, University of Natural Resources and Life Sciences, Vienna, Austria

Insect Pest Control Laboratory, Joint FAO/IAEA Programme of Nuclear Techniques in Food and Agriculture, Vienna, Austria

Sergio M. Ovruski LIEMEN, División Control Biológico de Plagas, PROIMI Biotecnología, CCT NOA Sur-CONICET, San Miguel de Tucumán, Tucumán, Argentina

Tommaso Pantezzi Center for Technology Transfer, Fondazione Edmund Mach, San Michele all'Adige (TN), Italy

Elena M. Rhodes University of Florida, Gainesville, FL, USA

Cesar Rodriguez-Saona Rutgers University, Chatsworth, NJ, USA

Giorgio De Ros Center for Technology Transfer, Fondazione Edmund Mach, San Michele all'Adige (TN), Italy

Fabiana Sassù Department of Forest and Soil Sciences, Boku, University of Natural Resources and Life Sciences, Vienna, Austria

Marc F. Schetelig Department of Insect Biotechnology in Plant Protection, Justus-Liebig-University Giessen, Institute for Insect Biotechnology, Giessen, Germany

Maxwell J. Scott Department of Entomology and Plant Pathology, North Carolina State University, Raleigh, NC, USA

Rady Shawer Department of Plant Protection, Faculty of Agriculture (Saba Basha), University of Alexandria, Alexandria, Egypt

Christian Stauffer Department of Forest and Soil Sciences, Boku, University of Natural Resources and Life Sciences, Vienna, Austria

Lorena Suárez Dirección de Sanidad Vegetal, Animal y Alimentos (DSVAA), San Juan, Argentina

Programa de Control y Erradicación de Mosca de los Frutos de San Juan (ProCEM-San Juan), San Juan, Argentina

Spencer S. Walse USDA-ARS, San Joaquin Valley Agricultural Sciences Center, Parlier, CA, USA

Xingeng Wang USDA ARS Beneficial Insects Introduction Research Unit, Newark, DE, USA

Ying Yan Department of Insect Biotechnology in Plant Protection, Justus-Liebig-University Giessen, Institute for Insect Biotechnology, Giessen, Germany

Chapter 1
Introduction to *Drosophila suzukii* Management

Flávio Roberto Mello Garcia

Abstract *Drosophila suzukii* Matsumura (Diptera: Drosophilidae), or the Spotted Wing Drosophila (SWD), belongs to the most important invasive pestiferous species. This chapter presents the main concepts necessary for the management of SWD in an area-wide integrated pest management approach. Morphological aspects and the basis for the correct identification of *D. suzukii* are also presented.

Keywords Identification · Morphology · Area-wide · Integrated pest management

1.1 Impacts and Management of *Drosophila suzukii*

Drosophila suzukii Matsumura (Diptera: Drosophilidae), or the Spotted Wing Drosophila (SWD), belongs to the most important invasive pestiferous species. The SWD is polyphagous, and its preferred hosts are fine epicarp fruits, such as strawberry (*Fragaria* spp.) (Rosaceae), blackberry (*Rubus* spp.) (Rosaceae), blueberry (*Vaccinium* spp.) (Ericaceae), and raspberry (*Rubus* spp.) (Rosaceae) (Wollmann et al. 2020).

The SWD originated in Asia and was first detected in 1980 outside its original environment (Hawaii) and later in other Hawaiian Islands; however, no damage was recorded (O'Grady et al. 2002). In 2008, SWD was detected simultaneously in California (USA) (Hauser 2011) and in many European localities (Calabria et al. 2012). In South America, the species was detected in Brazil (Deprá et al. 2014), and years after being detected in Brazil, it was detected in the other countries of South America, in Argentina (Santadino et al. 2015), Chile (Rojas et al. 2019), and in Uruguay (González et al. 2015). Its rapid dispersion among countries demonstrated the ability of this species to adapt to newly invaded areas, where high losses were recorded in agriculture (De Ros et al. 2013). There are still environmentally adequate areas with potential for *D. suzukii* occurrence in Oceania and Africa, although

F. R. M. Garcia (✉)
Departamento de Ecologia, Zoologia e Genética, Instituto de Biologia, Universidade Federal de Pelotas, Pelotas, RS, Brazil
e-mail: flavio.garcia@ufpel.edu.br

F. R. M. Garcia (ed.), *Drosophila suzukii Management*,
https://doi.org/10.1007/978-3-030-62692-1_1

there has been no record of species on these continents. Recently, this species has been detected in Africa (Boughdad et al. 2020). The models indicate that the environmental conditions in these areas are prone to the establishment of these species in case of future invasion (dos Santos et al. 2017).

The dispersion and colonization of SWD in invaded areas are due to human activity through the international trade of infested fruit, being the main means of dispersion of the species across several continents (Westphal et al. 2008). In addition, in general, exotic species in a new territory suffer less impacts from natural enemies, which leads to an increase and distribution of invasive species (Keane and Crawley 2002; Roy et al. 2011).

SWD can cause at least four types of damage:

- Cultural damages—the damaged fruits are discharged for industrial processing or for *in natura* consumption; infested orchards are depreciated; fruits lose their commercial value;
- Economic damages—orchards become economically inefficient, leading to investment losses and to the insolvency of producers; commercialization of fruits in the external market becomes inviable; foreign currencies surpluses are no longer generated; fruits from infested areas do not reach profitable prices in internal markets; amount of collectable taxes and duties decline.
- Social damages—cause the growing impoverishment of people, the rural exodus, and unemployment.
- Environmental damages—loss of biodiversity and ecosystem processes. The establishment of *D. suzukii* in new areas can cause disruption of genetic isolation and interfere in the dynamics of natural systems, causing premature extinction of species. Thus, *D. suzukii* can dominate large areas after having replaced autochthonous species and changing the trophic chains

Therefore, it is of fundamental importance to manage *D. suzukii* population in an Integrated Pest Management (IPM) approach to avoid or minimize the impacts. IPM seeks the use of methods aimed at keeping pest populations at densities below a level of economic damage, and these methods are represented by several control techniques, which can be used in a combined manner, as long as there is harmonization of this integrated use (Moino Júnior 2000). In the context of IPM, a combination of methods that have additive or synergistic effects may be applied that could result in elective management of SWD (Alkema et al. 2019).

1.2 Biological and Ecological Aspects of *Drosophila suzukii*

Drosophila suzukii is a species r strategist. In general, a species r strategist has small size, high mobility, tendency to immigration, fast sexual ripening, early start of reproduction, production of a great offspring by individual, and short generations (Garcia 2014). A female of *D. suzukii* produces from 200 to 500 eggs throughout its life, depending on several factors such as nutrition and temperature (Cini et al. 2012). Females may deposit more than one egg per fruit, and several females may live on the same fruit. The egg lasts approximately 24 h, and the larva remains within it, feeds, and develops from 3 to 13 days. The pupa usually remains within the fruits or in the soil for 3–43 days (Cini et al. 2012).

The first step toward the correct management of *D. suzukii* is the knowledge of its morphological characteristics, which avoids errors of identification of the species (Souza and Carvalho 2000) since there are many species of Drosophilidae. This family has more than 4000 species distributed throughout the world, of which the genus *Drosophila* has more than 1500 species (Marshall 2012). The SWD is a species belonging to the subgenus *Sophophora*; this subgenus contains 344 species organized in nine species groups (Yuzuki and Tidon 2020). The SWD is a species belonging the *melanogaster* group. This group contains almost 200 species (Yuzuki and Tidon 2020) and is very diverse in the Eastern and Afro-tropical region and has species that have invaded all the biogeographical regions of the world, such as *Drosophila melanogaster* Meigen and *Drosophila simulans* Sturtevant (Poppe et al. 2014).

The SWD deposits eggs inside the fruits. The eggs have approximately 0.65 mm and two respiratory filaments on one of two ends (Fig. 1.1). The larvae pass through three instars (L1, L2, and L3). Larvae (Fig. 1.2) measure about of 0.67 mm, 2.13 mm, and 3.94 in the L1, L2, and L3 instars, respectively (Walsh and Bolda 2011). The pupae (Fig. 1.3) have two extensions on one of the terminal faces, the size varies according to sex, the pupa of the males is greater (±3.18 mm in length) than that of females (±2.90 mm in length) (Walsh and Bolda 2011). The immature stages of *D. suzukii* are very similar to those of other species of *Drosophila*, thus being the identification made mainly by the adult.

Adults of *D. suzukii* are small flies, males are between 2.3 and 3.0 mm in length and females are between 2.5 and 4.0 mm in length. The male can be identified by the presence of large subdistal black spot (Fig. 1.4) and by protarsus with two small rows of 3–4 peg-like setae forming a sex comb (Fig. 1.5). Females do not have spots on the wings (Fig. 1.6), and present a large serrated oviscapt (Fig. 1.7) (Yuzuki and Tidon 2020).

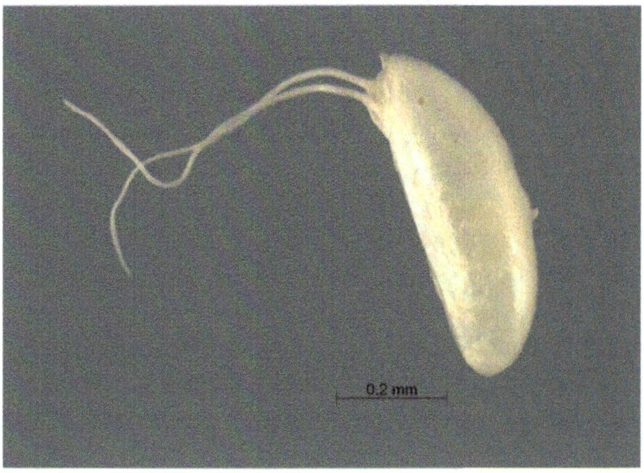

Fig. 1.1 *Drosophila suzukii* eggs

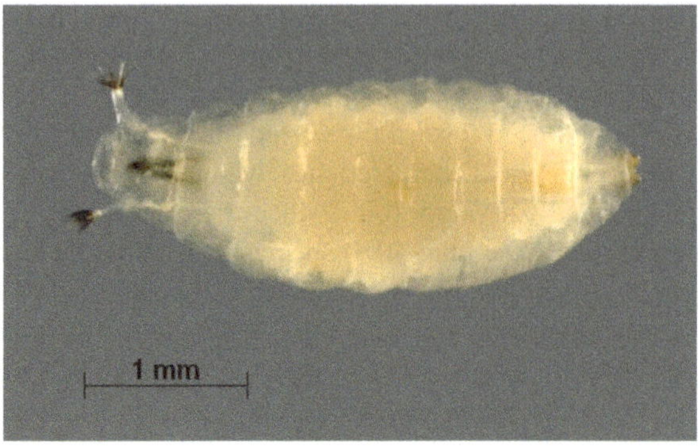

Fig. 1.2 Larva of *Drosophila suzukii*

Fig. 1.3 *Drosophila suzukii* pupa

1.3 The Book

This book was organized on the basis of an Integrated Area-Wide Pest Management approach (AW-IPM). AW-IPM is a coordinated, sustainable, and preventive approach that targets pest populations in all areas, including non-commercial urban areas (Hendrichs et al. 2007). According to Hendrichs et al. (2007), the AW-IPM approach has four components: (1) Basic research; (2) Modeling and developing new methods; (3) Feasibility studies and regulation, and (4) Pilot and operational AW-IPM programs. Thus, this book attempts to cover in its 14 chapters the various aspects that allow the management of *D. suzukii* in an AW-IPM approach.

Chapter 2 covers the economic impacts of *D. suzukii* and its control programs for this species, providing information on the cost of these control measures, what is

Fig. 1.4 *Drosophila suzukii* male with large subdistal black spot

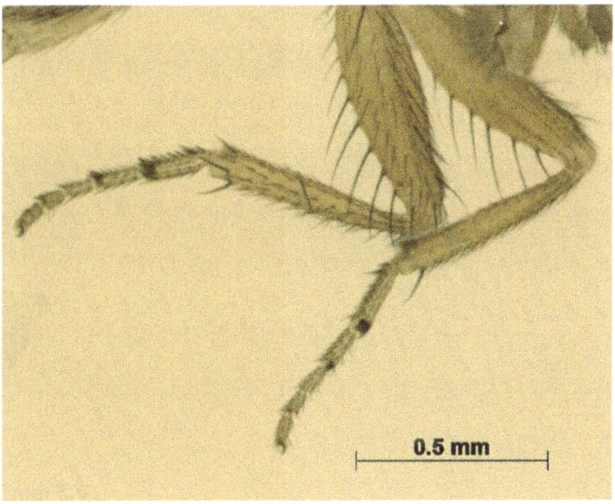

Fig. 1.5 Male protarsus of *Drosophila suzukii* with two small rows of 3–4 peg-like setae forming a sex comb

very important for farmers and technicians to be able to choose a better cost-effective control method. The basis for the regulation necessary to avoid the dispersion of *D. suzukii* between countries and localities is laid down in Chap. 3 by appropriate regulatory measures.

Chapter 4 discusses the biological, ecological, and behavioral aspects of *D. suzukii*, dealing with host plants, natural enemies, and the effects of abiotic factors on populations of this species around the world.

Fig. 1.6 *Drosophila suzukii* female

Fig. 1.7 The large serrated oviscapt of *Drosophila suzukii*

The basis for the management of *D. suzukii* in the most recently invaded region, Latin America, in an AW-IPM approach are dealt with in Chap. 5. The methods necessary for the adoption of management of this species suitable for the region are presented. Chapter 6 presents the best traps and attractive to monitor the populations of SWD, detailing the influence of different variables on the monitoring of the species.

The control method most used for SWD chemical control is treated in Chap. 7. The work highlights the best synthetic and natural insecticides in the control of the species. Biological control is covered in Chap. 8. The biological control includes a complex of entomophagous organisms, such as parasitoids and predators, and the entomopathogens such as bacteria, nematodes, and fungi; numerous researches have been developed all over the world seeking biotic mortality agents that are effective against the SWD (Garcia et al. 2017).

In Chap. 9, the use of Sterile Insect Technique (SIT) in the control of *D. suzukii* is discussed. The SIT is considered a type of autocotida control, where the pest is used for its own control, this technique meets the current requirements of the fruit-importing countries and the consumer internal market (Dias and Garcia 2014). This technique is important for both suppression and eradication of pest.

In Chap. 10, methods including the Tet-Off and Tet-On, Erythromycin-Off, Biotin-On, Vanillic-acid regulated, Phloretin-Off, Bile acid-Off, and the Quinic-acid systems for expression controls are covered. In addition, systems that work on stimuli based on light and temperature are discussed.

The use of physical barriers allows producers to suppress the damage caused by *D. suzukii*. Network barriers are a promising alternative or companion to pesticides for conventional and organic fruit producers. This issue is covered in Chap. 11.

Management of SWD in organic berry crops, including sanitation techniques, composting and the use of synthetic and organic mulches, canopy management (pruning), which have provided different levels of control for *D. suzukii* populations, is covered in Chap. 12.

The use of postharvest treatment in the control of *D. suzukii* is covered in Chap. 13. The postharvest treatment is often the simplest means to overcome an insect-related trade barrier, and several treatment options that provide quarantine security against *D. suzukii* are available.

Finally, Chap. 14 intended to compile the information presented in previous chapters highlighting the advances in SWD management and to project how research should advance to the management of this pest in a sustainable way in the future. This chapter deals how should be the *D. suzukii* management in the Agriculture 4 context.

The texts of the chapters were carefully prepared by 35 experts, experienced researchers from renowned research and teaching institutions in nine countries, to offer a global approach with the latest information and the best scientific and practical evidence in the management of *D. suzukii*.

Acknowledgments I would like to thank the Brazilian National Council for Scientific and Technological Development (CNPq) research productivity grant and the International Atomic Energy Agency (IAEA) for the project funding entitled evaluating the use of Sterile Insects and pupal parasitoids to managing *Drosophila suzukii* in greenhouse, which I am the coordinator.

References

Alkema JT, Dicke M, Wertheim B (2019) Context-dependence and the development of push-pull approaches for integrated Management of *Drosophila suzukii*. Insects 10(12):454. https://doi.org/10.3390/insects10120454

Boughdad A, Haddi K, El Bouazzati A et al (2020) First record of the invasive spotted wing Drosophila infesting berry crops in Africa. J Pest Sci. https://doi.org/10.1007/s10340-020-01280-0

Calabria G, Máca J, Bächli G et al (2012) First records of the potential pest species *Drosophila suzukii* (Diptera: Drosophilidae) in Europe. J Appl Entomol 136:139–147

Cini A, Ioratti C, Anfora G (2012) A review of the invasion of *Drosophila suzukii* in Europe and a draft research agenda for integrated pest management. Bull Insectol 65:149–160

De Ros G, Anfora G, Grassi A et al (2013) The potential economic impact of *Drosophila suzukii* on small fruits production in Trentino (Italy). IOBC/WPRS Bull 93:317–332

Deprá M, Poppe JL, Schmitz HJ et al (2014) The first records of the invasive pest *Drosophila suzukii* in the south American continent. J Pest Sci 87:379–383. https://doi.org/10.1007/s10340-014-0591-5

Dias NP, Garcia FRM (2014) Fundamentos da técnica do inseto estéril (TIE) para o controle de moscas-das-frutas (Diptera, Tephritidae). O Biológico 76:57–62

dos Santos LA, Mendes MF, Kruger AP et al (2017) Global potential distribution of *Drosophila suzukii* (Diptera, Drosophilidae). PLoS One 12:e0174318. https://doi.org/10.1371/journal.pone.0174318

Garcia FRM (2014) Zoologia Agrícola: manejo ecológico de pragas. Editora Rígel, Porto Alegre

Garcia FRM, Wollmann J, Krüger AP et al (2017) Biological control of *Drosophila suzukii* (Diptera, Drosophilidae): state of the art and prospects. In: Davenport L (ed) Biological control: methods, applications and challenges. Nova Science Publishers, New York, pp 1–27

González G, Mary AL, Goñi B (2015) *Drosophila suzukii* (Matsumura) found in Uruguay. Dros Inf Serv 98:103–107

Hauser M (2011) Historic account of the invasion of *Drosophila suzukii* (Matsumura) (Diptera: Drosophilidae) in the continental United States, with remarks on their identification. Pest Manag Sci 67:1352–1357. https://doi.org/10.1002/ps.2265

Hendrichs J, Kenmore P, Robinson A, Vreysen MJ (2007) Area-wide integrated pest management (AW-IPM): principles, practice and prospects in. In: Vreysen MJ, Robinson A, Hendrichs J (eds) Area-wide control of insect pests: from research to field implementation. Springer, Netherlands, pp 3–33

Keane RM, Crawley MJ (2002) Exotic plant invasions and the enemy release hypothesis. Trends Ecol Evol 17(6):164–170

Marshall S (2012) Flies the natural history & diversity of Diptera. Firefly Books. Ontario, 616 p

Moino Júnior AM (2000) Introdução ao manejo integrado de pragas. Editora da Universidade Federal de Lavras, Lavras

O'Grady PM, Beardsley JW, Pereira WD (2002) New records for introduced Drosophilidae (Diptera) in Hawaii. Bishop Museum Occasional Papers 69:34–35

Poppe JL, Schimitz HJ, Grimaldi D et al (2014) High diversity of Drosophilidae (Insecta, Diptera) in the Pampas biome of South America, with descriptions of new Rhinoleucophenga species. Zootaxa 3779:215–245

Rojas E, Andrade J, Concha C et al (2019) Manual de reconocimiento. Estados de desarrollo de *Drosophila suzukii* (Diptera: Drosophilidae) y otras especies del género, comunes en el sur de Chile. Ministerio de Agricultura, Servicio Agrícola y Ganadero, Santiago

Roy HE, Lawson HJ, Sschörogge K et al (2011) Can the enemy release hypothesis explain the success of invasive alien predators and parasitoids? BioControl 56:451–468

Santadino MV, Riquelme Virgala MB, Ansa MA et al (2015) First record of *Drosophila suzukii* (Diptera: Drosophilidae) in blueberry (*Vaccinium* spp.) from Argentina. Rev Soc Entomol Arg 74:183–185

Souza B, Carvalho CF (2000) Bases morfológicas para o reconhcimento de insetos-praga e inimigos naturais. Editora da Universidade Federal de Lavras, Lavras

Walsh DB, Bolda MP, Goodhue RE et al (2011) *Drosophila suzukii* (Diptera: Drosophilidae): invasive pest of ripening soft fruit expanding its geographic range and damage potential. J Integr Pest Manag 2(1):G1–G7. https://doi.org/10.1603/IPM10010

Wollmann J, Schlesener DC, Mendes SR et al (2020) Infestation index of *Drosophila suzukii* (Diptera: Drosophilidae) in small fruit in southern Brazil. Arq Inst Biol 87:e0432018. https://doi.org/10.1590/1808-1657000432018

Westphal MI, Browne M, Mackinnon K et al (2008) The link between international trade and the global distribution of invasive alien species. Biol Invasions 10:391–398

Yuzuki K, Tidon R (2020) Identification key for drosophilid species (Diptera, Drosophilidae) exotic to the Neotropical region and occurring in Brazil. Rev Bras Entomol 64:e2019100. https://doi.org/10.1590/1806-9665-rbent-2019-100

Chapter 2
Recent Trends in the Economic Impact of *Drosophila suzukii*

Giorgio De Ros, Alberto Grassi, and Tommaso Pantezzi

Abstract As *D. suzukii* was spreading almost globally, during the last decade the economic implications of the invasion have been examined by a rather heterogeneous body of research. Our review point out that the focus of these studies gradually moved from raising awareness about the magnitude of the damage, particularly on the industry, to aiding the decision-making process with regard to the control strategies progressively developed. The chapter then presents an update of an estimate carried out in the early phase of the invasion about the pest's impact on soft fruit industry in the province of Trento (Italy). On the basis of conservative assumptions, the economic impact on soft fruit production in the province of Trento amounts to 9% of the potential revenues, mostly due to costly control measures. Our estimates show that a decade after its introduction, *D. suzukii* keeps having a substantial negative impact on berry and cherry production.

Keywords Integrated pest management · Control strategies · Yield losses · Evaluation · Soft fruits · Economics

2.1 Introduction

Measuring the economic impact of an invasive pest like *Drosophila suzukii* Matsumura, also known as Spotted Wing Drosophila (SWD) in the USA, is important for both public and private stakeholders. This information, as highlighted by Born et al. 2005, can serve to raise awareness about the magnitude of the problem for a specific industry or for society as a whole, or to support decision-making activities regarding the choice of a control strategy. Indeed, as was previously the case with the economic literature on terrestrial invasive species (Olson 2006), over

G. De Ros (✉) · A. Grassi · T. Pantezzi
Center for Technology Transfer, Fondazione Edmund Mach,
San Michele all'Adige (TN), Italy
e-mail: giorgio.deros@fmach.it

F. R. M. Garcia (ed.), *Drosophila suzukii Management*,
https://doi.org/10.1007/978-3-030-62692-1_2

time a few economic studies approached the *D. suzukii* invasion along two[1] lines: one focused on assessing the pest's economic impact, and a second one aimed at comparing the economic viability of different control strategies.

In the last decade, *D. suzukii* has been rapidly spreading across the North-American mainland and Europe (Asplen et al. 2015), after being first detected in 2008 in both California (Bolda et al. 2010) and Spain (Calabria et al. 2012). More recently, it settled in South America (Andreazza et al. 2017). Given the pest's tolerance of a wide range of climatic conditions, with temperature and humidity being the main environmental variables affecting its distribution (Winkler et al. 2020), researchers found a potential risk for further invasions in areas of the Australian and African continents (Dos Santos et al. 2017).

Berries, particularly raspberries and strawberries, and sweet cherries have shown the greatest host potential (Bellamy et al. 2013), although a wide range of cultivated and wild soft-skinned fruits can serve as host (Cini et al. 2012). Female *D. suzukii* oviposits in intact, ripening soft-skinned fruits; larval feeding causes the fruit to collapse around the oviposition site and makes it unmarketable. Favoured by a high daily or seasonal dispersal rate (Tait et al. 2018), and not significantly limited by natural enemies in the newly invaded areas (Miller et al. 2015), it caused substantial damage in many cropping situations around the world (Asplen et al. 2015).

As *D. suzukii* was spreading almost globally, academic research, extension services as well as industry operators tested and developed a range of agricultural practices and equipment to complement insecticides in controlling the pest.

This included attractant-based traps for pest monitoring and control (Hampton et al. 2014), field sanitation and rapid harvest (Leach et al. 2018), exclusion netting (Leach et al. 2016; Kuesel et al. 2019), pruning (Schöneberg et al. 2020), etc. They can be combined together with insecticides (Beers et al. 2011; Bruck et al. 2011; Van Timmeren and Isaacs 2013; Shawer et al. 2018) in various management programmes, each consisting of a different mix of "ingredients" and each potentially ready for a technical and economical evaluation as well.

The remainder of this chapter is organised as follows. The first, following section reviews the studies about the economic impact of *D. suzukii* in different contexts in order to establish how this field of studies contributed to deepen the knowledge about the pest and the tools to control it. The second and third sections present an update of a previous estimate of the economic impact of *D. suzukii* in the early phase of the pest's invasion in the province of Trento (De Ros et al. 2015). After the 2009 sightings (Grassi et al. 2011), in 2011 *D. suzukii* caused significant damage to cherry and berry production in the area. Since then, the *D. suzukii*'s infestation in the province and its associated impacts have been studied in some detail.

[1]The review of Olson (2006) singled out a third line of research, concerning the relationship between terrestrial invasive species and international trade and trade policies, but to the best of the authors' knowledge, there is no published study about *Drosophila suzukii* invasion and international trade.

2.2 Studies on the Economic Evaluation of a *Drosophila suzukii*'s Invasion

The economic literature about the *D. suzukii*'s invasion consists of about a dozen research studies. Four of them (Bolda et al. 2010; De Ros et al. 2013; Benito et al. 2016; DiGiacomo et al. 2019), mostly carried out in the early phase of the invasion, estimate the economic impact of *D. suzukii* taking into account just the potential or actual damage caused by the pest. Other studies (Goodhue et al. 2011; De Ros et al. 2015; Farnsworth et al. 2017; Mazzi et al. 2017) are similarly focused on impact assessment, but include an economic evaluation of the control costs as well.

The remaining studies (Del Fava et al. 2017; Fan et al. 2020; Yeh et al. 2020), all carried out when the pest was established in the newly invaded areas and a set of control management measures already developed, are not strictly impact assessments. Their objective is, rather, to compare, and in the more recent studies to model the growers' responses to the pressure exerted by *D. suzukii*.

2.2.1 *Economic Impact Assessments*

The first study about the economic impact of *D. suzukii* (Bolda et al. 2010) focused on five crops, strawberries, blueberries, raspberries, blackberries and cherries in California, Oregon and Washington. Bolda et al. (2010) estimated the loss of production according to two hypotheses: yield losses of 20% for all crops in all the three State, and maximum reported yield losses per crop in each State. Results vary from $US 511.3 million to $US 421.5 million in potential damage. The study then charted the revenue losses as a function of yield losses in three California Counties with a relevant raspberry and strawberry production.

A subsequent research (Goodhue et al. 2011) refined the early estimate by taking into account a positive price response to the yield losses and by providing information regarding the cost of managing *D. suzukii* infestations. Combining estimates about yield losses based on field observations in California's Central Coastal region and estimates about own-price elasticity from prior studies, Goodhue et al. (2011) assessed a decrease in the California berry industry revenues by 37% for raspberries and by 20% for processed strawberries, while no significant yield losses were observed for fresh strawberries. The study also calculated the cost per hectare of insecticide treatments and the overall cost of a specific conventional pest management programme for raspberries.

The first economic impact assessment carried out outside the Pacific Coast of the USA considered the strawberry, blueberry, raspberry, blackberry and cherry production in the Province of Trento, Italy (De Ros et al. 2013). Estimates were based on the weekly yield losses reported through grading by a marketing co-op in 2011, when the area experienced an early infestation of *D. suzukii*. The temporal aspects of production and infestation were taken into consideration, but not the pest control costs. Estimated revenue losses amounted to € 3.3 million, that is to say, around 13% of potential revenue. In a follow-up study on the same region and crops (De Ros et al. 2015), updated information was provided for the period 2011–2013, and the costs and benefits of an integrated pest management programme were calculated. The study compared the scenarios before and after the implementation of an integrated pest management programme, when revenue losses decreased to € 1.55 million, a mere 7% of the industry's potential output.

A few years after the first record of the introduction of *D. suzukii* in South America (Deprá et al. 2014), another study evaluated the potential revenue losses for peaches and figs in Brazil (Benito et al. 2016). On the basis of yield loss estimates from prior studies, Benito et al. (2016) assessed potential revenue losses of up to $US 21.4 million for peaches and $US 7.8 million for figs.

A study focused on *D. suzukii*'s impact on the California raspberry industry a few years after the early outbreak (Farnsworth et al. 2017) maintains the methodological approach of Goodhue et al. (2011), providing updated information on yield losses and management practices. The analysis concerns the economic impact on both conventional and organic raspberry cultivation. While conventional producers benefited over time by chemical management programmes developed thanks to academic research and industry experience, organic ones must rely on labour-intensive management practices like more frequent harvesting, field sanitation and construction and maintenance of attractant-based traps. Combining the effects of a production decrease on market prices and yield loss estimate resulting from field trials, the study shows that conventional producers experienced almost no change in revenues in the 2012–2014 period, while organic growers experienced losses of up to −5.5% in the same period.

A research carried out in Switzerland on sweet cherry production (Mazzi et al. 2017) departs to some extent from the previous literature on the economic impact of *D. suzukii*, focusing on the costs of the implemented measures. The yield losses experienced by the growers, which measures (visual checks, monitoring traps, field sanitation insecticides, etc.) were adopted and the associated level of satisfaction were surveyed by means of a questionnaire submitted to local growers in 2015 and 2016. Then the cost of every measure was estimated. The results show a cost of Swiss Francs (CHF[2]) 1857 per hectare for a strategy encompassing all the considered measures. The study finally simulated four scenarios of increasing fruit infestation and the associated management costs.

A recent assessment of the economic impact of *D. suzukii* on the Minnesota raspberry industry (DiGiacomo et al. 2019) is characterised by a similar choice of data source: data on yield losses were derived from a survey submitted to local growers. The results, collected in November 2017, showed a median 20% yield loss reported by the surveyed growers. Applying this value to ex-ante production estimates, the study calculated yearly revenue losses of $US 2.35 million. DiGiacomo et al. (2019) indicate in the adoption of less aggressive pest control strategies a

[2]At the end of 2016, the $US to CHF was approximately 1.01.

possible explanation for the difference between their estimate and the assessment of Farnsworth et al. (2017) concerning the California raspberry industry.

2.2.2 Economic Evaluations of Control Management Programmes

Del Fava et al. (2017) compared two control strategies adopted among the berry producers in the province of Trento, Italy, in the period 2014–2015, when pest pressure significantly increased. A first strategy utilised insecticides, mass trapping and cultural measures, the other relied mainly on exclusion netting. The study incorporated an estimate of the societal costs linked to the use of insecticides, in terms of agricultural workers' exposure, consumers' exposure through residues on foods, and damage to ecosystems, adding extra costs for an amount equal to the cost of the insecticides. After submitting the data to a probabilistic sensitivity analysis, they found that the strategy based on exclusion netting had a probability of more than 85% of being more profitable.

Two other studies carried out very recently in the USA (Fan et al. 2020; Yeh et al. 2020) make use of some of the methodological tools refined by economic research to inform the selection of the optimal control management of invasive species (Epanchin-Niell and Hastings 2010; Dana et al. 2014; Buyuktahtakin and Haight 2018).

Fan et al. (2020) developed a Bayesian State-Space model to compare alternative *D. suzukii* control strategies in blueberry production in conditions of observational uncertainty. Twelve strategies were tested on data collected from blueberry farms located in New York State: from a strictly calendar-based management to decreasingly intensive monitoring-based management types. The cost function is formulated in order for the environmentally conscious growers to consider the environmental and health effects of insecticides. Results indicate that there are little private incentives to adopt monitoring-based strategies.

The study of Yeh et al. (2020) considered the economic impact of *D. suzukii* on the wild blueberry industry in Maine from two different angles. It first estimated the economic effect at a State level on the basis of some hypotheses regarding yield reduction and incremental levels of insecticide applications. Then it used the Monte Carlo method on a farm-level optimisation model to simulate the decision-making process in the weeks when the fruits are about to mature. Three options were analysed: no action, early harvest, insecticide application. The estimated economic impact on the Maine wild blueberry industry can amount to up to $US 6.8 million in the worst-case scenario, while optimal management depends on the level of infestation and on market prices.

2.3 An Update of the Economic Impact of *Drosophila suzukii* on Soft Fruit Production in the Province of Trento, Italy

2.3.1 Background

The Autonomous Province of Trento, commonly known as Trentino, is a mountain area in the North East of Italy where soft fruit production is an important agricultural activity. Between 2000 and 2010, the cultivated areas almost doubled in size and the output increased accordingly, from about € 15 to about € 30 million per year. Overall, this made soft fruits a viable alternative to the traditional, and commercially mature, apple production (Fig. 2.1). Farms usually do not specialise in a single crop, but grow strawberries and one or more other berry crops.

After *D. suzukii* introduced in the area, a monitoring system was established. Hundreds of fruits' samples[3] were taken every year in the field and then submitted to a microscopic inspection in order to detect the presence of eggs or larvae.

The percentage of infected fruits resulting from the microscopic inspection show the progress of the pressure of *D. suzukii* in the province. After an increase between

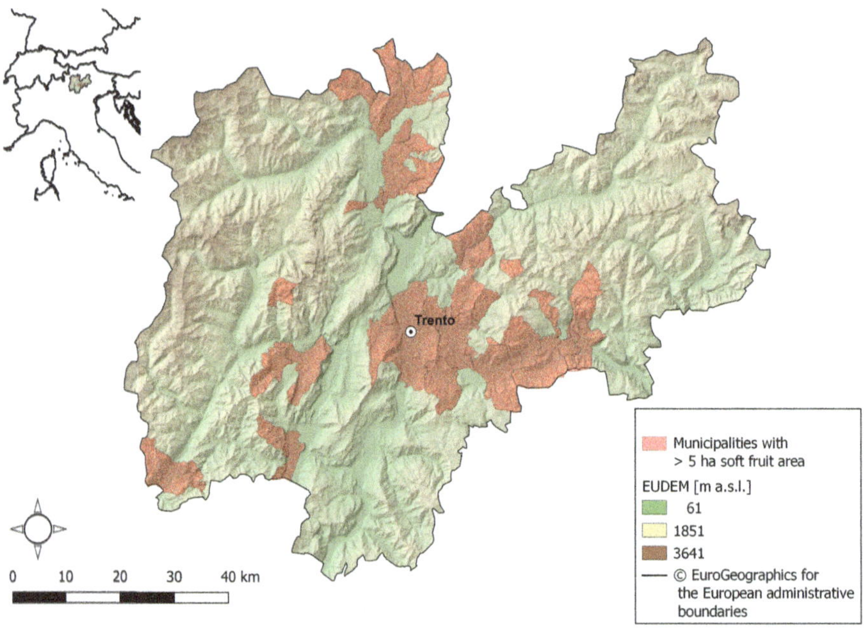

Fig. 2.1 Soft fruit-producing areas in the province of Trento, Italy. Map created using QGIS software v. 3.10

[3]A total of 561 samples of about 50 fruits were taken in 2012, 685 in 2013, 698 in 2014, 972 in 2015, 978 in 2016, 1151 in 2017, 1083 in 2018, 451 in 2019 when the diffusion of exclusion netting hampered the collection of many samples.

2012 and 2015, the pest's pressure basically stabilised thereafter, with the 2017 season, characterised by low temperatures in spring and above-average temperatures in summer, being the exception (Fig. 2.2).

A pest's control strategy including additional insecticides application, mass trapping and field sanitation was first implemented in 2012 (De Ros et al. 2015). It further evolved in three directions: adopting the exclusion netting technology on a progressively larger scale, discontinuing the maintenance of attractant-based traps and increasing the frequency of harvesting.

Exclusion netting consists in the installation around the field, or the planted row, of nets with a mesh fine enough to prevent the entrance of *D. suzukii* and other pests. In the province of Trento, it is becoming a mainstream practice for blueberry, cherry and blackberry. In fact, according to private communication with representatives of Producers' Organisations, in 2019 exclusion netting was implemented in 54%, 47% and 36% of the whole cultivated area, respectively, of blueberry, cherry and blackberry. In 2014, this technology was implemented in 35% of the whole cultivated area for blueberries and 11% for blackberries (Del Fava et al. 2017). A further increase can be expected in the future since the purchase of anti-insect nets is currently eligible for incentives under the EU-funded Operational Programmes for the fruit sector, and the expense for growers may be reduced up to 40%. Strawberry tunnels are not suitable for exclusion netting since they require more frequent access for applying treatments for other pests. Moreover, according to the same source, only 9% of raspberry area is managed with exclusion netting. This may be explained with the relative success of local raspberry growers in reducing the commercial

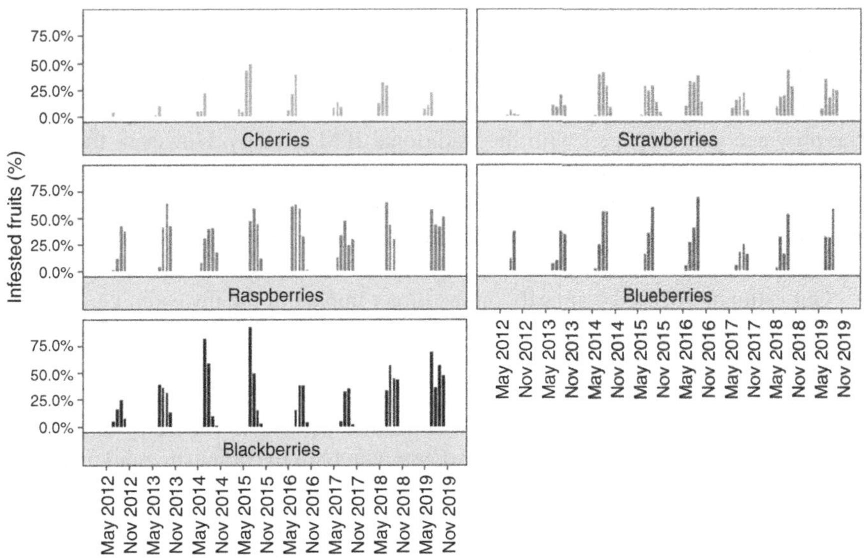

Fig. 2.2 *Drosophila suzukii*'s pressure on various soft fruits in the province of Trento, Italy (2012–2019)

damage by means of insecticide applications and frequent harvesting, similarly with what was reported by Farnsworth et al. (2017) for California raspberry industry.

During the early phase of *D. suzukii*'s invasion, mass trapping used to be a key component of the pest control strategy. But, as pest pressure was increasing, local growers grew sceptical about this practice as either control or monitoring tool. This is consistent with what was found out among the Swiss cherry producers: surveyed about their satisfaction with the effectiveness of single pest management's measures only 15% of mass-trapping users declared themselves satisfied with this practice (Mazzi et al. 2017). In short, the use of traps is now neglected in the province of Trento. Instead, in the area not managed with exclusion netting producers shortened the harvest intervals in order to reduce as much as possible the amount of ripe fruit that could be become targeted by *D. suzukii*.

2.3.2 Methodological Issues

According to a framework proposed by Born et al. (2005), four "cornerstones" have to be taken into account in the economic evaluation of biological invasions: the evaluated strategy, the field of application, the range of impact to be considered and the type of data.

In this update, we considered the pest's control measures currently adopted in the province of Trento, for the five fruit crops mostly affected by *D. suzukii* (cherries, strawberries, raspberries, blackberries and blueberries). For strawberries and raspberries, the control strategy consists of applications of pesticides complemented by field sanitation and frequent harvesting, while, for cherries, blueberries and blackberries, we hypothesised a control strategy relying on exclusion netting with one application of insecticides before the installation of the nets. It is a simplifying assumption since there are still relevant cherry, blueberry and blackberry areas in the province still managed with the traditional IPM strategy. However, the trends regarding the adoption of exclusion netting indicate that it will soon become the standard measure for these crops. On the other hand, it is unclear if the lack of interest towards the exclusion netting showed until now by the raspberry growers could change in the near future.

Our estimate focused primarily on the direct impact of the invasion, i.e. the specific effects on the host crops in terms of yield losses and associated control costs. Differently from what was done in two studies carried out in California (Goodhue et al. 2011; Farnsworth et al. 2017), we did not take into account indirect market effects like price changes and shifts in consumer demands. The berry and cherry production of the province is sold almost completely in the domestic market and, as Italy is a net importer of these products, we assume that a decrease of local supply could be easily substituted without significant effects on market prices.

As it was done in two recent studies about the economic impact of *D. suzukii* (Del Fava et al. 2017; Fan et al. 2020), we attempted to take into account the wider social costs of strategies, particularly those connected with the external, social and environmental costs associated with insecticides. On the basis of the assessment of Pimentel and Burgess (2014), who calculated conservatively that a $US 10 billion yearly investment in pesticide control corresponds to an almost equal amount of health and environmental costs, we estimated these costs as equal to the total expenses for insecticides purchase.

With regard to the type of data, we used as much as possible business data, i.e. the results of the grading performed by a soft fruit marketing co-op through visual inspection, to assess the revenue losses. Since the infestation is not always visible from the outside, these data are a conservative representation of the immediate commercial damage.[4] If maggots develop in the interval between grading and consumption, the co-op may have to endure hidden extra-costs for managing the consumer complaints or dissatisfaction. Instead, if the time interval between harvest, delivery and consumption is very short as it can happen for fresh fruit, there will be no impact on revenues. The same is true when cold treatments can stop the development of larvae (Saeed et al. 2020). As in our previous study, the reason we opted for this data source was that growers base their pest management decisions on this type of data. Since co-op grading data do not include the control strategy adopted on the farms, we estimated yield losses with exclusion netting using the results of field trials carried out in 2015, a year with high pest pressure levels.

In order to have an *ex ante* reference, i.e. the revenue generated by the industry in standard condition, we combined data from different sources. The output was estimated according to the average yield per hectare in the period 2007–2009, namely the years prior to the detection of considerable damage to cultivated berries in the area, used as a baseline also in our previous study on the impact of *D. suzukii* in the area (De Ros et al. 2015, Table 1). Data on the cultivated area in 2019 were obtained by the Provincial Paying Agency in Agriculture. For product prices at farm gate, we calculated the average of the available data for the most recent 3-year period: 2015–2017. Accordingly, the potential revenue for the mentioned fruit crops in Trentino can be calculated at around € 30.25 million (Table 2.1), almost half of which comes from strawberries.

[4] For a comparison between these data and the results from the microscopic inspection of sampled fruits, see Del Fava et al. (2017, figure 3).

Table 2.1 Revenue potential of soft fruit production in the province of Trento, Italy

Crop	Area (Ha)	Yield (t)	Price (€/kg)	Potential revenue (€)
Cherry	295	1828	3.07	5,604,409
Strawberry	152	4948	2.94	14,531,544
Raspberry	74	532	5.16	2,746,785
Blueberry	132	999	4.62	4,617,360
Blackberry	41	687	4.00	2,745,952
TOTAL				30,246,050

2.3.3 *Analytical Technique*

We estimated the additional costs and revenue losses associated with the invasion of *D. suzukii* in the province of Trento according to a partial budget approach. Originally designed to assess the economic effects of a management option at a farm business level, partial budgeting is also used for basic economic analysis of these effects on a wider scale, regional or national.

In a thorough discussion about the analytical techniques available for estimating the impact of an invasive pest, Soliman et al. (2010) highlighted the strength of partial budgeting in its simplicity and transparency. The fundamental shortcoming of this method is that it is not suitable for measuring the indirect impacts of the invasion. In fact partial budgeting does not consider the effects on market prices and the related welfare changes, or the repercussions on other sectors of the economy, or the long-term effects (e.g. the possible pickers' shortage envisaged by Farnsworth et al. (2017) as a consequence of the labour-intensive practices needed for pest control). However, when mainly direct, host-specific impacts are at stake, as appears to be the case for *D. suzukii* in the province of Trento, the strengths of partial budgeting, easily replicable in many situations, outweigh its limitations.

2.4 Estimates

2.4.1 *Revenue Losses*

In order to estimate the revenue losses with the adoption of exclusion netting, we utilised data from field trials carried out by the Edmund Mach Foundation in different cherry, raspberry, blueberry and blackberries areas managed with exclusion netting during 2015, well representative of high pest pressure levels in the province of Trento. Overall 17,197 fruits were sampled and then submitted to microscopic inspection which detected 693 fruits, i.e. 4.03%, with *D. suzukii*'s damage. These data offer a general overview of the effectiveness of exclusion netting in average operating conditions, even though, being based on microscopic inspection, they presumably overestimate the potential commercial damage.

For the crops managed with the integrated strategy, namely strawberry and raspberry, we considered the results of the marketing co-op grading through visual inspection in 2015: 1.50% for strawberries and 2.21% for raspberries.[5]

According to the data used, the direct impacts of *D. suzukii* on revenues would amount to little more than 800,000 €, i.e. about 2.6% of the estimated revenue potential (Table 2.2). This result is consistent with the recent assessment of

[5] In 2015 the co-op grading detected *D. suzukii*'s damage in 9.09% of inspected cherries, in 11.31% of blueberries and in 6.51% of blackberries.

Table 2.2 Estimated *Drosophila suzukii*-induced revenue losses for soft fruit production in the province of Trento, Italy

Crop	Potential revenue (€)	Yield reduction (%)	Estimated revenue losses (€)
Cherry	5,604,409	4.03	225,857.7
Strawberry	14,531,544	1.50	217,973.2
Raspberry	2,746,785	2.21	60,703.9
Blueberry	4,617,360	4.03	183,079.6
Blackberry	2,745,050	4.03	110,661.9
TOTAL	30,246,050		801,276.3

Table 2.3 Estimated cost per hectare of insecticides used against *Drosophila suzukii* in the province of Trento, Italy

		Application rate		
Crop	Treatment	ml/hl	hl/ha	Cost (€/ha)
Cherry	Phosmet	150	10	71.6
Strawberry	Deltamethrin	50	10	56.2
	Spinosad	25	8	296.2
Raspberry	Deltamethrin	50	10	56.2
	Spinetoram	300	8	222.8
Blueberry	Phosmet	150	10	71.6
Blackberry	Spinetoram	300	8	222.8

Farnsworth et al. (2017) for the California raspberry industry, but significantly lower than the estimate of DiGiacomo et al. (2019) for the Minnesota wild raspberry industry.

2.4.2 Control Costs

2.4.2.1 Chemical Control

Insecticides are a crucial element of a pest management strategy. According to an estimate on the USA (Pimentel 2014), the amount of insecticides applied against invasive pest insects generates costs for $US 500 million per year.

Also in the case of *D. suzukii*, for the producers managing the invasion implies an increase in the use of insecticides and in application costs. In order to assess the costs of insecticides, reference is made to the guidelines of the local Extension Service about the specific insecticide treatments for *D. suzukii* and their application rates. Only one application of insecticides is estimated for cherry, blueberry and blackberry since we assumed that these crops are exclusively managed with exclusion netting. The material costs are evaluated on the basis of market prices for insecticides (Table 2.3).

The application costs are evaluated according to a standard level of mechanisation, i.e. a tractor and a mounted sprayer with 300 and 100 h of annual use for the tractor and for the sprayer, respectively. Taking into account also the operator labour, the hourly costs amount to 44.3 €/h. Hence, assuming an average of 2 h/ha for application, each application is estimated to have costs of 88.6 €/ha, an intermediate value between the higher costs recently calculated for sweet cherry production in Switzerland (Mazzi et al. 2017) and the lower application costs estimated for raspberry industry in California (Farnsworth et al. 2017).

2.4.2.2 Labour Management Costs

The most onerous *D. suzukii*'s control measure, besides chemical control, is the shortening of harvest intervals. More frequent harvests affect costs as, given the reduced quantity of fruit to be picked in every harvest, pickers' efficiency decreases. As highlighted by Farnsworth et al. (2017) for the California raspberry industry, this translates to an increased need for workforce. According to private conversations with growers, the increase of workforce needed to harvest ranges from 20 to 30%. In order to estimate the associated costs, we utilised the data of the yield per hectare for the 3-year period 2007–2009 assumed as baseline. On the basis of an average picker's productivity of 15 kg/h for strawberries and 5 kg/h for raspberries, we first estimated an amount of 2170 h/ha and 1440 h/ha for harvesting strawberries and raspberries, respectively. Assuming a conservative estimate of a 20% labour increase, with the minimum hourly rate for pickers according to the collective contracts for workers, i.e. 8.5 €/h, we obtained additional costs of 3689 €/ha for strawberries and 2448 €/ha for raspberries (Table 2.4).

In addition, during harvest it is necessary to remove and destroy every rejected piece of fruit since every fallen or damaged fruit can become a source of infestation. This practice requires additional work that can be estimated at around 1 h per hectare for each harvest day. Total field clean-up costs are calculated taking into consideration the harvest periods: 75 days for everbearing strawberries, which currently constitutes 90% of strawberry area in the province, and strawberries. At the minimum hourly rate for pickers, i.e. 8.5 €/h, we thus estimated additional costs of 637.5 €/ha for strawberries and raspberries.

Table 2.4 Estimated *Drosophila suzukii*'s control costs per hectare in the province of Trento, Italy

Crop	Insecticides (€/ha)	Insecticides application (€/ha)	Additional harvesting (€/ha)	Field sanitation (€/ha)	Nets and material (€/ha)	Exclusion netting management (€/ha)	TOTAL (€/ha)
Cherry	71.6	88.6			1000.0	625.0	1785.2
Strawberry	352.4	265.8	3689.0	637.5			4944.7
Raspberry	279.0	265.8	2448.0	637.5			3630.3
Blueberry	71.6	88.6			1700.0	625.0	2485.2
Blackberry	222.8	88.6			300.0	625.0	1236.4

2.4.2.3 Exclusion Netting Costs

The adoption of exclusion netting implies both material and labour costs. Material costs consist of the purchase of anti-insect nets with a mesh not larger than 1.00 mm and, if there are not pre-positioned tunnels or rain covers in the planting, a metal structure to support the nets. Labour is required for the installation of the nets at the completion of flowering and their removal at the end of the growing season.

We considered the full material costs, even though the exclusion netting is currently eligible for public incentives in the province of Trento. For cherry cultivation two types of netting are available: the single-row system, where nets are hung on every row of plants, and the less expensive so-called mono-block system, where the nets have to be placed on the sides of the planting. We assumed an average investment of 10,000 €/ha with an economic life expectancy of 10 years, i.e. 1000 €/ha each year. For blackberry crops, which generally are cultivated in tunnels, the nets have to be applied only to the sides and to the ends of the tunnels, so we calculated an investment of 3000 €/ha, i.e. 300 €/ha each year. Since the implementation of the exclusion netting for blueberry involves also the positioning of a metal structure to support the nets and enough plastic material to cover not only the sides but also the top, we assumed an investment of 17,000 €/ha, i.e. a yearly cost of 1700 €/ha.

According to private communications with growers, the labour management of the nets requires an average of 50 h/ha. At the minimum hourly rate for a specialised agricultural worker, i.e. 12.5 €/h, that means additional costs for 625 €/ha.

In spite of the relevant material and labour costs, according to our estimate exclusion netting is relatively less expensive in comparison to the strategy relying on insecticides application and labour-intensive cultural practices. In fact, while *D. suzukii*'s control costs for cherries, blueberries and blackberries, that we assumed to be wholly managed with exclusion netting, do not overcome 2485.2 €/ha, strawberries and raspberries require control costs of 4944.7 €/ha and 3630.3 €/ha, respectively (Table 2.4).

2.4.3 *Economic Impact of* Drosophila suzukii *on the Soft Fruit Production in the Province of Trento*

In order to estimate the economic impact of *D. suzukii*, we multiplied the management costs per hectare for the area cultivated with the different crops. Then we combined the results with the estimate of potential revenue losses (Table 2.5).

Our results show that the control strategies adopted in the province of Trento seem effective in reducing the direct impacts of *D. suzukii* on revenues, even with high levels of pest's pressure. The pest's management costs are however remarkable and almost reach at industry level € 2 million. Thus, the economic impact on producers amount to € 2.73 million, more than 9% of the potential revenues of soft fruit production in the province of Trento.

Table 2.5 Estimated economic impact of *Drosophila suzukii* on soft fruit production in the province of Trento, Italy

Crop	Area (ha)	Revenue losses (€)	Management costs (€/ha)	Management costs (€)	Social costs (€/ha)	Social costs (€)
Cherry	295	225,857.7	1785.2	526,634.0	71.6	21,122.0
Strawberry	152	217,973.2	4944.7	751,594.4	352.4	53,564.8
Raspberry	74	60,703.9	3630.3	268,642.2	279.0	20,646.0
Blueberry	132	183,079.6	2485.2	328,046.4	71.6	9451.2
Blackberry	41	110,661.9	1236.4	50,692.4	222.8	9134.8
Total		801,276.3		1,925,609.4		113,918.8

If one considers even the social impacts of control strategies, like the environmental and social costs deriving from pesticides, the effects are even greater. On the basis of the estimate of Pimentel and Burgess (2014), we did attempt to take into account the external costs of insecticides used to control the *D. suzukii*'s invasion and added extra costs for an amount equal to the material costs for the purchase of insecticides. These costs amount to about 114,000 €, so the overall economic impact of *D. suzukii*'s in the province of Trento can be assessed in € 2.84 million.

2.5 Conclusions

In this chapter we reviewed the economic literature on *D. suzukii*. It is a relatively heterogeneous body of research, utilising different methods, different data sources, with a focus on different crops and different areas. Over time it moved its focus from raising awareness about the magnitude of the effects, particularly on the industry, to aiding the decision-making process. This happened as the invasion passed from the introduction to the establishment and spread phases and as a set of control strategies became available to producers. Recently few studies approached the economic evaluation of farm responses through bio-economic models using Bayesian methods to address the uncertainty surrounding both the pest effect and the result of control measures.

We also updated a previous impact assessment for the five most affected crops in the province of Trento. On the basis of conservative assumptions, the economic impact on soft fruit production in the province of Trento amounts to 9% of the potential revenues, mostly due to costly control measures. Our estimates show that, a decade after its introduction, *D. suzukii* keeps having a substantial negative impact on berry and cherry production. Indirectly they also point out that there is still a need for accurate empirical impact assessments investigating the costs of control besides the damage caused by the pest.

In our opinion, there are not only geographical areas but also additional aspects that deserve attention and analysis with regard to the economic impacts of *D. suzukii*. One is the increasing variety of producer responses to pest pressure as shown by the six clusters of Swiss grape growers each implementing different combinations of various measures as noted in a recent study (Knapp et al. 2019). Another is the use of data from structured surveys to growers, using an approach pioneered by the North Carolina State Extension (Burrack et al. 2013), possibly combined with information gathered by extension specialists. In addition, an evaluation of the social costs and the public preferences for different control strategies is critical (Hanley and Roberts 2019).

Acknowledgements We thank the Association of Fruit and Vegetable Producers of Trentino (APOT) and Gianluca Savini, of Sant'Orsola Farmers Co-operative, for providing us with the data on the current implementation of the exclusion netting and the data about the soft fruit grading, Pietro Molfetta of Provincial Agricultural Paying Agency for the data about soft fruit cultivation in the province of Trento. We also wish to thank our colleagues Sandro Conci and Sergio Franchini for the useful discussions.

References

Andreazza F, Bernardi D, dos Santos RSS et al (2017) *Drosophila suzukii* in southern neotropical region: current status and future perspectives. Neotrop Entomol 46:591–605. https://doi.org/10.1007/s13744-017-0554-7

Asplen MK, Anfora G, Biondi A et al (2015) Invasion biology of spotted wing Drosophila (*Drosophila suzukii*): a global perspective and future priorities. J Pest Sci 88:469–494. https://doi.org/10.1007/s10340-015-0681-z

Beers EH, Van Steenwyk RA, Shearer PW et al (2011) Developing *Drosophila suzukii* management programs for sweet cherry in the western United States. Pest Manag Sci 67:1386–1395. https://doi.org/10.1002/ps.2279

Bellamy DE, Sisterson MS, Walse SS (2013) Quantifying host potentials: indexing postharvest fresh fruits for spotted wing Drosophila, Drosophila suzukii. PLoS One 8:e61227. https://doi.org/10.1371/journal.pone.0061227

Benito NP, Lopes-da-Silva M, dos Santos RSS (2016) Potential spread and economic impact of invasive *Drosophila suzukii* in Brazil. Pesqui Agropecu Bras 51:571–578. https://doi.org/10.1590/S0100-204X2016000500018

Bolda MP, Goodhue RE, Zalom FG (2010) Spotted wing Drosophila: potential economic impact of a newly established pest. Agric Resour Econ Updat Univ California Giannini Found 13:5–8. https://doi.org/10.1016/j.jff.2015.04.027

Born W, Rauschmayer F, Bräuer I (2005) Economic evaluation of biological invasions—a survey. Ecol Econ 55:321–336. https://doi.org/10.1016/j.ecolecon.2005.08.014

Bruck DJ, Bolda M, Tanigoshi L et al (2011) Laboratory and field comparisons of insecticides to reduce infestation of Drosophila suzukii in berry crops. Pest Manag Sci 67:1375–1385. https://doi.org/10.1002/ps.2242

Burrack HJ, Smith JP, Pfeiffer DG et al (2013) Using volunteer-based networks to track *Drosophila suzukii* (Diptera: Drosophilidae) an invasive pest of fruit crops. J Integr Pest Manag 3:1–5. https://doi.org/10.1603/ipm12012

Buyuktahtakin E, Haight R (2018) A review of operations research models in invasive species management: state of the art, challenges, and future directions. Ann Oper Res 271:357–403

Calabria G, Máca J, Bächli G et al (2012) First records of the potential pest species *Drosophila suzukii* (Diptera: Drosophilidae) in Europe. J Appl Entomol 136:139–147. https://doi.org/10.1111/j.1439-0418.2010.01583.x

Cini A, Ioriatti C, Anfora G (2012) A review of the invasion of *Drosophila suzukii* in Europe and a draft research agenda for integrated pest management. Bull Insectology 65:149–160

Dana ED, Jeschke JM, García-De-Lomas J (2014) Decision tools for managing biological invasions: existing biases and future needs. Oryx 48:56–63. https://doi.org/10.1017/S0030605312001263

De Ros G, Anfora G, Grassi A, Ioriatti C (2013) The potential economic impact of *Drosophila suzukii* on small fruits production in Trentino (Italy). IOBC-WPRS Bull 91:317–321

De Ros G, Conci S, Pantezzi T, Savini G (2015) The economic impact of invasive pest *Drosophila suzukii* on berry production in the province of Trento, Italy. J Berry Res 5:89–96. https://doi.org/10.3233/JBR-150092

Del Fava E, Ioriatti C, Melegaro A (2017) Cost–benefit analysis of controlling the spotted wing drosophila (*Drosophila suzukii* (Matsumura)) spread and infestation of soft fruits in Trentino, northern Italy. Pest Manag Sci 73:2318–2327. https://doi.org/10.1002/ps.4618

Deprá M, Poppe JL, Schmitz HJ et al (2014) The first records of the invasive pest *Drosophila suzukii* in the south American continent. J Pest Sci 87:379–383. https://doi.org/10.1007/s10340-014-0591-5

DiGiacomo G, Hadrich J, Hutchison WD et al (2019) Economic impact of spotted wing Drosophila (Diptera: Drosophilidae) yield loss on Minnesota raspberry farms: a grower survey. J Integr Pest Manag 10:0–5. https://doi.org/10.1093/jipm/pmz006

Dos Santos LA, Mendes MF, Krüger AP et al (2017) Global potential distribution of *Drosophila suzukii* (Diptera, Drosophilidae). PLoS One 12:1–13. https://doi.org/10.1371/journal.pone.0174318

Epanchin-Niell RS, Hastings A (2010) Controlling established invaders: integrating economics and spread dynamics to determine optimal management. Ecol Lett 13:528–541. https://doi.org/10.1111/j.1461-0248.2010.01440.x

Fan X, Gómez MI, Atallah SS, Conrad JM (2020) A Bayesian state-space approach for invasive species management: the case of spotted wing Drosophila. Am J Agric Econ 00:1–18. https://doi.org/10.1002/ajae.12028

Farnsworth D, Hamby KA, Bolda M et al (2017) Economic analysis of revenue losses and control costs associated with the spotted wing Drosophila, *Drosophila suzukii* (Matsumura), in the California raspberry industry. Pest Manag Sci 73:1083–1090. https://doi.org/10.1002/ps.4497

Goodhue RE, Bolda M, Farnsworth D et al (2011) Spotted wing drosophila infestation of California strawberries and raspberries: economic analysis of potential revenue losses and control costs. Pest Manag Sci 67:1396–1402. https://doi.org/10.1002/ps.2259

Grassi A, Giongo L, Palmieri L et al (2011) *Drosophila (Sophophora) suzukii* (Matsumura) (Diptera: Drosophilidae), new pest of soft fruits in Trentino (North-Italy) and in Europe. IOBC/WPRS Bull 70:121–128

Hampton E, Koski C, Barsoian O et al (2014) Use of early ripening cultivars to avoid infestation and mass trapping to manage *Drosophila suzukii* (Diptera: Drosophilidae) in *Vaccinium corymbosum* (Ericales: Ericaceae). J Econ Entomol 107:1849–1857. https://doi.org/10.1603/ec14232

Hanley N, Roberts M (2019) The economic benefits of invasive species management. People Nat 1:124–137. https://doi.org/10.1002/pan3.31

Knapp L, Mazzi D, Finger R (2019) Management strategies against *Drosophila suzukii*: insights into Swiss grape growers choices. Pest Manag Sci 75:2820–2829. https://doi.org/10.1002/ps.5397

Kuesel R, Hicks DS, Archer K et al (2019) Effects of fine-mesh exclusion netting on pests of blackberry. Insects 10:1–12. https://doi.org/10.3390/insects10080249

Leach H, Van Timmeren S, Isaacs R (2016) Exclusion netting delays and reduces Drosophila suzukii (Diptera: Drosophilidae) infestation in raspberries. J Econ Entomol 109:2151–2158. https://doi.org/10.1093/jee/tow157

Leach H, Moses J, Hanson E et al (2018) Rapid harvest schedules and fruit removal as non-chemical approaches for managing spotted wing Drosophila. J Pest Sci 91:219–226. https://doi.org/10.1007/s10340-017-0873-9

Mazzi D, Bravin E, Meraner M et al (2017) Economic impact of the introduction and establishment of *Drosophila suzukii* on sweet cherry production in Switzerland. Insects 8:1–13. https://doi.org/10.3390/insects8010018

Miller B, Anfora G, Buffington M et al (2015) Resident parasitoids associated with *Drosophila suzukii* and their seasonal occurrence in two small fruit production regions. Biol Control 68:255–263

Olson LJ (2006) The economics of terrestrial invasive species: a review of the literature. J Agric Resour Econ 35:178–194. https://doi.org/10.1017/S1068280500010145

Pimentel D (2014) Pesticides applied for the control of invasive species in the United States. In: Abrol DP (ed) Integrated pest management: current concepts and ecological perspective. Academic, San Diego, pp 111–123

Pimentel D, Burgess M (2014) Environmental and economic costs of the application of pesticides primarily in the United States. In: Pimentel D, Peshin R (eds) Integrated pest management: pesticide problems, vol 3. Springer, Dordrecht, pp 47–71

Saeed N, Tonina L, Battisti A, Mori N (2020) Postharvest short cold temperature treatment to preserve fruit quality after *Drosophila suzukii* damage. Int J Pest Manag 66:23–30. https://doi.org/10.1080/09670874.2018.1539531

Schöneberg T, Arsenault-Benoit A, Taylor CM et al (2020) Pruning of small fruit crops can affect habitat suitability for *Drosophila suzukii*. Agric Ecosyst Environ 294:106860. https://doi.org/10.1016/j.agee.2020.106860

Soliman T, Mourits MCM, Oude Lansink AGJM, van der Werf W (2010) Economic impact assessment in pest risk analysis. Crop Prot 29:517–524. https://doi.org/10.1016/j.cropro.2009.12.014

Shawer R, Tonina L, Tirello P et al (2018) Laboratory and field trials to identify effective chemical control strategies for integrated management of Drosophila suzukii in European cherry orchards. Crop Prot 103:73–80. https://doi.org/10.1016/j.cropro.2017.09.010

Tait G, Grassi A, Pfab F et al (2018) Large-scale spatial dynamics of *Drosophila suzukii* in Trentino, Italy. J Pest Sci 91:1213–1224. https://doi.org/10.1007/s10340-018-0985-x

Van Timmeren S, Isaacs R (2013) Control of spotted wing Drosophila, *Drosophila suzukii*, by specific insecticides and by conventional and organic crop protection programs. Crop Prot 54:126–133. https://doi.org/10.1016/j.cropro.2013.08.003

Winkler A, Jung J, Kleinhenz B, Racca P (2020) A review on temperature and humidity effects on *Drosophila suzukii* population dynamics. Agric For Entomol 22:179–192. https://doi.org/10.1111/afe.12381

Yeh DA, Drummond FA, Gómez MI, Fan X (2020) The economic impacts and management of spotted wing Drosophila (*Drosophila suzukii*): the case of wild blueberries in Maine. J Econ Entomol 113:1–8. https://doi.org/10.1093/jee/toz360

Chapter 3
Drosophila suzukii: Pathways and Pathway Management by Regulation

Jocelyn A. Berry

Abstract For geographically isolated countries, where Spotted Wing Drosophila (SWD) or *Drosophila suzukii* remains a quarantine pest, regulating entry pathways is of paramount importance. Countries such as New Zealand and Australia reacted to the establishment of *D. suzukii* in the United States by imposing emergency measures on imported host material. However, under the International Plant Protection Convention (IPPC 2015), emergency measures cannot be imposed indefinitely and must be justified by risk analysis within a reasonable timeframe. This chapter reviews the invasion history and pathways of *D. suzukii* and discusses justification for imposing measures, host determination in a scientific context and host categorisation in a regulatory context.

Keywords Host status · *Drosophila suzukii* · Regulation · Pathway · Invasion history

3.1 Introduction

The worldwide distribution of *Drosophila suzukii* Matsumura (SWD) has expanded rapidly over the last 20 years. The species is considered to be native to Western Asia (dos Santos et al. 2017) and was described from Japan by Matsumura in 1931. It had, however, been noted as a pest of cherries in that country in 1916 (Kanzawa 1935) and is widespread and abundant there (Tamada 2009). SWD is also widespread and abundant in Korea (Lee 1964) and China (Kai et al. 1993). Its wider distribution in Asia includes India (as the subspecies *indicus*; Singh and Bhatt 1988); Myanmar, Taiwan, Nepal and Thailand (Toda 1991), Pakistan (Amin ud Din et al. 2005) and Far East Russia (Calabria et al. 2010).

SWD was first reported outside Asia in Hawaii on the island of Oahu in 1980, followed by several other Hawaiian Islands (Hauser 2011). More recently it has

J. A. Berry (✉)
Ministry for Primary Industries, Wellington, New Zealand
e-mail: jo.berry@mpi.govt.nz

F. R. M. Garcia (ed.), *Drosophila suzukii Management*,
https://doi.org/10.1007/978-3-030-62692-1_3

invaded the Americas and Europe. In continental North America, it was first detected in California in 2008, though the response to the detection was delayed due to misidentification (Hauser 2011). It was subsequently reported from South America (Deprá et al. 2014). It was first confirmed present in Europe from Italy in 2009 and then in Spain, France and Germany (EPPO 2020). EPPO (the European Plant Protection Organization) (2020) has now reported this species to be present in many European countries. In addition, it has the potential to colonise agricultural crops in countries in Oceania and Africa (Dos Santos et al. 2017). Table 3.1 summarises the invasion history of this fly.

3.2 Important Invasion Pathways

The lifecycle of SWD is very similar to that of the economically important true fruit flies (Tephritidae). All lay their eggs and feed internally as immatures within fruit, or very occasionally in flowers. Pupation usually occurs outside the host, most usually in soil. As such, by far the most important pathway for international (non-natural) spread of these pests is as immature stages within the commodities they have a biological relationship with. This is in contrast to "contaminating" or "hitchhiking" pests such as brown marmorated stink bug, *Halyomorpha halys* Stal (Hemiptera: Pentatomidae), spotted lantern fly, *Lycorma delicatula* (White) (Hemiptera: Fulgoridae), and various species of tramp ants, which move in most part on inanimate pathways (e.g. containers, vehicles and machinery).

Since adults feed on dropped, spoiled and fermented fruit (Walsh et al. 2011), they may be attracted to fruit in packhouses; however, these stages are highly mobile and are likely to move off hosts when disturbed. Depending on transit conditions, late instar larvae may pupate and emerge as adults in transit.

Since SWD are mainly associated with fruiting bodies, they are unlikely to move in association with nursery stock (also called plants for planting), another high-risk pathway for a variety of important plant pests and diseases. However, like tephritids, larval stages may leave the host and pupate in a substrate such as soil. Although Kanzawa (1939) considered that "pupation in the fruit seems to be the norm," Bolda (2009) reared many flies from strawberry and raspberry and found they always exited the fruit to pupate. Presumably, like tephritids, in natural conditions SWD larvae may drop to the ground to pupate in soil. Therefore, the movement of fruiting host plants with attached soil may be a pathway for SWD for countries that do not regulate the entry of soil or growing media associated with host trees. Some countries (such as New Zealand and Australia) have strict measures in place to regulate the movement of soil, either in association with plants for planting or as a contaminant. These measures are aimed at reducing the likelihood of a wide ranges of pests and pathogens (in particular), not specifically tephritids.

Although there are a few records of flowers as SWD hosts, the available evidence does not suggest that the commercial cut flower pathway is an important one for the

Table 3.1 The invasion history of *Drosophila suzukii*: from its native Asia to Oceania, the Americas, Europe and the Middle East

Country	Year	Means of detection and location/host of first records	Reference
Oceania			
USA—Hawaii	1980	Collected in forest reserve	Kaneshiro (1983)
North America			
USA—Mainland (California)	2008	Field-grown raspberry	Hauser (2011)
Canada (British Colombia)	2009		Hauser (2011)
Mexico (Michoacán)	2011		Lasa and Tadeo (2015)
South America			
Brazil	2013	Rearing commercial fruit; banana baiting	Vilela and Mori (2014); Deprá et al. (2014)
Uruguay	2013	Baiting and field-grown blueberry	EPPO 2016/113
Argentina	2014	Field-grown raspberry	EPPO 2018/002
Chile	2017	Trapping in blackberry bushes	EPPO 2017/182
Europe			
Spain	2008	Trapping in pine forest with *Rubus* undergrowth	Calabria et al. (2010)
Italy	2008	Malaise traps[a]	Cini et al. (2012)
France	2009	Field-grown strawberry and cherry	EPPO (2020)
Slovenia	2010	Field-grown raspberry and grape	EPPO (2020)
Croatia	2010	Trapping in raspberry, peach and grapevine	EPPO (2020)
Germany	2011	Trapping	EPPO (2020)
Belgium	2011	Private garden	EPPO (2020)
Austria	2011	Field-grown raspberry, kiwifruit and elder	EPPO (2020); Asplen et al. (2015)
Switzerland	2011	Vinegar traps in strawberries, raspberries, blueberries and cherry orchards	EPPO (2020); Asplen et al. (2015)
Portugal	2012	Raspberries in commercial greenhouse	EPPO (2020)
Netherlands	2012	Apple cider vinegar and wine traps in wilderness areas and private gardens	EPPO (2020)
United Kingdom	2012	Raspberry and blackberry experimental plots	EPPO (2020)
Hungary	2012	Survey (apple cider vinegar trap in road area)	EPPO (2020)
Bosnia and Herzegovina	2013	Trapping	EPPO (2020)
Montenegro	2013	Tephri traps	Asplen et al. (2015)
Romania	2013	Trapping in wild blackberry	EPPO (2020)

(continued)

Table 3.1 (continued)

Country	Year	Means of detection and location/host of first records	Reference
Serbia	2014	Fruit survey of commercial blackberry, fig, raspberry and grapes	EPPO (2020)
Sweden	2014	Trapping	EPPO (2020)
Ukraine	2014	Survey	Lavrinienko et al. (2017)
Turkey	2014	Strawberry crops	EPPO (2020)
Czech Republic	2014	Apple cider vinegar trap in garden	EPPO (2020)
Slovak Republic	2014	Trapping in farmland	EPPO (2020)
Ireland	2015	Pheromone trapping in commercial nurseries	EPPO (2020)
Cyprus	2016	Trapping in commercial crops	EPPO (2020)
Middle East			
Israel	2020	Private garden	EPPO (2020)

[a]Note EPPO reports the first occurrence as 2009 in field-grown raspberries

movement of SWD. The fly has only ever been recorded from two species of old or fallen flowers (DAFF 2013).

The most important pathway for international movement is therefore as eggs or larvae inside fruiting bodies of host commodities (fruit and some vegetables). Host commodities may be moved across borders as part of commercial trade (by land, sea or air), carried by passengers, moved in craft such as yachts or cruise ships and sent by mail. In the United States and Europe, the first records were on the coast close to seaports (Deprá et al. 2014), providing circumstantial evidence of the role of commercial trade.

SWD has been detected (or at least reported) relatively infrequently in international interceptions:

- at the New Zealand border, it has been detected twice in commercial fruit consignments; once in 2012 as dead larvae in nectarines from the USA, and once in 2019 as live larvae in citrus from the USA (discussed below). Live larvae have also been identified once in 2019 in blueberries carried by an air passenger (MPI Internal database);
- EPPO reports only one detection, in sour cherry (*Prunus cerasus*) fruit exported from the Lebanon and detected in France in 2018;
- SWD has also been detected at the Australian, Canadian, Japanese, South Korean, and mainland USA borders. The detections at the Canadian border were from the USA in 2009 (live larvae in blackberries) and Italian grapes (in 2016, with no viability information). The detection at the Japanese border was on fresh fruit from Mexico in 2017. There are no further details available about the detections in Australia and the USA (Rebecca Turner, Scion, pers. comm.).

However, many drosophilids are not identified to species level when detected at international borders, and it is likely that the low detection rate does not reflect the true level of movement in international trade of fresh produce. For geographically isolated countries, where this fly remains a quarantine pest, identifying and regulating entry pathways are of paramount importance. Reducing the likelihood of SWD establishing in new regions via the pathway of imported commercial hosts is relatively straightforward compared to, for example, managing hitchhikers.

Prerequisites for successful management of this pathway are effective treatments, appropriate regulation and compliance with such regulation. Where commercial trade is managed effectively and consistently, the non-commercial movement of host commodities (intentional or non-intentional) may be the pathway that SWD is most likely to move on.

3.3 Regulation

The global trade in plants and plant products provides many pathways for invasive species to enter and establish in new geographical regions. New trade routes are opening, and the type and volume of trade on existing routes are changing and generally increasing. The goal of phytosanitary practices is to reduce the damage inflicted by invasive species that may be associated with these commodities by regulating their movement. For signatories of the International Plant Protection Convention (IPPC), National Plant Protection Organisations (NPPOs) are the competent and legally responsible bodies for regulatory plant protection, and each country has the sovereign right to set its own level of protection (IPPC 2015). However, these restrictions are a primary impediment to global trade, which is a key and growing component of most economies. Their imposition also benefits domestic providers (Heather and Hallman 2007). For these reasons, signatories to the International Plant Protection Convention (IPPC) are required to impose the least restrictive conditions to achieve their desired level of protection (IPPC 1997).

NPPOs have therefore to deal with conflicting demands, on the one hand the need to minimise the damage caused by invasive organisms, on the other to facilitate global trade, or at least not to impede it. NPPOs can only impose measures that are scientifically justified (IPPC 1997); however, this can lead to a number of areas of contention. One of the more important areas of contention is the determination of host status.

3.4 Host Status

Follett and Neven (2006) use a definition for a fruit fly host as "a fruit or vegetable onto which an insect deposits eggs, the eggs hatch into larvae, and the larvae feed and develop to form viable pupae from which adults emerge." If the insect cannot completely develop to form viable adults, the plant is considered to be a non-host.

Determining whether a particular plant species fits this definition can often be a difficult and controversial problem for regulatory authorities. Host status has often been historically conferred on commodities found with the pest in or on it in the field. According to ISPM 37 (2016), host status may still be determined from historical production records or from trade, or from interception data indicating natural infestations. However, there are many problems associated with historical data, and these have sometimes led to quarantine regulations that have little or no objective validity (Armstrong 1994). Common problems include misidentification of pests (especially immatures) or of hosts, contamination of samples, transcription errors or misreading of sample numbers. A number of records where an adult has been collected on a plant probably refer to casual associations rather than host records.

Science-based decision-making processes represent the only acceptable mechanism to resolve any dispute on this matter (Aluja and Mangan 2008). ISPM 37 (2016) provides guidelines for the determination of host status of fruit to tephritid fruit flies and describes three categories of host status of fruit to fruit flies for regulatory applications:

- Natural hosts: plant species or cultivars that have been scientifically found to be infested by the target fruit fly species under natural conditions and are able to sustain their development to viable adults;
- Conditional hosts: plant species or cultivars that are not natural hosts but have been experimentally demonstrated to be infested by the target fruit fly species and are able to sustain its development to viable adults as concluded from semi-natural field conditions (field cages, etc.);
- Non-hosts: plant species or cultivars that have not been found to be infested by the target fruit fly species or are not able to sustain its development to viable adults under natural conditions or under semi-natural field conditions.

Natural hosts range from poor to good (Aluja and Mangan 2008). Other terminology is used to describe this continuum, for example "main" and "other," "favoured," "preferred" and "occasional."

In common with tephritids (and many other pests and pathogens), the accurate determination of regulatory hosts of SWD can be problematic. The fly is still expanding its geographical range and novel hosts that may not be predictable are being encountered (Lee et al. 2015). Its relatively recent invasive status means that SWD lacks the long history of potentially dubious host records that is common for some pest species, and may cause problems with quarantine regulation. However, the lack of historical records also meant that, in the early stages of its invasions, NPPOs lacked guidance on which commodities to regulate. Although emergency measures can be imposed on imported host material, for signatories of the IPPC these measures cannot be imposed indefinitely, and must be justified by risk analysis within a reasonable timeframe. Risk analysis is a science-based process that provides the rationale for determining appropriate phytosanitary measures for a particular pest or commodity for a specified area. Pests can be either regulated or not and the risk analysis process assists with determining whether a pest fits either of these two categories and the strength of phytosanitary measures, if any, that should be taken in response to it.

3.5 Assigning Host Status to Plants Associated with *Drosophila suzukii*—A Regulatory Perspective

SWD is reported to infest around 30 families of host plants, with thin-skinned berries and stone fruits being particularly susceptible, both as wild hosts and as cultivated crops. Unlike other species of *Drosophila*, which infest overripe and damaged fruit, SWD shows a preference for ripening or ripe fruit (Asplen et al. 2015). In the original Japanese description of host biology, Kanzawa (1935) stated that the fly "prefers to infest and develop in slightly under ripe perfect fruit," whereas if the preferred host stages were unavailable, it was able to infest damaged or rotten fruit. SWD is able to exploit this unusual (for species of *Drosophila*) niche because it can penetrate fruit skin with its serrated ovipositor, which is lacking in nearly all other species in this very speciose genus.

Some fruits such as raspberries and strawberries appear to be preferred hosts, while others are unsuitable unless they are damaged or overripe, and SWD is not regarded as a significant pest of these crops (Asplen et al. 2015). Fruit penetration force is one potential measure of host susceptibility (although host attractiveness is also likely to depend on other factors, such as soluble sugar content or acidity), and SWD is unlikely to be able to penetrate fruit with very thick or tough skin (Burrack et al. 2013). However, if the skin is damaged (mechanically or by other means such as fungal infection, or if it breaks down after ripening), adults may be able to oviposit in the fruit. If it is a physiological host, the fruit may support development to emergence of adults. The reported host range on intact, undamaged fruit is thus much narrower than on all fruit regardless of its physical condition, though it may be difficult to tell from literature records whether intact fruit can be hosts.

Although host status is clearly a continuum in nature, in the regulatory world clear, defensible and consistent binary decisions need to be made through the process of risk analysis. According to ISPM 37 (2016) definitions, a non-host is one that "is not able to sustain development to viable adults under natural or semi-natural field conditions." For this category of commodities, there is no justification for requiring phytosanitary measures. Conversely, the imposition of measures is clearly justified for demonstrated natural hosts. In the case of SWD, damaged or overripe hosts may be infested under natural conditions, and may be able to sustain development of larvae to viable adults. However, infestation depends on the condition of the fruit: the skin must be damaged or softened by ripening or infection to allow oviposition to take place. ISPM 37 (2016) includes "*taking into consideration the conditions in which the commodity is known to be traded, such as physiological condition, cultivar and stage of maturity*" under general requirements for determining host status. Conditional non-host status in quarantine protocols can also specify absence of relevant injury (Heather and Hallman 2007). While risk analysis and consequent categorisation of hosts often makes the implicit (or explicit) assumption that fruit for export is ripe, healthy and commercially produced, it may be necessary for NPPOs to additionally specify that the skin of potential host commodities is intact and is free of splits and cuts.

An example that illustrates the relevance of this requirement is the recent detection by New Zealand border officials of SWD larvae in damaged imported orange (*Citrus sinensis*) fruit.[1] Intact orange fruit is generally not considered to be a host of SWD, as known reports are all from fallen and likely overripe fruit (Price and Nagle 2009; Plant Health Australia 2019). Commercially produced fruit is therefore considered to be very unlikely to host *Drosophila suzukii*, and consequently oranges imported into New Zealand do not currently require specific measures for SWD.[2] The larvae detected in imported oranges were alive and were subsequently reared to adults. This demonstrates that orange is a physiological host as the fly is able to complete its development.

3.6 Regulatory Action in Response to Global Spread of SWD

3.6.1 New Zealand

In response to the late 2008 detection in California, New Zealand's NPPO (then Ministry of Agriculture and Fisheries, now Ministry for Primary Industries) imposed emergency mitigation measures in May and June 2010 requiring cold treatment or methyl bromide fumigation of host fruit of SWD exported from the USA to New Zealand. New Zealand finalised a PRA for host material from the USA in 2012. Additional regulation is managed through systems such as the MPI Emerging Risk System (Reed 2014), a semi-automated system that monitors changes in distribution and host records of plant and animal pests and pathogens. Alerts received through this system allow, for example, the updating of import requirements from newly invaded countries, or the regulation of newly reported hosts.

MPI has a list of approved treatments for SWD that must be applied before export to manage the likely pathways of entry for hosts from non-Pest Free Areas. These include cold disinfestation, methyl bromide fumigation, SO_2/CO_2 fumigation AND cold disinfestation, and SO_2/CO_2 fumigation AND methyl bromide fumigation.

3.6.2 Australia

Biosecurity Australia (now the Department of Agriculture) introduced emergency measures on high-risk commercial fruit in 2010 in response to reports of SWD affecting commercial crops in western North America in March 2010. An

[1] https://www.stuff.co.nz/business/farming/111977242/fruit-fly-larvae-found-on-us-orange-imports-on-hold; https://www.kvh.org.nz/vdb/document/104784.

[2] https://www.mpi.govt.nz/dmsdocument/1147/send.

all-pathways Pest Risk Analysis identifying import pathways and proposing management options was finalised in 2013.

Management options proposed by the Australian NPPO for SWD include area freedom, or a systems approach for fruit, or application of either methyl bromide or SO_2/CO_2 fumigation (commodity dependent) followed by cold treatment.

3.6.3 *European Plant Protection Organisation (EPPO)*

SWD was put on the EPPO A2 quarantine pest list in 2011 and is currently still on it. EPPO A2 pests are locally present in the EPPO region, and the list is reviewed every year by the Working Party on Phytosanitary Regulations and approved by Council.

3.7 Conclusions

- SWD can move internationally on a limited number of pathways; therefore, regulating its movement in trade is relatively straightforward compared to, for example, managing hitchhiking pests.
- The key to controlling invasions is an understanding of biology and pathways and the ability to translate this into appropriate regulation that fits into the IPPC framework.
- The main regulatory problem is defining and predicting hosts, including conditional host status. A precautionary stance by countries for which SWD is a quarantine pest is acceptable, but only for a limited time.

References

Aluja M, Mangan RL (2008) Fruit fly (Diptera: Tephritidae) host status determination: critical conceptual, methodological, and regulatory considerations. Annu Rev Entomol 53(1):473–502

Amin ud Din M, Mazha K, Haque S et al (2005) A preliminary report on *Drosophila* fauna of Islamabad (Capital, Pakistan). DIS 88:6–7. http://www.ou.edu/journals/dis/DIS88/Research/R1.pdf. Accessed May 2020

Armstrong JW (1994) Commodity resistance to infestation by quarantine pests. In: Sharp JL, Hallman GJ (eds) Quarantine treatments for pests and food plants. Westview, Boulder, pp 199–211. 290 pp

Asplen MKG, Anfora A, Biondi A et al (2015) Invasion biology of spotted wing Drosophila (*Drosophila suzukii*): a global perspective and future priorities. J Pest Sci 88(3):469–494

Bolda M (2009) Update on the cherry vinegar fly, *Drosophila suzukii*, now known as the spotted wing Drosophila. Strawberries and Caneberries. https://ucanr.edu/blogs/blogcore/postdetail.cfm?postnum=1483

Burrack HJ, Fernandez GE, Spivey T et al (2013) Variation in selection and utilization of host crops in the field and laboratory by *Drosophila suzukii* Matsumura (Diptera: Drosophilidae), an invasive frugivore. Pest Manag Sci 69(10):1173–1180

Calabria G, Máca J, Bächli G et al (2010) First records of the potential pest species *Drosophila suzukii* (Diptera: Drosophilidae) in Europe. J Appl Entomol 136:139–147

Cini A, Ioratti C, Anfora G (2012) A review of the invasion of *Drosophila suzukii* in Europe and a draft research agenda for integrated pest management. Bull Insectology 65:149–160

DAFF (2013) Final pest risk analysis report for *Drosophila suzukii*. Department of Agriculture, Fisheries and Forestry Biosecurity, Canberra

Deprá M, Poppe JL, Schmitz HJ et al (2014) The first records of the invasive pest *Drosophila suzukii* in the south American continent. J Pest Sci 87:379–383

dos Santos LA, Mendes MF, Kruger AP et al (2017) Global potential distribution of *Drosophila suzukii* (Diptera, Drosophilidae). PLoS One 12(3):e0174318. https://doi.org/10.1371/journal.pone.0174318

EPPO (2020) *Drosophila suzukii* (DROSSU) reporting service articles. EPPO Global Database. https://gd.eppo.int/taxon/DROSSU/reporting

Follett PA, Neven LG (2006) Current trends in quarantine entomology. Annu Rev Entomol 51:359–385

Hauser M (2011) A historic account of the invasion of *Drosophila suzukii* (Matsumura) (Diptera: Drosophilidae) in the continental United States, with remarks on their identification. Pest Manag Sci 67(11):1352–1357

Heather NW, Hallman GJ (2007) Pest management and phytosanitary trade barriers. In: Pest management and phytosanitary trade barriers. CAB International, Oxfordshire, pp 1–13

IPPC (1997) International plant protection convention. Rome, IPPC, FAO. https://www.ippc.int/static/media/files/publications/en/2013/06/06/1329129099_ippc_2011-12-01_reformatted.pdf

IPPC (2015) Establishing a national plant protection organization: A guide to understand the principal requirements for establishing an organization to protect national plant resources from pests. Capacity Development 09/2015. Rome, IPPC, FAO. https://www.standardsfacility.org/sites/default/files/PG_350_Manual_Establishing_of_an_NPPO_guide_0.pdf

ISPM 37 (2016) Determination of host status of fruit to fruit flies (Tephritidae). IPPC, Food and Agricultural Organisation, Rome

Kai H, Zhang W-X, Carson HL (1993) The Drosophilidae (Diptera) of Hainan Island (China). Pac Sci 47(4):319–327

Kaneshiro KY (1983) Drosophila (Sophophora) suzukii (Matsumura). Proc Hawaii Entomol Soc 24:179. https://scholarspace.manoa.hawaii.edu/bitstream/10125/11150/24_2-3proc.pdf

Kanzawa T (1935) Research into the fruit-fly *Drosophila suzukii* Matsumura (preliminary report). Yamanashi Prefecture Agricultural Experiment Station Report

Kanzawa T (1939) Studies on *Drosophila suzukii* Mats. Kofu, Yamanashi Agricultural Experimental Station. 49 pp

Lasa R, Tadeo E (2015) Invasive drosophilid pests *Drosophila suzukii* and *Zaprionus indianus* (Diptera: Drosophilidae) in Veracruz. Mexico Fla Entomol 98:987–988

Lavrinienko A, Kesäniemi J, Watts PC et al (2017) First record of the invasive pest *Drosophila suzukii* in Ukraine indicates multiple sources of invasion. J Pest Sci 90(2):421–429

Lee TJ (1964) Taxonomy, and geographical distribution of Drosophilidae (Diptera) in Korea. Reprinted from the Chungang University Theses Collection vol. 9, pp 424–459

Lee JC, Dreves AJ, Cave AM et al (2015) Infestation of wild and ornamental noncrop fruits by *Drosophila suzukii* (Diptera: Drosophilidae). Ann Entomol Soc Am 108(2):117–129

Plant Health Australia (2019) Spotted wing drosophila host list. Hort innovation project MT17005: improving the biosecurity preparedness of Australian horticulture for the exotic spotted wing drosophila (*Drosophila suzukii*). https://www.horticulture.com.au/growers/help-your-business-grow/research-reports-publications-fact-sheets-and-more/mt17005/

Price JF, Nagle CA (2009) Spotted wing Drosophila new in Florida berry culture. IFAS Extension, University of Florida

Reed C (2014) The emerging risks system. Surveillance 41(2):3–4. Ministry for Primary Industries, Wellington, New Zealand

Singh BK, Bhatt M (1988) A preliminary report on the Drosophilidae of the Kumaun region with description of two new species and three new records. Orient Insects 22:147–161

Tamada T (2009) Current trends of blueberry culture in Japan. Acta Hort 810:109–116

Toda MJ (1991) Drosophilidae (Diptera) in Myanmar (Burma) VII. The *Drosophila melanogaster* species-group, excepting the *D. montium* species sub-group. Orient Insects 25:69–94

Vilela CR, Mori L (2014) The invasive spotted-wing Drosophila (Diptera, Drosophilidae) has been found in the city of Sao Paulo (Brazil). Rev Bras Entomol 58(4):371–375

Walsh D, Bolda M, Goodhue R et al (2011) *Drosophila suzukii* (Diptera: Drosophilidae): invasive pest of ripening soft fruit expanding its geographic range and damage potential. J Integr Pest Manag 106(2):289–295

Chapter 4
The Biology and Ecology of *Drosophila suzukii* (Diptera: Drosophilidae)

Daniel S. Kirschbaum, Claudia F. Funes, María J. Buonocore-Biancheri, Lorena Suárez, and Sergio M. Ovruski

Abstract Preference of egg laying in ripe (or in ripening process), healthy, soft, small fruits (berries and stone fruits) still attached to the plant makes *Drosophila suzukii* (spotted wing *Drosophila*, SWD) a severe, harmful pest, capable of producing huge food and economic losses to Asia, Europe, and America. The high degree of polyphagy shown by SWD either in their native region, Eastern and Southeastern Asia, or in the invaded continents, besides its seasonal phenotypic plasticity that allows it adaptations to adverse thermal periods, greatly favors SWD establishment and dispersion into new environments. In view of its high-damaging profile and rapid adaptability to settle in different world regions, several authors have focused on the study of biological and ecological features of this invasive pestiferous insect. In this regard, the chapter provides summarized information on life cycle, oviposition preference, larva feeding effects on fruits, crop and non-crop host plants world-

D. S. Kirschbaum (✉)
INTA Estación Experimental Agropecuaria Famaillá, Tucumán, Argentina

Universidad Nacional de Tucumán, Facultad de Agronomía y Zootecnia, Tucumán, Argentina
e-mail: kirschbaum.daniel@inta.gob.ar

C. F. Funes
INTA Estación Experimental Agropecuaria Famaillá, Tucumán, Argentina
e-mail: funes.claudia@inta.gob.ar

M. J. Buonocore-Biancheri
LIEMEN, División Control Biológico de Plagas, PROIMI Biotecnología, CCT NOA Sur-CONICET, Avda. Belgrano y Pje. Caseros, T4001MVB San Miguel de Tucumán, Tucumán, Argentina

L. Suárez
Dirección de Sanidad Vegetal, Animal y Alimentos (DSVAA), San Juan, Argentina

Programa de Control y Erradicación de Mosca de los Frutos de San Juan (ProCEM-San Juan), San Juan, Argentina
e-mail: ldcsuarez@sanjuan.gov.ar

S. M. Ovruski
LIEMEN, División Control Biológico de Plagas, PROIMI Biotecnología, CCT NOA Sur-CONICET, San Miguel de Tucumán, Tucumán, Argentina
e-mail: sovruski@conicet.gov.ar

F. R. M. Garcia (ed.), *Drosophila suzukii Management*,
https://doi.org/10.1007/978-3-030-62692-1_4

wide range, host preference, continental dispersal, seasonal biology, population dynamic, thermal susceptibility, reproductive behavior, interspecific and intergeneric interactions with other frugivorous dipterans, and relationships with natural enemies in native and introduced regions. A deep knowledge of the relationships between SWD and ecological components of newly invaded landscapes is critical for designing effective, environmental-friendly, SWD management strategies.

Keywords Spotted wing Drosophila · Biological invasions · Host range · Fruit pests · Seasonal biology · Natural enemies

4.1 Introduction

Drosophila suzukii (Matsumura) (Diptera: Drosophilidae), commonly known as the spotted wing *Drosophila* (SWD), is a major invasive global pest of small, soft, and stone fruits. Originally from Eastern and Southeastern Asia, SWD causes significant economic damage to fruit production in invaded regions of Europe and the American continent (Lee et al. 2019). From the earliest records on a specific invaded continent, SWD was characterized by its rapid range expansion within fruit-growing areas (Asplen et al. 2015). Based on this fact, and taking into account the very damaging profile of SWD, several authors have been giving special attention to biological and ecological features of SWD to understand its adaptation and dispersal strategies to invade and settle in different world regions. Thus, this chapter reviews over a large number of publications and provides summarized information on: (1) life cycle, oviposition preference, and feeding effects; (2) crop and non-crop host plants worldwide range, host preference and continental dispersal; (3) seasonal biology and population dynamics; (4) adult and immature stages thermal susceptibility; (5) mating behavior and sex ratio; (6) interspecific and intergeneric interactions with other frugivorous dipterans; and (7) relationships with natural enemies in native and introduced regions.

4.2 Life Cycle, Oviposition Preference, and Feeding Effects

Two exclusive features of SWD make it an economically dangerous pest: its preference for healthy, ripening fruit and the powerful, sclerosed, and serrated ovipositor of the female, which penetrates the fruit epidermis causing physical damage. Frequently, these oviposition wounds provide access to secondary fruit-feeding organisms, both insects (i.e., other frugivorous drosophilids) and pathogens (fungi and bacteria), which together cause additional losses (Walsh et al. 2011). Eggs develop into larvae within the fruit mesocarp, causing its rapid softening and rotting, bringing as consequence enormous yield losses, with significantly negative economic impact. Nevertheless, SWD may also lay eggs in wounded and/or fermenting fruit (Kienzle et al. 2020). The SWD may thrive well in fallen fruit, when the preferred fruit stages are unavailable or scarce (Bal et al. 2017; Kienzle et al.

2020). As a result, SWD individuals can emerge from eggs laid on berries either intact or injured (Mazzetto et al. 2020; Kienzle et al. 2020).

SWD females lay eggs with an erratic distribution pattern across fruits, which can be explained by random visits of females to fruits and the production of clutches of mostly single eggs (Mitsui et al. 2006; Schlesener et al. 2017). SWD females tend to oviposit at softer parts of the fruit, which imply that they examine fruits very carefully to locate the most suitable parts for egg laying (Kinjo et al. 2014). Tests carried out with different *Vaccinium* spp. cultivars showed more egg laying in softer fruit blueberry varieties than in firmer fruit varieties (Kinjo et al. 2013). A SWD female may lay 20–419 eggs in a lifetime, depending on the environmental conditions (Hamby et al. 2016). Once the larva hatched, it feeds inside fruit pulp. SWD larvae develop through three instars; the mature larva tends mainly to migrate out of the fruit to pupate in the soil (Woltz and Lee 2017; Lee et al. 2019). A total of 13–14 days is needed for the fly to develop from egg to adult at 22 °C (egg stage duration 1.4 day, larval stage 6 days, and pupal stage 6 days) (Tochen et al. 2014). Adults of both summer and winter morphotypes may live up to 30–179 days in the lab when provided food depending on temperature (Shearer et al. 2016; Rendon et al. 2019; Stockton et al. 2019). Females have a short pre-ovipositional period, during which they mostly feed. After first mate, 1- to 5-day-old females start laying eggs under standard lab conditions (Hamby et al. 2016).

4.3 Host Plants and Continental Dispersal

The SWD has a broad crop and non-crop host plants range, mainly throughout Asia, Europe, and America (Asplen et al. 2015), and with potential for adaptation and establishment in Oceania and Africa (Dos Santos et al. 2017). Soft-skinned and stone fruits such as berries and cherries are highly susceptible to infestation (Mitsui et al. 2010; Cini et al. 2012; Bellamy et al. 2013; Burrack et al. 2013; Lee et al. 2015). In addition to cultivated host species, the SWD can develop in both native and exotic fruit of wild and ornamental non-crop hosts. This occurs in various habitats, such as forests, forest edges, meadows, hedges of agricultural areas, urban gardens and parks, whereby this insect may switch among host plants and environments when ripe fruits are available (Lee et al. 2015; Asplen et al. 2015; Kenis et al. 2016). This feature, besides other biological factors, greatly facilitates its dispersion in invaded fruit-growing regions (Lee et al. 2011a, b, 2012; Cini et al. 2012, 2014; Kenis et al. 2016).

4.3.1 Asia

A total of 58 host plant species, belonging to 11 families, were recorded in Asia (Table 4.1). This complete list includes crop and non-crop fruit species from south-eastern Asia, including China, Japan, South Korea (the region where SWD is native), and Turkey. Four species were cited only at the generic level, such as *Prunus* sp., *Rubus* sp., *Morus* sp., and *Vaccinium* sp. About 16% of the host plant species are cultivated, while the majority are non-crop fruits, upon which SWD multiplies. Most of the SWD host plant species belongs to the Rosaceae (53.4%), in which 55% and 26% belong to *Prunus* L. and *Rubus* L., respectively. All of the host plant species cited for Asia had naturally SWD-infested fruits (Table 4.1). Additionally, two of them, *Prunus donarium* Siebold (Mitsui et al. 2006) and *P. avium* L. (Kasuya et al. 2013), were infested under lab conditions as well. The SWD has mainly caused economic damage in cherry (*Prunus tomentosa* Thunb. and *P. avium*), blueberry (*Vaccinium* spp.), wax-myrtle (*Myrica rubra* Sieb. Et Zucc.), and autumn olive (*E. umbellate* Thunberg) crops (Asplen et al. 2015).

4.3.2 Europe

From the first report of *D. suzukii* in Europe during 2008 in both Spain (Calabria et al. 2012) and Italy (Cini et al. 2012), the SWD widely spread in 6 years across the continent (Asplen et al. 2015). Severe damage in several small fruit crops, such as sweet cherry (*Prunus avium*), sour cherry (*Prunus cerasus* L.), strawberry (*Fragaria x ananassa* Duch.), raspberry (*Rubus idaeus* L.), blackberry (*Rubus fruticosus* aggr.), and blueberry (*Vaccinium* spp.), was mainly recorded at numerous European locations (Cini et al. 2012; Weydert and Mandrin 2013; Asplen et al. 2015; Weydert et al. 2016; Mazzi et al. 2017). SWD infestations have also been reported in cultivated peach (*Prunus persica* (L.) Stokes), apricot (*Prunus armeniaca* L.), plum (*Prunus domestica* L.), apple (*Malus domestica* (L.) Borkh.), and fig (*Ficus carica* L.), although without economically significant damage (Grassi et al. 2011; Weydert and Mandrin 2013; Asplen et al. 2015; Kenis et al. 2016). Similarly, infestations in grape (*Vitis vinefera* L.) varieties were recorded in both Italy (Grassi et al. 2011; Cini et al. 2012; Kenis et al. 2016) and Germany (Asplen et al. 2015). Although grape cultivars with soft-skinned berries are more susceptible for egg laying (Mazzetto et al. 2020), successful SWD adult development is limited (Asplen et al. 2015). Late-ripening grape varieties were mainly infested by SWD in Italy, probably because of an increase in the number of flies and/or the reduction of alternative hosts (Mazzetto et al. 2020). In fact, surrounding wild vegetation suitable for SWD oviposition increased the capture of flies inside vineyards as well as in other crops (Poyet et al. 2015; Kenis et al. 2016; Mazzetto et al. 2020).

In Europe, 126 species of crop and non-crops fruits belonging to 27 families were recorded as SWD hosts (Table 4.2). From a total of SWD host species cited in Table 4.2, 91 were found producing natural infestations, while the remaining 35 were infested only under lab conditions. However, in 13 of these lab-infested fruit species, *D. suzukii* laid eggs, but there was no adult emergence (Table 4.2). The highest percentage of host species (41%) belongs to the Rosaceae family, with *Prunus* as the genus with more host species (24%), followed by *Cotoneaster* Medik. (14%) and *Rubus* (12%). Caprifoliaceae, Ericaceae, Solanaceae, Adoxaceae, and Cornaceae also have numerous SWD host species (Table 4.2). Several non-crop host plants such as *Rubus* spp., *Sambucus* spp., *Prunus* spp., *Lonicera* spp., *Arbutus*

Table 4.1 Crop and non-crop host species of *Drosophila suzukii* recorded in Asia

Plant family	Plant species	Host plant status		Fruit-collection country[a]	Infestation type		Source
		Crop	Non-crop		Nat.	Lab.	
Adoxaceae	*Sambucus adnata* Wallich ex Candolle	–	Yes	CN	Yes	–	Giorgini et al. (2019)
	Sambucus williamsii Hance	–	Yes	CN	Yes	–	Girod et al. (2018a, b, c)
	Viburnum dilatatum Thunberg	–	Yes	JP	Yes	–	Mitsui et al. (2010), Lee et al. (2015)
Caprifoliaceae	*Lonicera maacki* (Rupr.) Herder	–	Yes	CN	Yes	–	Girod et al. (2018a, b, c)
Coriariaceae	*Coriaria nepalensis* Wall.	–	Yes	CN	Yes	–	Girod et al. (2018a, b, c)
Cornaceae	*Alangium platanifolium* (Sieb. et Zucc.)	–	Yes	JP	Yes	–	Mitsui et al. (2010), Lee et al. (2015)
	Aucuba japónica Thunberg	–	Yes	JP	Yes	–	Mitsui et al. (2010), Lee et al. (2015)
	Cornus controversa Hemsl. ex Prain	–	Yes	JP	Yes	–	Mitsui et al. (2010), Lee et al. (2015)
	Cornus kousa F. Buerger ex Miq.	–	Yes	JP	Yes	–	Mitsui et al. (2010), Lee et al. (2015)
Ebenaceae	*Diospyros kaki* Thunberg	–	Yes	JP	Yes	–	Mitsui et al. (2010), Lee et al. (2015)
Elaeagnaceae	*Elaeagnus multiflora* Thunberg	–	Yes	JP	Yes	–	Sasaki and Sato (1995), Lee et al. (2015)
	Elaeagnus umbellate Thunberg	Yes	–	KR	Yes	–	Asplen et al. (2015)
Ericaceae	*Gaultheria adenothrix* (Miq.) Maximovich	–	Yes	JP	Yes	–	Mitsui et al. (2010), Lee et al. (2015)
	Vaccinium smallii A. Gray	–	Yes	JP	Yes	–	Kasuya et al. (2013)
	Vaccinium ovalifolium Sm.	–	Yes	JP	Yes	–	Kasuya et al. (2013)
	Vaccinium sp.	Yes	–	CN	Yes	–	Asplen et al. (2015)
		–	Yes	JP	Yes	–	Sasaki and Sato (1995), Mitsui et al. (2010), Girod et al. (2018a, b, c)
Moraceae	*Morus alba* L.	–	Yes	JP	Yes	–	Lee et al. (2015)
	Morus australis Poiret (= *bombycis*)	–	Yes	JP	Yes	–	Lee et al. (2015)
	Morus bombycis Koidz	–	Yes	JP	Yes	–	Sasaki and Sato (1995), Mitsui et al. (2010)
	Morus sp.	–	Yes	JP	Yes	–	Lee et al. (2015), Girod et al. (2018a, b, c)

(continued)

Table 4.1 (continued)

Plant family	Plant species	Host plant status		Fruit-collection country[a]	Infestation type		Source
		Crop	Non-crop		Nat.	Lab.	
Myricaceae	*Myrica rubra* Sieb. Et Zucc.	Yes	–	JP	Yes	–	Asplen et al. (2015)
		–	Yes	CN	Yes	–	Girod et al. (2018a, b, c)
Phytolaccaceae	*Phytolacca americana* L.	–	Yes	JP	Yes	–	Sasaki and Sato (1995), Lee et al. (2015)
Rosaceae	*Cerasus jamasakura* (Siebold ex Koidz.)	–	Yes	JP	Yes	–	Kasuya et al. (2013)
	Eriobotrya japónica (Thunbergerg) Lindley	–	Yes	JP	Yes	–	Lee et al. (2015)
	Fragaria x *ananassa* Duch.	Yes	–	TR	Yes	–	Orhan et al. (2016); Efil (2018)
	Fragaria moupinensis Cardot	–	Yes	CN	Yes	–	Giorgini et al. (2019)
	Malus domestica (Borkh.)	Yes	–	KR	Yes	–	Daane et al. (2016)
	Malus pumila Miller	–	Yes	JP	Yes	–	Lee et al. (2015)
	Princepia utilis Royle	–	Yes	CN	Yes	–	Girod et al. (2018a, b, c)
	Prunus armeniaca L.	–	Yes	JP	Yes	–	Lee et al. (2015)
	Prunus avium L.	Yes	–	JP	Yes	–	Mitsui and Kimura (2010)
		–	Yes	JP	–	Yes	Kasuya et al. (2013)
		–	Yes	JP	Yes	–	Sasaki and Sato (1995), Lee et al. (2015)
		Yes	–	CN	Yes	–	Asplen et al. (2015)
	Prunus buergeriana Miquel	–	Yes	JP	Yes	–	Sasaki and Sato (1995), Lee et al. (2015)
	Prunus cerasoides BuCN-Ham. Ex d. Don	–	Yes	CN	Yes	–	Girod et al. (2018a, b, c)
	Prunus cerasus L.	–	Yes	JP	Yes	–	Lee et al. (2015)
	Prunus donarium Siebold	–	Yes	JP	Yes	–	Kasuya et al. (2013), Lee et al. (2015)
		–	Yes	JP	–	Yes	Mitsui et al. (2006)
	Prunus japónica Thunberg	–	Yes	JP	Yes	–	Lee et al. (2015)
	Prunus mahaleb L.	–	Yes	JP	Yes	–	Lee et al. (2015)

	Prunus nipponica Matsumura	–	Yes	JP	Yes	–	Mitsui et al. (2010), Lee et al. (2015)
	Prunus padus L.	–	Yes	JP	Yes	–	Girod et al. (2018a, b, c)
	Prunus persica (L.) Batsch	–	Yes	JP	Yes	–	Sasaki and Sato (1995), Lee et al. (2015)
	Prunus salicina Lindley *(= P. triflora)*	–	Yes	JP	Yes	–	Lee et al. (2015)
	Prunus sargentii Rehder	–	Yes	JP	Yes	–	Lee et al. (2015)
	Prunus serrulata Lindl.	–	Yes	JP	Yes	–	Girod et al. (2018a, b, c)
	Prunus sp.	–	Yes	JP	Yes	–	Mitsui et al. (2010)
	Prunus tomentosa Thunb.	Yes	–	KR	Yes	–	Asplen et al. (2015)
	Prunus yedoensis Matsumura	–	Yes	JP	Yes	–	Sasaki and Sato (1995), Lee et al. (2015)
	Rubus crataegifolius Bunge	–	Yes	JP	Yes	–	Mitsui et al. (2010), Lee et al. (2015)
	Rubus ellipticus Sm.	–	Yes	CN	Yes	–	Girod et al. (2018a, b, c)
	Rubus foliosus Weihe	–	Yes	CN	Yes	–	Giorgini et al. (2019)
	Rubus microphyllus L.f.	–	Yes	JP	Yes	–	Mitsui et al. (2010), Lee et al. (2015)
	Rubus niveus Thunb.	–	Yes	CN	Yes	–	Giorgini et al. (2019)
	Rubus parvifolius L. *(= triphyllus)*	–	Yes	JP	Yes	–	Sasaki and Sato (1995), Lee et al. (2015)
	Rubus sp.	–	Yes	CN	Yes	–	Girod et al. (2018a, b, c)
		Yes	–	KR	Yes	–	Daane et al. (2016)
	Rubus ulmifolius Schott	Yes	–	KR	Yes	–	Daane et al. (2016)
Solanaceae	*Solanum lycopersicum* L.	–	Yes	JP	Yes	–	Lee et al. (2015)
	Solanum nigrum L.	–	Yes	CN	Yes	–	Girod et al. (2018a, b, c)
Styracaceae	*Styrax japonicas* Siebold & Zuccarini	–	Yes	JP	Yes	–	Mitsui and Kimura (2010)
Taxaceae	*Torreya nucifera* (L.) Siebold & Zuccarini	–	Yes	JP	Yes	–	Mitsui et al. (2010), Lee et al. (2015)
Vitaceae	*Vitis vinifera* L.	Yes	–	KR	Yes	–	Asplen et al. (2015)

[a]ISO Country Codes: CN = China, JP = Japan, KR = Republic of Korea (South Korea), TR = Turkey

Table 4.2 Crop and non-crop host species of *Drosophila suzukii* recorded in Europe

Plant family	Plant species	Host plant status		Fruit-collection country[a]	Infestation type		Source
		Crop	Non-crop		Nat.	Lab.	
Actinidiaceae	*Actinidia arguta* (Siebold & Zucc.) Planch. ex Miq.	Yes	–	AT	Yes	–	Asplen et al. (2015)
	Actinidia chinensis Planch.	Yes	–	IT	Yes	–	Kenis et al. (2016)
Adoxaceae	*Sambucus ebulus* L.	–	Yes	IT	Yes	–	Kenis et al. (2016)
		–	Yes	FR	–	Yes	Poyet et al. (2015)
	Sambucus nigra L.	–	Yes	IT	Yes	–	Grassi et al. (2011), Kenis et al. (2016)
		–	Yes	ES	Yes	Yes	Arnó et al. (2016)
		–	Yes	NL, CH	Yes	–	Kenis et al. (2016)
		–	Yes	FR	–	Yes	Poyet et al. (2015)
	Sambucus racemosa L.	–	Yes	IT, NL, CH	Yes	–	Kenis et al. (2016)
		?	?	CH	–	Yes	Baroffio et al. (2014)
	Sambucus spp.	Yes	Yes	AT, DE, NL	Yes	–	Asplen et al. (2015)
	Viburnum lantana L.	–	Yes	IT	Yes	–	Kenis et al. (2016)
		?	?	CH	–	Yes	Baroffio et al. (2014)
	Viburnum opulus L.	?	?	CH	–	Yes	Baroffio et al. (2014)
		–	Yes	FR	–	Yes[b]	Poyet et al. (2015)
	Viburnum rhytidophyllum Hemsl.	–	Yes	NL	Yes	–	Kenis et al. (2016)
Aquifoliaceae	*Ilex aquifolium* L.	–	Yes	FR	–	Yes[b]	Poyet et al. (2015)
Araceae	*Arum italicum* Mill.	–	Yes	NL	Yes	–	Kenis et al. (2016)
	Arum maculatum L.	–	Yes	FR	–	Yes	Poyet et al. (2015)
		?	?	CH	–	Yes	Baroffio et al. (2014)
Araliaceae	*Hedera hélix* L.	–	Yes	FR	–	Yes[b]	Poyet et al. (2015)

Asparagaceae	*Asparagus officinalis* L.	–	Yes	FR	–	Yes[b]	Poyet et al. (2015)
	Ruscus aculeatus L.	–	Yes	FR	–	Yes[b]	Poyet et al. (2015)
Berberidaceae	*Mahonia aquifolium* (Pursh) Nutt.	–	Yes	NL	–	Yes	Kenis et al. (2016)
		–	Yes	FR	–	Yes	Poyet et al. (2015)
	Mahonia x media Brickell	–	Yes	FR	–	Yes	Poyet et al. (2015)
	Mahonia ES	–	Yes	IT	Yes	–	Kenis et al. (2016)
Caprifoliaceae	*Bryonia cretica* (Jacq.)	–	Yes	ES	–	Yes	Arnó et al. (2016)
	Lonicera alpigena L.	–	Yes	IT	Yes	–	Kenis et al. (2016)
		?	?	CH	–	Yes	Baroffio et al. (2014)
	Lonicera caerulea L.	–	Yes	IT	Yes	–	Kenis et al. (2016)
	Lonicera caprifolium L.	–	Yes	NL	Yes	–	Kenis et al. (2016)
	Lonicera ferdinandii Franch.	–	Yes	NL	Yes	–	Kenis et al. (2016)
	Lonicera nigra L.	–	Yes	IT	Yes	–	Kenis et al. (2016)
	Lonicera nitida E. H. Wilson	–	Yes	NL	–	Yes	Kenis et al. (2016)
		–	Yes	FR	–	Yes	Poyet et al. (2015)
	Lonicera periclymenum L.	?	?	CH	–	Yes	Baroffio et al. (2014)
	Lonicera ES	–	Yes	IT	Yes	–	Grassi et al. (2011)
		?	?	NL	Yes	–	Kenis et al. (2016)
	Lonicera xylosteum L.	–	Yes	IT	Yes	–	Kenis et al. (2016)
		–	Yes	FR	–	Yes	Poyet et al. (2015)
		?	?	CH	–	Yes	Baroffio et al. (2014)
	Symphoricarpos albus (L.)	–	Yes	NL	Yes	–	Kenis et al. (2016)
		–	Yes	FR	–	Yes	Poyet et al. (2015)
	Symphoricarpos x chenaultii Rehder	–	Yes	FR	–	Yes[b]	Poyet et al. (2015)

(continued)

Table 4.2 (continued)

Plant family	Plant species	Host plant status		Fruit-collection country[a]	Infestation type		Source
		Crop	Non-crop		Nat.	Lab.	
Cornaceae	*Cornus alba* L.	–	Yes	NL	Yes	–	Kenis et al. (2016)
	Cornus kousa Hance	–	Yes	NL	Yes	–	Kenis et al. (2016)
	Cornus mas L	–	Yes	IT, NL, CH	Yes	–	Kenis et al. (2016)
		?	?	CH	–	Yes	Baroffio et al. (2014)
	Cornus sanguinea L.	–	Yes	IT, NL,	Yes	–	Kenis et al. (2016)
		?	?	CH	–	Yes	Baroffio et al. (2014)
		–	Yes	FR	–	Yes[b]	Poyet et al. (2015)
	Cornus sericea L.	–	Yes	FR	–	Yes	Poyet et al. (2015)
Dioscoreaceae	*Tamus communis* L.	–	Yes	IT, CH	Yes	–	Kenis et al. (2016)
Elaeagnaceae	*Elaeagnus* x *ebbingei* (hybrid)	–	Yes	NL	Yes	–	Panel et al. (2018)
	Hippophae rhamnoides L.	Yes	Yes	CH	Yes	Yes	Kenis et al. (2016), Baroffio et al. (2014)
		–	Yes	FR	–	Yes	Poyet et al. (2015)
Ericaceae	*Arbutus unedo* L.	–	Yes	ES	Yes	–	Gabarra et al. (2012), Arnó et al. (2012)
		–	Yes	IT	Yes	–	Kenis et al. (2016)
	Gaultheria procumbes L.	–	Yes	FR	–	Yes[b]	Poyet et al. (2015)
	Gaultheria x wisleyensis M. & M.	–	Yes	NL	Yes	–	Kenis et al. (2016)
	Vaccinium corymbosum L.	Yes	–	IT	Yes	–	Grassi et al. (2011)
	Vaccinium myrtilloides Michx.	Yes	Yes	NL	Yes	–	Kenis et al. (2016)
	Vaccinium myrtillus L.	Yes	Yes	IT	Yes	Yes	Grassi et al. (2009, 2011), Kenis et al. (2016)
		Yes	Yes	CH	Yes	–	Kenis et al. (2016)
	Vaccinium oldhamii Miquel.	Yes	Yes	NL	Yes	–	Kenis et al. (2016)
	Vaccinium praestans Lamb.	–	Yes	NL	Yes	–	Kenis et al. (2016)
	Vaccinium uliginosum L.	?	?	FR	–	Yes	Poyet et al. (2015)
	Vaccinium vitis-idea L.	–	Yes	NL	Yes	–	Kenis et al. (2016)

Garriaceae	*Aucuba japonica* Thunb.	–	Yes	FR	–	Yes	Poyet et al. (2015)
Liliaceae	*Polygonatum multiflorum* (L.) All	–	Yes	NL, CH	Yes	–	Kenis et al. (2016)
Melanthiaceae	*Paris quadrifolia* L.	—	Yes	IT, CH	Yes	–	Kenis et al. (2016)
Moraceae	*Ficus carica* L.	Yes	–	IT	Yes	–	Grassi et al. (2011), Kenis et al. (2016)
		Yes	–	CH	Yes	–	Kenis et al. (2016)
	Morus nigra L.	–	Yes	IT	–	Yes	Grassi et al. (2011)
	Morus ES	–	Yes	FR	–	Yes	Poyet et al. (2015)
Onagraceae	*Fuchsia* ES	?	?	FR	–	Yes	Poyet et al. (2015)
Phytolaccaceae	*Phytolacca americana* L.	–	Yes	IT, NL, CH	Yes	–	Kenis et al. (2016)
		–	Yes	FR	–	Yes	Poyet et al. (2015)
		?	?	CH	–	Yes	Baroffio et al. (2014)
	Phytolacca esculenta Van Houtte	–	Yes	NL	Yes	–	Kenis et al. (2016)
Rhamnaceae.	*Frangula alnus* Mill.	–	Yes	IT	Yes	–	Grassi et al. (2009), Kenis et al. (2016)
		–	Yes	NL	Yes	–	Kenis et al. (2016)
		–	Yes	CH	Yes	–	Kenis et al. (2016)
		–	Yes	FR	–	Yes	Poyet et al. (2015)
	Rhamnus cathartica L	–	Yes	NL	Yes	–	Kenis et al. (2016)
		–	Yes	FR	–	Yes[b]	Poyet et al. (2015)
Rosaceae	*Amelanchier lamarckii* F.G. Schr.	–	Yes	NL	Yes	–	Kenis et al. (2016)
	Amelanchier ovalis Medik.	–	Yes	IT	Yes	–	Kenis et al. (2016)
	Cotoneaster bullatus Boiss.	?	?	FR	–	Yes	Poyet et al. (2015)
	Cotoneaster franchetii Boiss.	–	Yes	NL	Yes	–	Kenis et al. (2016)
	Cotoneaster horizontalis DecNL	?	?	CH	–	Yes	Baroffio et al. (2014)
		–	Yes	FR	–	Yes[b]	Poyet et al. (2015)
	Cotoneaster lacteus W.W. Smith	–	Yes	IT	Yes	–	Kenis et al. (2016)
	Cotoneaster rehderi Pojark.	–	Yes	NL	Yes	–	Kenis et al. (2016)
	Cotoneaster salicifolius Franch.	–	Yes	FR	–	Yes[b]	Poyet et al. (2015)

(continued)

Table 4.2 (continued)

Plant family	Plant species	Host plant status		Fruit-collection country[a]	Infestation type		Source
		Crop	Non-crop		Nat.	Lab.	
	Cotoneaster x *watereri* Exell.	?	?	FR	–	Yes	Poyet et al. (2015)
	Crataegus chrysocarpa Ashe	–	Yes	NL	Yes	–	Kenis et al. (2016)
	Crataegus monogyna Jacq.	–	Yes	NL	Yes	–	Kenis et al. (2016)
		–	Yes	FR	–	Yes[b]	Poyet et al. (2015)
	Duchesnea indica (Andr.) Focke	–	Yes	IT, NL, CH	–	Yes	Kenis et al. (2016)
		?	?	FR	–	Yes	Poyet et al. (2015)
	Eriobotrya japonica (Thunb.) Lindl.	–	Yes	IT	Yes	–	Kenis et al. (2016)
	Fragaria x ananassa Duch.	Yes	–	IT	Yes	–	Grassi et al. (2011)
		Yes	–	ES	Yes	Yes	Sarto and Sorribas (2011), Gabarra et al. (2012), Arnó et al. (2016)
		Yes	–	FR	Yes	–	Weydert and Mandrin (2013)
	Fragaria vesca L.	–	Yes	IT, NL, CH	Yes	–	Kenis et al. (2016)
		–	Yes	FR	–	Yes	Poyet et al. (2015)
		?	?	CH	–	Yes	Baroffio et al. (2014)
		–	Yes	DE	–	Yes	Gong et al. (2016)
	Malus baccata (L.) Borkh.	–	Yes	NL	Yes	–	Kenis et al. (2016)
	Malus domestica Borkh.	Yes	–	FR	Yes	–	Weydert and Mandrin (2013)
	Photinia beauverdiana C. K. Schn.	–	Yes	NL	Yes	–	Kenis et al. (2016)
	Photinia villosa (Thunb.) DC.	–	Yes	NL	Yes	–	Kenis et al. (2016)

	Photinia prunifolia Lindl.	–	Yes	NL	Yes	–	Kenis et al. (2016)
	Prunus armeniaca L.	Yes	–	IT	Yes	–	Grassi et al. (2011), Kenis et al. (2016)
		Yes	–	FR	Yes		Weydert and Mandrin (2013)
	Prunus avium L.	Yes	Yes	IT	Yes	–	Grassi et al. (2011), Cini et al. (2012)
		Yes	Yes	NL, CH	Yes	–	Kenis et al. (2016)
		Yes	Yes	FR	Yes	Yes	Weydert and Mandrin (2013), Poyet et al. (2015), Weydert et al. (2016)
		Yes	–	CH	Yes	–	Mazzi et al. (2017)
	Prunus cerasifera Ehrh	–	Yes	IT	Yes	–	Kenis et al. (2016)
	Prunus cerasus L.	Yes	Yes	IT	Yes	–	Kenis et al. (2016)
		Yes	–	DE	Yes	–	Asplen et al. (2015)
	Prunus domestica L.	Yes	–	CH	Yes	–	Kenis et al. (2016)
		Yes	–	HU	Yes	–	Asplen et al. (2015)
	Prunus laurocerasus L.	–	Yes	IT	–	Yes	Grassi et al. (2011)
		–	Yes	IT, NL, CH	Yes	–	Kenis et al. (2016)
	Prunus lusitanica L.	–	Yes	IT	Yes	–	Kenis et al. (2016)
			Yes	FR	–	Yes	Poyet et al. (2015)

(continued)

Table 4.2 (continued)

Plant family	Plant species	Host plant status		Fruit-collection country[a]	Infestation type		Source
		Crop	Non-crop		Nat.	Lab.	
	Prunus mahaleb L.	–	Yes	ES	Yes	–	Arnó et al. (2016)
		–	Yes	IT	Yes	–	Kenis et al. (2016)
		–	Yes	FR	–	Yes	Poyet et al. (2015)
	Prunus padus L.	–	Yes	NL, CH	–	Yes	Kenis et al. (2016)
		–	Yes	FR	–	Yes[b]	Poyet et al. (2015)
	Prunus persica (L.) Stokes	Yes	–	FR	Yes	–	Weydert and Mandrin (2013)
		Yes	–	HU	Yes	–	Asplen et al. (2015)
	Prunus serotina Ehrh.	–	Yes	FR	Yes	Yes	Poyet et al. (2013, 2015)
		–	Yes	NL	Yes	–	Kenis et al. (2016)
	Prunus spinosa L.	–	Yes	IT, NL, CH	–	Yes	Kenis et al. (2016)
		–	Yes	FR	–	Yes	Poyet et al. (2015)
		–	Yes	DE	Yes	–	Asplen et al. (2015)
	Pyracantha coccinea M. Roem.	–	Yes	FR	–	Yes[b]	Poyet et al. (2015)
	Pyracantha ES	–	Yes	NL	Yes	–	Kenis et al. (2016)
	Pyrus calleryana Decne	–	Yes	FR	–	Yes	Poyet et al. (2015)
	Ribes nigrum L.	–	Yes	FR	–	Yes[b]	Poyet et al. (2015)
	Ribes rubrum L.	Yes	–	NL, CH	–	Yes	Kenis et al. (2016)
		–	Yes	FR	–	Yes	Poyet et al. (2015)
	Ribes sanginium Pursh.	–	Yes	FR	–	Yes	Poyet et al. (2015)
	Rosa acicularis Lindl.	–	Yes	NL	Yes	–	Kenis et al. (2016)

	Rosa canina L.	–	Yes	ES	–	Yes	Arnó et al. (2016)
		–	Yes	NL	Yes	–	Kenis et al. (2016)
	Rosa glauca Pourr.	–	Yes	NL	Yes	–	Kenis et al. (2016)
	Rosa pimpinellifolia L.	–	Yes	NL	Yes	–	Kenis et al. (2016)
	Rosa rugosa Thunb.	–	Yes	NL	Yes	–	Kenis et al. (2016)
	Rubus caesius L.	–	Yes	IT, NL	Yes	–	Kenis et al. (2016)
	Rubus fruticosus aggr.	Yes	Yes	IT, NL, CH	Yes	–	Cini et al. (2012), Kenis et al. (2016)
		Yes	Yes	FR	Yes	Yes	Weydert and Mandrin (2013), Poyet et al. (2015), Weydert et al. (2016)
		Yes	–	AT, DE	Yes	–	Asplen et al. (2015)
	Rubus idaeus L.	Yes	Yes	IT	Yes	–	Grassi et al. (2011), Kenis et al. (2016)
		Yes	–	ES	Yes	–	Arnó et al. (2012, 2016)
		Yes	Yes	NL, CH	Yes	–	Kenis et al. (2016)
		Yes	Yes	FR	Yes	Yes	Weydert and Mandrin (2013), Poyet et al. (2015), Weydert et al. (2016)
		?	?	H	–	Yes	Baroffio et al. (2014)
		Yes	–	AT, DE, HU	Yes	–	Asplen et al. (2015)
	Rubus phoenicolasius Maxim.	Yes	–	CH	Yes	–	Kenis et al. (2016)
	Rubus saxatilis L.	–	Yes	IT	Yes	–	Kenis et al. (2016)
	Rubus ulmifolius Schott	–	Yes	ES	Yes	–	Arnó et al. (2016)
	Sorbus aria (L.)	–	Yes	IT	Yes	–	Kenis et al. (2016)
	Sorbus aucuparia L.	–	Yes	NL	Yes	–	Kenis et al. (2016)
Rubiaceae	*Rubia tinctorum* L.	–	Yes	FR	–	Yes[b]	Poyet et al. (2015)
Santalaceae	*Viscum album* L.	–	Yes	FR	–	Yes	Poyet et al. (2015)

(continued)

Table 4.2 (continued)

Plant family	Plant species	Host plant status		Fruit-collection country[a]	Infestation type		Source
		Crop	Non-crop		Nat.	Lab.	
Solanaceae	*Atropa belladona* L.	–	Yes	FR	–	Yes	Poyet et al. (2015)
	Lycium barbarum L.	Yes	Yes	IT	Yes	–	Kenis et al. (2016)
	Physalis alkekengi L.	–	Yes	FR	–	Yes	Poyet et al. (2015)
	Solanum chenopodioides Lam.	–	Yes	ES	–	Yes	Arnó et al. (2016)
	Solanum dulcamara L.	– – –	Yes Yes –	ES NL, CH FR	– Yes –	Yes – Yes	Arnó et al. (2016) Kenis et al. (2016) Poyet et al. (2015)
	Solanum dulcamara L. *f. littorale*	–	Yes	FR	–	Yes	Poyet et al. (2015)
	Solanum nigrum L	– – – ?	Yes Yes Yes ?	ES NL, CH FR CH	– Yes – –	Yes – Yes Yes	Arnó et al. (2016) Kenis et al. (2016) Poyet et al. (2015) Baroffio et al. (2014)
	Solanum villosum Mill. (= *S. luteum*)	–	Yes	ES	Yes	–	Gabarra et al. (2012), Arnó et al. (2012)
Taxaceae	*Taxus baccata* L	– – ?	Yes Yes ?	IT, NL, CH FR CH	Yes – –	– Yes Yes	Kenis et al. (2016) Poyet et al. (2015) Baroffio et al. (2014)
Thymelaeaceae	*Daphne mezereum* L.	–	Yes	IT	Yes	–	Kenis et al. (2016)
Vitaceae	*Parthenocissus inserta* (A. Kern.) FrIT	–	Yes	FR	–	Yes[a]	Poyet et al. (2015)
	Parthenocissus quinquefolia (L.)	–	Yes	IT, CH	Yes	Yes	Grassi et al. (2011), Cini et al. (2012), Baroffio et al. (2014)
	Vitis vinifera L.	Yes Yes	– –	IT DE	Yes Yes	Yes –	Grassi et al. (2011), Cini et al. (2012), Kenis et al. (2016) Asplen et al. (2015)

[a]AT =Austria, CH = Switzerland, DE = Germany, ES = Spain, FR = France, HU = Hungary, IT = Italy, NL = Netherlands
[b]Fruit with *D. suzukii* eggs but there was no adult emergence

unedo L., and *Frangula alnus* Mill. (Table 4.2) have been constantly found throughout Europe with high natural infestation rates, and SWD switch between these hosts depending on the fruit seasonality (Arnó et al. 2012, 2016; Asplen et al. 2015; Poyet et al. 2015). Those host plants could be potential resources for SWD population increase in Europe (Asplen et al. 2015; Arnó et al. 2016; Kenis et al. 2016). Green fruits of the wild host *Rubus ulmifolius* Schott collected at an altitude of 1273 m in Spain were highly infested (Arnó et al. 2016), which shows that SWD can be found within a broad range of altitude.

4.3.3 North America

SWD was first recorded on berry crops in California (USA) in 2008 and from that year it was dispersed through the soft and stone fruits-producing regions of mainland United States and Canada (Table 4.3) (Asplen et al. 2015). The most important economically affected crops in these countries include blueberries, raspberries, blackberries, strawberries, and cherries (Lee et al. 2011a, b; Bellamy et al. 2013). In 2011, SWD was detected in traps in Mexico (García-Cancino et al. 2015), but then it was found infesting *Psidium guajava* L. (Myrtaceae) crops (Lasa et al. 2017). A total of 85 crop and non-crop fruit species belonging to 21 families were recorded as SWD hosts in North America (Canada, United States, and Mexico) (Table 4.3). Four host plant species were only recorded at the generic level, e.g., *Sambucus* sp., *Lonicera* sp., *Morus* sp., and *Rubus* sp. The 93% of recorded SWD host plant species was naturally infested, while six fruit species were infested only under lab conditions. Most SWD host species are non-crop fruits, while 23% are cultivated species. The main SWD host plant family is Rosaceae, which includes 27% of the total fruit species recorded as hosts. Both *Prunus* and *Rubus* species represented 39% and 31%, respectively, of all Rosaceae species. Other important plant families with numerous SWD host species are Caprifoliaceae, Cornaceae, Adoxaceae, and Ericaceae (Table 4.3).

4.3.4 South America

SWD was first recorded in Brazil in 2013 (Deprá et al. 2014), and then in other South American countries, such as Uruguay (Gonzales et al. 2015), Argentina (Santadino et al. 2015), and Chile (Morales 2020). In 7 years, 31 crop and non-crop fruit species belonging to 10 families were recorded as SWD hosts in South America (Table 4.4). Two host plant species were only recorded at the generic level, such as *Butia* sp. and *Rubus* sp. All host fruit species recorded in Table 4.4 were naturally infested by SWD. Most SWD host species (58%) are crop fruits of economic importance. Damage caused by SWD in *Rubus idaeus* L., *R. fruticosus, R. ulmifolius* Schott, *Vaccinium corymbosum* L., *V. myrtillus* L., and *Fragaria* x *ananassa* has

Table 4.3 Crop and non-crop host species of *Drosophila suzukii* recorded in North America

Plant family	Plant species	Host plant status		Fruit-collection	Infestation type		Source
		Crop	Non-crop	US state or country[a]	Nat.	Lab.	
Adoxaceae	*Sambucus* sp.	–	Yes	US-Mi	Yes	–	Lee et al. (2015)
	Sambucus canadensis L.	Yes	–	US-Mn	Yes	–	Sward (2017)
		–	Yes	US-Sd	Yes	–	Agbaba (2017)
	Sambucus nigra L.	–	Yes	US-Or	Yes	–	Lee et al. (2015)
	Viburnum lantana L.	–	Yes	US-Sd	Yes	–	Agbaba (2017)
Berberidaceae	*Berberis aquifolium* Pursh	–	Yes	US-Or	Yes	–	Lee et al. (2015)
Buxaceae	*Sarcococca confusa* Sealy	–	Yes	US-Or	Yes	–	Lee et al. (2015)
Cactaceae	*Opuntia streptacantha* Lem.	–	Yes	US-Ca	Yes	--	Wang et al. (2019a, b)
Cannabaceae	*Celtis occidentalis* L.	–	Yes	US-Sd	Yes	–	Agbaba (2017)
Caprifoliaceae	*Lonicera* sp.	–	Yes	US-Mi	Yes	–	Lee et al. (2015)
	Lonicera caerulea L.	–	Yes	US-Or	Yes	–	Lee et al. (2015)
	Lonicera japonica Thunberg	–	Yes	US-Va	Yes	–	Shrader (2017)
	Lonicera morrowii A. Gray	–	Yes	US-Mn	Yes	–	Sward (2017)
		–	Yes	US-Ny	Yes	–	Elsensohn and Loeb (2018)
	Lonicera tatarica L.	–	Yes	US-Va	Yes	–	Shrader (2017)
		–	Yes	US-Mn	Yes	–	Sward (2017)
		–	Yes	US-Sd	Yes	–	Agbaba (2017)
	Symphoricarpos albus (L.) S.F. Blake	–	Yes	US-Or	Yes	–	Lee et al. (2015)
		–	Yes	US-Sd	Yes	–	Agbaba (2017)
Basellaceae	*Basella alba* L.	–	Yes	US-Sd	Yes	–	Diepenbrock and McPhie (2018)
Cornaceae	*Cornus alternifolia* L.	–	Yes	US-Sd	Yes	–	Agbaba (2017)
	Cornus amomum Miller	–	Yes	US-Mi	Yes	–	Lee et al. (2015)
		–	Yes	US-Ny	Yes	–	Elsensohn and Loeb (2018)
	Cornus foemina Miller	–	Yes	US-Mi	Yes	–	Lee et al. (2015)
	Cornus kousa Hance	–	Yes	US-Or	Yes	–	Lee et al. (2015),

	Cornus racemosa Lam.	–	Yes	US-Mn	Yes	–	Sward (2017)
		–	Yes	US-Sd	Yes	–	Agbaba (2017)
		–	Yes	US-Ny	Yes	–	Elsensohn and Loeb (2018)
	Cornus sericea L.	–	Yes	US-Or	Yes	–	Lee et al. (2015)
		–	Yes	US-Sd	Yes	–	Agbaba (2017)
Ebenaceae	*Diospyros kaki* L.	Yes	–	US-Ca	Yes	–	Wang et al. (2019a, b)
Elaeagnaceae	*Hippophae rhamnoides* L.	–	Yes	CA	–	Yes	Little et al. (2017)
Ericaceae	*Elaeagnus umbellata* Thunberg	–	Yes	US-Mi	Yes	–	Lee et al. (2015)
	Vaccinium corymbosum L.	Yes	–	US-Or	Yes	Yes	Dreves (2008), Lee et al. (2011a, b)
		Yes	–	US-Mn	Yes	–	Sward (2017)
		Yes	–	CA	–	Yes	Little et al. (2017)
	Vaccinium macrocarpon Aiton	Yes	Yes	US-Wi	Yes	Yes	Steffan et al. (2013)
	Vaccinium vitis-idaea L.	–	Yes	CA	–	Yes	Little et al. (2017)
Lauraceae	*Lindera benzoin* (L.) Blume	–	Yes	US-Mi	Yes	–	Lee et al. (2015)
Lythraceae	*Punica granatum* L.	Yes	–	US-Ca	Yes	–	Wang et al. (2019a, b)
Moraceae	*Ficus carica* L.	Yes	Yes	US-Ca	Yes	–	Lee et al. (2015), Wang et al. (2019a, b)
		Yes	–	US-Ca	Yes	–	Yu et al. (2013)
	Morus spp.	–	Yes	US-Ca	Yes	–	Yu et al. (2013)
	Morus alba L.	–	Yes	US-Sd	Yes	–	Agbaba (2017)
	Morus alba x rubra	–	Yes	US-Ca	Yes	–	Yu et al. (2013)
	Morus nigra L.	–	Yes	US-Or	–	Yes	Lee et al. (2015)
	Morus rubra L.	–	Yes	US-Fl	–	Yes	Lee et al. (2015)
Myrtaceae	*Eugenia uniflora* L.	–	Yes	US-Fl	Yes	–	Lee et al. (2015)
	Psidium guajava L.	Yes	–	MX	Yes	–	Lasa et al. (2017)

(continued)

Table 4.3 (continued)

Plant family	Plant species	Host plant status		Fruit-collection	Infestation type		Source
		Crop	Non-crop	US state or country[a]	Nat.	Lab.	
Phytolaccaceae	*Phytolacca americana* L.	–	Yes	US-Mi	–	Yes	Lee et al. (2015)
		–	Yes	US-Va	Yes	–	Shrader (2017)
		–	Yes	US-Ny	Yes	–	Elsensohn and Loeb (2018)
Rhamnaceae	*Frangula purshiana* (de Candolle) A. Gray	–	Yes	US-Or	Yes	–	Lee et al. (2015)
	Rhamnus cathartica L.	–	Yes	US-Mn	Yes	–	Sward (2017)
		–	Yes	US-Sd	Yes	–	Agbaba (2017)
		–	Yes	US-Ny	Yes	–	Elsensohn and Loeb (2018)
Rosaceae	*Aronia melanocarpa* (Mitchx.) Elliot	–	Yes	US-Sd	Yes	–	Agbaba (2017)
	Cotoneaster lacteus W.W. Smith	–	Yes	US-Or	Yes	–	Lee et al. (2015)
	Duchesnea indica (Andr.) Focke	–	Yes	US-Va	–	Yes	Shrader (2017)
	Eriobotrya japonica (Thunb.) Lindl.	–	Yes	US-Fl	Yes	–	Wang et al. (2019a, b)
	Fragaria x ananassa Duch.	Yes	–	US-Ca	Yes	–	Bolda (2008)
		Yes	–	US-Or	Yes	–	Lee et al. (2011a, b)

	Malus domestica Borkh.	Yes	–	US-Ca	Yes	–	Wang et al. (2019a, b)
	Prunus armeniaca L.	Yes	–	US-Ca	Yes	–	Wang et al. (2019a, b)
	Prunus avium L.	Yes	Yes	US-Or	Yes	–	Lee et al. (2011a, b, 2015)
		–	Yes	US-Va	Yes	–	Shrader (2017)
		Yes	–	US-Ca	Yes	Yes	Hauser (2011), Stewart (2015), Wang et al. (2019a, b)
		Yes	–	US-Ca,US-Wa	Yes	–	Beers et al. (2011)
	Prunus domestica L.	Yes	–	US-Ca	–	Yes	Wilson et al. (2013)
		Yes	–	US-Ca	Yes	–	Wang et al. (2019a, b)
	Prunus laurocerasus L.	–	Yes	US-Or	–	Yes	Lee et al. (2015)
	Prunus lusitanica L.	–	Yes	US-Or	Yes	–	Lee et al. (2015)
	Prunus persica (L.) Stokes	Yes	–	US-Ca	Yes	–	Wang et al. (2019a, b)
	Prunus serotina Ehrh.	–	Yes	US-Ny	Yes	–	Elsensohn and Loeb (2018)
	Prunus tomentosa Thunb.	–	Yes	US-Sd	Yes	–	Agbaba (2017)
	Prunus virginiana L.	–	Yes	US-Sd	Yes	–	Agbaba (2017)
		–	Yes	US-Ny	Yes	–	Elsensohn and Loeb (2018)
		–	Yes	CA	–	Yes	Little et al. (2017)
	Pyrus communis L.	Yes	–	US-Ca	Yes	–	Wang et al. (2019a, b)
	Rubus spp.	–	Yes	US-Va	Yes	–	Shrader (2017)
	Rubus allegheniensis Porter	–	Yes	US-Ny	Yes	–	Elsensohn and Loeb (2018)
	Rubus armeniacus Focke	–	Yes	US-Or	Yes	–	Lee et al. (2015)

(continued)

Table 4.3 (continued)

Plant family	Plant species	Host plant status		Fruit-collection US state or country[a]	Infestation type		Source
		Crop	Non-crop		Nat.	Lab.	
	Rubus fruticosus L.	Yes	–	US-Or	Yes	Yes	Lee et al. (2011a, b)
		Yes	–	US-Nc	Yes	–	Burrack et al. (2013), Swoboda-Bhattarai and Burrack (2016)
		Yes	–	MX	–	Yes	Lasa et al. (2017)
	Rubus idaeus L.	Yes	–	US-Or	Yes	Yes	Lee et al. (2011a, b)
		Yes	–	US-Mn	Yes	–	Sward (2017)
		Yes	–	US-Sd	Yes	–	Agbaba (2017)
		Yes	–	CA	Yes	–	Little et al. (2017)
		Yes	–	US-Ca	Yes	–	Bolda (2008), Hamby et al. (2014)
		Yes	–	US-Nc	Yes	–	Burrack et al. (2013), Swoboda-Bhattarai and Burrack (2016)
	Rubus occidentalis L.	–	Yes	US-Mn	Yes	–	Sward (2017)
		–	Yes	US-Ny	Yes	–	Elsensohn and Loeb (2018)
	Rubus spectabilis Pursh	–	Yes	US-Fl	Yes	–	Lee et al. (2015)
Rutaceae	*Citrus × sinensis* (L.) Osbeck	Yes	–	US-Ca	Yes	–	Wang et al. (2019a, b)
	Citrus reticulata Blanco	Yes	–	US-Ca	Yes	–	Wang et al. (2019a, b)
	Murraya paniculata (L.) Jack	–	Yes	US-Fl	Yes	–	Lee et al. (2015)
Solanaceae	*Solanum dulcamara* L.	–	Yes	US-Fl, US-Or	–	Yes	Lee et al. (2015)
		–	Yes	US-Ny	Yes	–	Elsensohn and Loeb (2018)
	Solanum lycopersicum L.	Yes	–	US-Fl	–	Yes	Lee et al. (2015)
Vitaceae	*Vitis vinifera* L.	Yes	–	US-Or	Yes	–	Dreves (2008), Lee et al. (2011a, b)
		–	Yes	US-Sd	Yes	Yes	Agbaba (2017)
		Yes	–	US-Va	–	Yes	Shrader (2017)

[a]**ISO Country code:** CA = Canada, MX = Mexico; US = United States. **US state code:** Ca = California, Fl = Florida, Mi = Michigan, Mn = Minnesota, Nc = North Carolina, Ny = New York, Or = Oregon, Sd = South Dakota, Va = Virginia, Wa = Washington; Wi = Wisconsin

Table 4.4 Crop and non-crop host species of *Drosophila suzukii* recorded in South America

Plant family	Plant species	Host plant status		Fruit-collection country[a]	Infestation type		Source
		Crop	Non-crop		Nat.	Lab.	
Arecaceae	*Butia* sp.	–	Yes	BR	Yes	–	Alexandre (2016)
Caricaceae	*Carica papaya* L.	–	Yes	BR[c]	Yes	–	Junior et al. (2018)
Ebenaceae	*Dispyros kaki* L.	?	?	UR	Yes	–	Lauyé (2017)
Elaeocarpaceae	*Aristotelia chilensis* (Molina) Stuntz	–	Ye	CL	Ye	–	Morales (2020)
Ericaceae	*Vaccinium ashei* Reade	Yes	–	UY[b]	Ye	–	Gonzales et al. (2015)
	Vaccinium corymbosum L.	Yes	–	BR	Yes	–	Vilela and Mori (2014), Bezerra Da Silva et al. (2019)
		Yes	–	AR	Yes	–	Santadino et al. (2015), Funes et al. (2018a, b)
		Yes	–	UY	Yes	–	Lauyé (2017)
		Yes	–	CL	Yes	–	SAG Servicio Agrícola y Ganadero (2019), Morales (2020)
Moraceae	*Ficus carica* L.	Yes	–	UY	Yes	–	Lauyé (2017)
Myrtaceae	*Acca sellowiana* Burret		?	UY	Yes	–	Lauyé (2017)
	Eugenia involucrata DC.	–	Yes	BR	Yes	–	Geisler et al. (2015)
	Eugenia uniflora L.	–	Yes	BR	Yes	–	Müller and Nava (2014), Nunes et al. (2014), Andreazza et al. (2015), Geisler et al. (2015)
	Luma apiculate (DC.) Burret	–	Yes	CL	Yes	–	SAG Servicio Agrícola y Ganadero (2019)
	Psidium cattleianum Sabine	– –	Yes Yes	UY BR	Yes Yes	– –	Lauyé (2017) Müller and Nava (2014), Andreazza et al. (2015), Junior et al. (2018)
	Psidium guajava L.	– –	Yes Yes	BR[d] AR[e]	Yes Yes	– –	Nunes et al. (2014), Andreazza et al. (2015), Andreazza et al. (2016c) Escobar et al. (2018)
	Ugni molinae Turcz	–	Yes	CL	Yes	–	SAG Servicio Agrícola y Ganadero (2019)

(continued)

Table 4.4 (continued)

Plant family	Plant species	Host plant status		Fruit-collection country[a]	Infestation type		Source
		Crop	Non-crop		Nat.	Lab.	
Rosaceae	*Eriobotrya japonica* (Thunb.)	–	Yes	BR	Yes	–	Andreazza et al. (2016b), Geisler et al. (2015)
	Fragaria x *ananassa* Duch.	Yes	–	BR	Yes	–	Deprá et al. (2014), Santos (2014), Nava et al. (2015), Andreazza et al. (2016b, c), Wollmann et al. (2016), Junior et al. (2018)
		Yes	–	BR	–	Yes	Schlesener et al. (2017)
		Yes	–	AR	Yes	–	Lochbaum (2017)
		Yes	–	CL	Yes	–	SAG Servicio Agrícola y Ganadero (2019)
	Malus domestica L.	Yes	–	BR	Yes	–	Oliveira et al. (2015)
	Prunus avium L.	Yes	–	AR	Yes	–	Lochbaum (2017)
		Yes	–	CL	Yes	–	Morales (2020)
	Prunus cerasifera Ehrh.	–	Yes	CL	Yes	–	Morales (2020)
	Prunus cerasus L.	Yes	–	CL	Yes	–	SAG Servicio Agrícola y Ganadero (2019), Morales (2020)
	Prunus persica L.	Yes	–	BR	Yes	–	Geisler et al. (2015), Borba et al. (2016)
	Pyrus communis L.	¿?	–	UY	Yes	–	Lauyé (2017)
	Rosa rubiginosa L.	–	Yes	CL	Yes	–	SAG Servicio Agrícola y Ganadero (2019)
	Rubus spp.	Yes	–	BR	Yes	–	Wollmann et al. (2016), Junior et al. (2018), Schlesener et al. (2018), Krüger et al. (2018a, b, 2019)
	Rubus L. subgen. *Rubus* Watson	Yes	–	AR	Yes	–	Dagatti et al. (2018)
	Rubus fruticosus L. agg.	Yes	–	AR	Yes	–	Lochbaum (2017), Funes et al. (2018a, b)
	Rubus idaeus L.	Yes Yes ¿? Yes	– – – –	AR BR UY CL	Yes Yes Yes Yes	– – –	Cichón et al. (2015), Escobar et al. (2018), Lochbaum (2017), Funes et al. (2018a, b), Lavagnino et al. (2018) Alexandre (2016) Lauyé (2017) SAG Servicio Agrícola y Ganadero (2019)
	Rubus ulmifolius Schott	Yes	–	UY	Yes	–	Lauyé (2017)
		–	Yes	CL	Yes	–	SAG Servicio Agrícola y Ganadero (2019), Morales (2020)

Rutaceae	*Citrus* x *sinensis* (L.) Osbeck	Yes	–	UY	Yes	–	Lauyé (2017)
Vitaceae	*Vitis vinifera* L.	Yes	–	UY	Yes	–	Lauyé (2017)
		Yes	–	BR	–	Yes	Andreazza et al. (2016a)

[a]ISO Country Codes: AR = Argentina, BR = Brazil; CL = Chile; UY = Uruguay
[b]Fruit rotting on the ground
[c]In damaged fruit, over-ripen or decayed
[d]Fruit with symptoms of infestation by several pests
[e]SWD was found in healthy, ripe fruit attached to the trees and in damaged fruit collected from the ground

been reported for Argentina, Brazil, Chile, and Uruguay (Table 4.4). All of these fruit species are primary SWD hosts (Deprá et al. 2014; Andreazza et al. 2016a; Lauyé 2017; Funes et al. 2018a, b; Bezerra Da Silva et al. 2019; SAG Servicio Agrícola y Ganadero 2019; Wollmann et al. 2016, 2019; Morales 2020).

Other cultivated plant species, such as *Malus domestica* L., *Prunus persica* L., *Prunus cerasus* L., *Pyrus communis* L., *Citrus sinensis* (L.) Osbeck, *Carica papaya* L., *Psidium guajava* L., *Ficus carica* L., and *Vitis vinifera* L. can be used by SWD as secondary hosts or reservoirs (Nunes et al. 2014; Oliveira et al. 2015; Geisler et al. 2015; Andreazza et al. 2015, 2016c, 2017; Borba et al. 2016; Lauyé 2017; Junior et al. 2018; Morales 2020). However, at least for some apple (Oliveira et al. 2015) and peach (Andreazza et al. 2017) cultivars, SWD females showed low oviposition frequencies, without economic damage in the fruit. Interestingly, *V. vinifera* varietal susceptibility studies carried out under lab conditions in Brazil showed that only some cultivars were vulnerable to SWD attack (Andreazza et al. 2016a). In field studies, the SWD has showed preference for different species within the genus *Rubus*, being *R. idaeus* much more attractive to SWD than *R. fruticosus* (Funes et al. 2018a). Furthermore, preference for determined *R. fruticosus* varieties has been detected for other frugivorous dipeterans (Funes et al. 2017), which could also be the case for the SWD.

Rosaceae is the plant family with the most SWD host species (45.2%) followed by Myrtaceae (23%) and Ericaceae (10%) (Table 4.4). Wild myrtaceous species (Table 4.4) are particularly alternative SWD hosts in South America (Müller and Nava 2014; Geisler et al. 2015; Alexandre 2016; Andreazza et al. 2015, 2016c; Lauyé 2017; SAG Servicio Agrícola y Ganadero 2019; Morales 2020). SWD pupae were usually recovered from healthy fruit, but there were some exceptions, e.g. SWD pupae were also found from damaged, over-ripen, or decayed *Vaccinium ashei* Reade fruit (Gonzales et al. 2015) or *P. guajava* fruit (Escobar et al. 2018), and from *Carica papaya* rotting fruit on the ground (Junior et al. 2018).

4.3.5 Africa and Oceania

SWD's potential for further invasions to Africa and Oceania is predicted, due to the environmental suitability of these areas for this species (Dos Santos et al. 2017). In Africa, SWD has been found on berry crops since 2013 in Morocco (North Africa) and on Reunion Island (EPPO European and Mediterranean Plant Protection Organization 2019). Only three crop fruit species were recorded as hosts of the SWD in this continent (Table 4.5). In Oceania, SWD was introduced into Hawaii in the 1980s (Kaneshiro 1983), and recorded in traps in the French Polynesia since 2017 (EPPO European and Mediterranean Plant Protection Organization 2019). Only two crop and two non-crop fruit species were cited as SWD hosts in Hawaii (Table 4.5).

Table 4.5 Crop and non-crop host species of *Drosophila suzukii* recorded in Africa and Oceania

Plant family	Plant species	Host plant status		Fruit-collection country, state or territory[a]	Infestation type		Source
		Crop	Non-crop		Nat.	Lab.	
AFRICA							
Ericaceae	*Vaccinium myrtillus* L.	Yes	–	FR-Ri	Yes	–	EPPO European and Mediterranean Plant Protection Organization (2019)
Rosaceae	*Fragaria* x *ananassa* Duch.	Yes	–	FR-Ri	Yes	–	EPPO European and Mediterranean Plant Protection Organization (2018, 2019); CABI EPPO Centre for Agricultural Bioscience International, European and Mediterranean Plant Protection Organization (2016)
	Rubus idaeus L.	Yes	–	MA, FR-Ri	Yes	–	EPPO European and Mediterranean Plant Protection Organization (2019)
OCEANIA							
Ericaceae	*Vaccinium reticulatum* (Smith)	–	Yes	US-Hi	Yes	–	Kaneshiro (1983); O'Grady et al. (2002); Magnacca et al. (2008); Hauser 2011; EPPO European and Mediterranean Plant Protection Organization (2014); Fraimout et al. (2015); CABI EPPO Centre for Agricultural Bioscience International, European and Mediterranean Plant Protection Organization (2016); Dos Santos et al. (2017); Koch et al. (2020)
Myrtaceae	*Psidium cattleianum* Sabine	Yes	–	US-Hi	Yes	–	Koch et al. (2020)
Rosaceae	*Rubus hawaiiensis* Gray	--	Yes	US-Hi	Yes	–	Koch et al. (2020)
	Rubus idaeus L.	Yes	–	US-Hi	Yes	–	Koch et al. (2020)

[a]MA = Morocco, FR-Ri = Reunion Island, US-Hi = Hawaii

4.4 Seasonal Biology and Population Dynamics

In temperate climate regions, such as Asia, Europe, and North America, adult SWD populations markedly decrease from winter onward (Kinjo et al. 2014; Wang et al. 2016a, b, c; Grassi et al. 2018). In Europe, SWD adult populations frequently reach peaks in spring and autumn, to decline again in the early winter (Weydert and Mandrin 2013; Asplen et al. 2015; Mazzetto et al. 2015; Zerulla et al. 2015; Arnó et al. 2016; Rossi-Stacconi et al. 2016; Zengin and Karaca 2019). This population dynamic is too similar to that reported in California, USA (Harris et al. 2014; Wiman et al. 2014; Wang et al. 2016a, b, c) where SWD populations increase in early spring, decrease during the hottest months of the year but increase again in autumn. Thus, cherries are usually the first commercial fruits available to SWD in Europe (Grassi et al. 2011; Cini et al. 2012; Poyet et al. 2015; Kenis et al. 2016; Mazzi et al. 2017). Damages to cherry crops during spring can be mainly explained by two SWD's biological abilities. Firstly, SWD adults can adapt to seasonal climate changes and survive harsh conditions through physiological and morphological adaptations (Shearer et al. 2016). These winter adaptations involve: (1) a reproductive diapause in which females have undeveloped ovaries and males produce very few sperm (Zerulla et al. 2015; Rossi-Stacconi et al. 2016; Shearer et al. 2016; Grassi et al. 2018), and (2) a phenotypic plasticity that allows a specific seasonal morphological change, which produces differences between a winter morphotype fly and a summer morphotype fly. The former has higher cold tolerance (10–15°C), darker pigmentation, and a larger body and longer wings than the latter (Asplen et al. 2015; Shearer et al. 2016; Fraimout et al. 2018; Stockton et al. 2018). Seasonal morphologies of SWD adults are unalterable, whereas reproductive dormancy is not; females can develop mature eggs when weather conditions are favorable (Rossi-Stacconi et al. 2016; Wallingford et al. 2016). Therefore, winter morphotype females that mated in autumn and overwintered may have mature eggs in early spring to infest the earliest ripening cherries and any other suitable non-crop fruits occurring at the same time (Panel et al. 2018), which facilitate the development of the summer morphotype SWD first generation, increasing pest population in late spring (Panel et al. 2018). Secondly, winter SWD survivors can use several early spring fruiting non-crop host plants for food and reproduction (Kenis et al. 2016; Grassi et al. 2018). Cases in point as *Viscum album* L. in Germany (Briem et al. 2016), *Hedera helix* L. in Italy (Grassi et al. 2018), and *Aucuba japonica* Thnb. in The Netherlands (Panel et al. 2018) are believed to be the earliest SWD reproductive hosts in early spring. Even though fitness of SWD adults emerging from these non-crop host species is poor, they could potentially attack the first available commercial crops (Grassi et al. 2018; Panel et al. 2018).

In subtropical climate regions like southern Brazil, a fruit-growing area invaded by SWD quite recently, SWD pattern of occurrence and population peaks were similar to those observed in berry-producing areas of the USA in late spring and mid-autumn (Wollmann et al. 2019). Probably, the growth of SWD population in autumn was determined by the occurrence of cooler conditions than in summer, with temperatures within a similar range than that of the fly's native region (Deprá et al. 2014). The gradual temperature increase in spring favored not only SWD activity but also triggered the onset of the fruiting season for both, commercial berry crops and non-crop fruits, such as strawberry guava (*Psidium cattleianum* Sabine) and Surinam cherry (*Eugenia uniflora* L) in southern Brazil (Wollmann et al. 2019). Although some SWD population peaks occurred during the spring-summer period, a significant decrease in the pest activity was detected at temperatures near 30°C, which are considered limitant for SWD oviposition (Wiman et al. 2014). In southern Brazil, SWD can survive at low population levels in alternative hosts such as loquat (*Eriobotrya japonica* Thunb.) even in winter, with thermal variations between 9 and 10.7°C (Wollmann et al. 2019). Alike in North America and Europe, SWD winter morphotypes were found in subtropical regions of South America, such as southern Brazil, demonstrating that seasonal phenotypic plasticity allows SWD to remain in the same habitat year after year. Similar SWD seasonal activity was found in blueberry crops in northwestern Argentina, another subtropical, South American fruit-growing region. High population peaks were detected in late spring, but the permanence of SWD in blueberry orchards was verified throughout the year, regardless of the presence of blueberry fruit in the area (Funes et al. 2018a). During the fruitless period, from early summer to mid-winter, there were low levels of SWD adult catches in traps (Funes et al. 2018b), which would indicate that SWD females use alternative host fruits, such as *P. guajava*, in wild vegetation areas between mid-summer and late-autumn (Escobar et al. 2018). SWD winter morphotypes were caught in peach crops in fruit-producing areas of the central-eastern region of Argentina as well (Gonsebatt et al. 2017).

4.5 Thermal Susceptibility

Seasonal temperature variations affect SWD population abundance (Wiman et al. 2014; Zengin and Karaca 2019), as well as daily temperature variations influence SWD adult dispersal between different habitats (Tait et al. 2020). In addition, temperature substantially influences SWD developmental, survival, and reproduction periods (Tochen et al. 2014). High temperatures would be responsible for SWD population summer decline in several temperate and subtropical regions (Arnó et al. 2016; Wollmann et al. 2019). SWD's activities are null or reduced when temperatures are out of the 10°–31°C range (Kimura 2004; Kinjo et al. 2014; Tochen et al. 2014; Zerulla et al. 2015; Rossi-Stacconi et al. 2016; Grassi et al. 2018). Although SWD adult detection is often difficult in Europe's and North America's winter, winter trapping has caught wild SWD individuals despite temperatures below 0°C (Hamby et al. 2016; Rossi-Stacconi et al. 2016; Thistlewood et al. 2018). This is a clear evidence of SWD's ability to overwinter locally (Rossi-Stacconi et al. 2016; Grassi et al. 2018; Stockton et al. 2018). Thus, these SWD winter morphotype adaptive features allow adult insects to absorb and retain heat throughout the winter (Kimura 1988; Tonina et al. 2016). Furthermore, in contrast with summer morphotypes, winter morphotypes may reduce water loss, improve immune function, and extend their longevity at cold temperatures (Stephens et al. 2015; Shearer et al. 2016; Toxopeus et al. 2016; Wallingford et al. 2016) through physiological changes that prevent ice crystals formation within the hemocoel, and fly death (Lee 1991; Shearer et al. 2016; Toxopeus et al. 2016).

Several studies on SWD thermal tolerance have used static and/or dynamic acclimation protocols in order to learn more about this biological capacity, taking into account closer approximations to natural temperature changes (Kimura 2004; Dalton et al. 2011; Kellermann et al. 2012; Jakobs et al. 2015; Stephens et al. 2015; Shearer et al. 2016; Wallingford et al. 2016; Stockton et al. 2018). In the static acclimation procedure, SWD individuals are kept at a single constant cool temperature for the duration of the acclimation period prior to testing, while a dynamic acclimation process involves gradually subjecting the insect to cold hardening temperatures (Stockton et al. 2018). Static acclimation procedure induces cold tolerance in SWD adults, which survived to temperatures below 0°C for up to 1 h and reached a lower lethal limit at −7.5°C (Jakobs et al. 2015). However, dynamic acclimation significantly improved SWD winter morphotype adults' survival since the survival rate was 50% after 72 h at −7.5°C (Stockton et al. 2018). Therefore, both acclimation and a suitable overwintering site may allow the survival of small founder populations of SWD winter morphotypes in the spring (Wallingford et al. 2016). Some models of *D. suzukii* potential distribution relate temperature with pest spread (Dos Santos et al. 2017). According to these models, annual mean temperature, maximum temperature of the warmest month (values >33°C), mean temperature of the coldest quarter (values <−10°C), and annual precipitation are the environmental variables with more influence on SWD distribution. Consequently, SWD potential geographical distribution range includes world subtropical regions with high rainfall throughout the year or during part of it.

There are no records on SWD overwintering as larva or pupa. Both immature stages have lesser cold tolerance compared to adults, and show little survival at temperatures below 5°C (Dalton et al. 2011; Stephens et al. 2015; Aly et al. 2017; Enriquez and Colinet 2017). Pupal survival drops noticeably below 0°C regardless of the dynamic acclimation procedure (Stockton et al. 2018).

Temperature is crucial for SWD development and establishment. According to studies carried out by Schlesener et al. (2020) in Brazil, for egg-to-adult survival, the lowest thermal threshold was 7.8 °C and optimum temperature was 23–25 °C. No SWD emergence occurred at 30–33 °C. The shortest egg-to-adult development time (10 days) was observed at 25–28 °C. The intrinsic rate of population increase was the highest at 23–25 °C. SWD body size is strongly affected by temperature, with 13°C and 28 °C causing the largest and the smallest body sizes, respectively. The annual number of generations ranged from 17.1 to 27.2 in cold and warm regions, respectively.

4.6 Reproduction

A high reproductive potential is one of the strongest reasons for the rapid worldwide spread of the polyphagous SWD. Courtship in drosophilids might involve aerial vibrations, visual displays, substrate-borne vibrations, and sex pheromones (Hamby et al. 2016). SWD sexual behavior is characterized by distinct facts of male courtship leading to female acceptance for mating. Furthermore, it has been reported that time of day and male age modulate *D. suzukii* mating activity (Revadi et al. 2015). The SWD displays characteristic courtship behavior, but to what extent the different sensory modalities are involved remains unclear. The importance of substrate-borne vibrations produced during male courtship has been demonstrated (Mazzoni et al. 2013), but the role of pheromonal and visual components in courtship remains unknown. Unlike most other drosophilids (e.g., *Drosophila melanogaster* Meigen), *D. suzukii* does not produce the male pheromone cis-vaccenyl acetate (Dekker et al. 2015). In this direction, female pheromones such as cuticular hydrocarbons (CHCs) were also studied in SWD since female sexual maturity is accompanied by a quantitative increase in these olfactory signals (Revadi et al. 2015). Despite of having been shown that CHCs were not needed to induce male courtship in *D. suzukii*, it turned out that these pheromones significantly increased male sexual behavior and that interfering in their naturally occurring ratios disrupts mate recognition (Revadi et al. 2015; Snellings et al. 2018).

Sex ratios are among the most basic of demographic parameters and provide an indication of both the relative survival of females and males and the future breeding potential of a population (Skalski et al. 2005). The production of males and females in a 1:1 ratio is usually the most prevalent evolutionary steady strategy, governed by frequency-dependent natural selection owing to competition for mates among individuals of the same sex (Sapir et al. 2008). SWD sex ratio is subjected to variations, depending on several factors. In SWD, sex ratio fluctuations throughout the year seem to be very conservative regardless the plant environment within the same region, but not between regions. In Italy (Gargani et al. 2015), in a trapping net implemented from January to December, sex ratio fluctuation patterns over time were almost the same in a vineyard than in a botanical garden, in Siena. However, in another location (Gosseto), the sex ratio fluctuation curve was different to that of Sienna, in three orchards surveyed (vineyard, apricot, and blueberry). Nevertheless, sex ratio fluctuation patterns were very similar among the three fruit crops. Besides, the sex ratios (expressed as percentage of females) varied between 0 and 50 in Siena, and from 0 to 100 in Grosseto. The highest sex ratios (biased to females) occurred at different seasons, depending on the location (winter in Siena, autumn in Grosseto). The lowest sex ratios (biased to males) were registered in spring. There was not SWD catches in none of the two locations in summer.

SWD sex ratio (% of females) decreased in parallel with host's fruit production (berries, and stone and pome fruits) from January to March, in semiarid high-elevation valleys with temperate summers and cold winters, in northwestern Argentina (CF Funes pers. comm.). As food supply slowed down, SWD sex ratios went from female-biased (January, plenty of fruit) to male-biased (March, scarce fruits). SWD sex ratio may vary between undamaged fruit collected from the plant and undamaged fruit lifted from the ground below the tree, as reported by Lasa et al. (2017), who worked in guava (*P. guajava*) in Mexico. In both cases, sex ratios were female-biased, varying within a range of 58–68%. Drummond et al. (2019) reported that across a 7-year survey, the relative abundances of sexes were slightly male-biased. Sex ratio (% of males) declined linearly between 2012 and 2018. The sex ratio of newly emerging adults remained more or less constant, with a mean of 0.98 males per female (Emiljanowicz et al. 2014).

No effect of egg density was observed on the sex ratio of SWD adults that emerged from attached pupae. Conversely, in detached pupae, sex ratio decreased linearly as egg density increased (Bezerra da Silva et al. 2019). A female-biased sex ratio for *D. suzukii* cultures probably results from the greater sensitivity of males to dietary ethanol. Without dietary alcohol, the sex ratio for *D. suzukii* stabilized near that of *D. melanogaster*, ~1:1 (Sampson et al. 2016).

4.7 Interspecific and Intergeneric Interactions with Other Frugivorous Dipterans

Selective pressure might have facilitated the evolution of *Drosophila* morphological traits such as a serrated ovipositor in *D. suzukii*. For instance, interspecific and intra-specific competitors could induce changes in SWD behavior of either female adult oviposition or larvae. The presence of competitors around fallen fruits on the ground probably suppressed *D. suzukii* oviposition on the fallen fruits and facilitated oviposition on non-fermenting substrates in nature, driving *D. suzukii* to use ripening fruits on the tree instead (Kidera and Takahashi 2020). Two species co-habiting similar niches would compete to either coexist or exclude the other. Interestingly, *D. suzukii* and *D. melanogaster* may have separate ecological niches, but *D. melanogaster* was observed ovipositing in fruit injuries caused by *D. suzukii* oviposition (Vilaire et al. 2011).

The presence of *D. melanogaster* in a substrate significantly reduced SWD emergence and egg laying; conversely, more SWD eggs were laid in blank media, not pre-inoculated with *D. melanogaster* (Shaw et al. 2018). Chemical cues or signs of previous oviposition by heterospecifics, such as *D. melanogaster*, can deter *D. suzukii* female from ovipositing in the same fruit (Shaw et al. 2018; Kidera and Takahashi 2020). This may be due to repellent olfactory cues such as trace amounts of the male *D. melanogaster* sex pheromone, cis-vaccenyl acetate, transferred to fruit during oviposition. In direct interspecific competition situations, *D. melanogaster* larvae have greater survival than *D. suzukii* larvae (Gao et al. 2018). This is due in part to higher tolerance to ethanol produced through decay and fermentation of fruit damaged by larval feeding and that higher levels of ethanol are produced in fruit containing *D. melanogaster* larvae than fruit containing *D. suzukii* larvae

(Sampson et al. 2016; Gao et al. 2018). Although SWD females prefer to oviposit in ripe fruit, they are able to shift ecological niches and use ripening fruit to avoid competitive pressures and reduce potential ethanol exposure to their larvae (Little et al. 2020).

Zaprionus indianus flies would not be capable of laying eggs in healthy fruit because they lack a serrated ovipositor, but they might take advantage of other dipeterans oviposition punctures for placing their own eggs (Lasa et al. 2017; Shrader et al. 2020). Guavas attached to the tree infested with *Z. indianus* were also infested with *D. suzukii*, *Anastrepha* spp., or both. Therefore, the infestation was possible because of previous injuries caused by the other two frugivorous dipterans (Lasa et al. 2017). Competition from *Z. indianus* limited *D. suzukii* numbers in interspecific laboratory studies (Shrader et al. 2020). When reared with grapes, *Z. indianus* also affected *D. suzukii* mortality, and increased developmental times to pupation and adult emergence, compared with the intraspecific SWD controls. Pupal volume was scarcely reduced, and it occurred only with the highest interspecific larval densities (Shrader et al. 2020). In laboratory trials with strawberry ripe fruits, there was a significant and positive interaction between damage caused by *D. suzukii* and infestation by *Z. indianus* (Bernardi et al. 2017).

4.8 Natural Enemies–*Drosophila suzukii* Relationships

Reviews on SWD natural enemies were published by Asplen et al. (2015), Haye et al. (2016), Garcia et al. (2017), and Lee et al. (2019), involving several parasitoid, predator, and entomopathogenic species. Parasitoids represent the most numerous and most studied group of SWD's natural enemies. In total, 30 parasitoid species associated with SWD were recorded throughout the world, either through field collections or through laboratory host specificity tests (Table 4.6). Nine of them are pupa parasitoids and 21 are larva parasitoids. The 40% of the parasitoid species recorded are Asian-native species, the SWD's origin region. The 75% of Asian parasitoid species belong to the genus *Asobara* Foerster (Braconidae, Alysiinae), such as *Asobara brevicauda* van Achterberg & Guerrieri, *A. elongata* van Achterberg & Guerrieri, *A. mesocauda* van Achterberg & Guerrieri, *A. triangulata* van Achterberg & Guerrieri, *A. unicolorata* van Achterberg & Guerrieri, *A. leveri* (Nixon), *A. japonica* Belokobylskij (Guerrieri et al. 2016), *A. pleuralis* (Ashmead) (Girod et al. 2018a), and *A. tabida* (Nees) (Mitsui et al. 2007), while 35% are eucoilines (Figitidae: Eucoilinae), such as *Ganaspis xanthopoda* (Ashmead) (Kasuya et al. 2013), *Ganaspis brasiliensis* (Ihering), and *Leptopilina japonica* Novkovic & Kimura (Daane et al. 2016; Girod et al. 2018a; Giorgini et al. 2019). However, the association between *G. xanthopoda* and *D. suzukii* is doubtful because *G. xanthopoda* individuals from Japan, reported by Mitsui et al. (2007) and Kasuya et al. (2013), have been assigned to *G. brasiliensis* by Bufington and Forshage (2016) and Nomano et al. (2017). In turn, *G. xanthopoda* individuals tested under lab conditions by Kacsoh and Schlenke (2012) would appear to be *G. brasiliensis* (Nomano et al. 2017).

Table 4.6 Country list of hymenopteran parasitoid species associated with *Drosophila suzukii*

Parasitoid Family	Parasitoid species	Host stage attacked	Parasitism conditions		Successful development on the host	Country[a]	Source
			Lab	Field			
Braconidae	*Asobara brevicauda* van Achterberg & Guerrieri	Larva	– –	Yes Yes	Yes Yes	CN KR	Guerrieri et al. (2016) Daane et al. (2016)
	Asobara citri Fisher	Larva	Yes	–	Yes	US	Kacsoh and Schlenke (2012), Poyet et al. (2013)
	Asobara elongata van Achterberg and Guerrieri	Larva	–	Yes	Yes	CN	Guerrieri et al. (2016)
	Asobara japonica Belokobylskij	Larva	– – – Yes Yes	Yes Yes Yes – –	Yes Yes Yes Yes Yes	JP CN KR US CH	Mitsui et al. (2007), Ideo et al. (2008), Guerrieri et al. (2016) Daane et al. (2016) Kacsoh and Schlenke (2012), Poyet et al. (2013), Wang et al. (2018a, a, b, 2020) Girod et al. (2018b, c)
	Asobara leveri (Nixon)	Larva	– –	Yes Yes	Yes Yes	CN KR	Guerrieri et al. (2016) Daane et al. (2016)
	Asobara mesocauda van Achterberg and Guerrieri	Larva	–	Yes	Yes	CN	Girod et al. (2018a)

	Asobara pleuralis (Ashmead)	Larva	–	Yes	Yes	CN	Girod et al. (2018a)
			Yes	–	No	JP	Nomano et al. (2015)
			Yes	–	No	US	Kacsoh and Schlenke (2012)
	Asobara tabida (Nees)	Larva	–	Yes	Yes	JP	Mitsui et al. (2007)
			Yes	–	No	JP	Nomano et al. (2015)
			Yes	–	No	CH	Knoll et al. (2017)
			Yes	–	No	US	Kacsoh and Schlenke (2012)
	Asobara triangulata van Achterberg & Guerrieri (likely *Asobara* sp. TK1)	Larva	—	Yes	Yes	JP	Nomano et al. (2015), Guerrieri et al. (2016),
			–	Yes	Yes	CN	Girod et al. (2018a)
	Asobara unicolorata van Achterberg & Guerrieri	Larva	—	Yes	Yes	CN	Guerrieri et al. (2016)
Diapriidae	*Trichopria drosophilae* Perkins	Pupa	Yes	–	Yes	CH	Knoll et al. (2017), Woltering et al. (2019)
			Yes	–	Yes	IT	Rossi-Stacconi et al. (2013), Mazzetto et al. (2016)
			–	Yes	Yes	US	Miller et al. (2015), Wang et al. (2016a, b, 2018b)
			Yes	–	Yes	US	Kaçar et al. (2017)
			Yes	–	Yes	FR	Chabert et al. (2012)
			–	Yes	Yes	ES	Gabarra et al. (2015)
			–	Yes	Yes	KR	Daane et al. (2016)
			–	Yes	Yes	CN	Giorgini et al. (2019)
			Yes	–	Yes	CN	Yi et al. (2020)
			–	Yes	Yes	MX	García-Cancino et al. (2015)
			Yes	–	Yes	MX	Gonzalez-Cabrera et al. (2019)
	Trichopria anastrephae Lima	Pupa	–	Yes	Yes	BR	Wollmann et al. (2016)
			Yes	–	Yes	BR	Schlesener et al. (2019), Krüger et al. (2019)
	Trichopria sp.1	Pupa	Yes	–	Yes	US	Kacsoh and Schlenke (2012)
	Trichopria sp.	Pupa	–	Yes	Yes	AR	Funes et al. (2019)

(continued)

Table 4.6 (continued)

Parasitoid Family	Parasitoid species	Host stage attacked	Parasitism conditions Lab	Parasitism conditions Field	Successful development on the host	Country[a]	Source
Figitidae	*Dieucoila* sp.	larva	–	Yes	Yes	AR	Funes et al. (2019)
	Ganaspis brasiliensis (Ihering)	Larva	–	Yes	Yes	CN	Girod et al. (2018a), Giorgini et al. (2019)
			–	Yes	Yes	JP	Nomano et al. (2017), Matsuura et al. (2018),
			Yes	–	Yes	CH	Girod et al. (2018a)
			–	Yes	Yes	KR	Girod et al. (2018b)
			Yes	–	Yes	US	Daane et al. (2016)
							Wang et al. (2018a)
	Ganaspis hookeri Crawford	Larva	–	Yes	Yes	AR	Lue et al. (2017)
	Ganaspis xanthopoda (Ashmead)[b]	Larva	–	Yes	Yes	JP	Mitsui et al. (2007), Kasuya et al. (2013)
			Yes	–	No	US	Kacsoh and Schlenke (2012), Poyet et al. (2013)
	Ganaspis sp.1	Larva	Yes	–	Yes	US	Kacsoh and Schlenke (2012)
	Ganaspis sp.	Larva	–	Yes	Yes	AR	Funes et al. (2019)
	Hexacola sp.	Larva	–	Yes	Yes	AR	Funes et al. (2019)
	Leptopilina boulardi Barbotin, Carton & Kelner-Pillault	Larva	–	Yes	Yes	AR	Garrido et al. (2018)
			–	Yes	Yes	MX	García-Cancino et al. (2015)
			Yes	–	Yes	MX	Gonzalez-Cabrera et al. (2019)
			–	Yes	Yes	BR	Wollmann et al. (2016)
			Yes	–	No	US	Kacsoh and Schlenke (2012)
			Yes	–	No	CH	Knoll et al. (2017)
			Yes	–	No	FR	Chabert et al. (2012)
			Yes	–	No	IT	Mazzetto et al. (2016)
	Leptopilina clavipes (Hartig)	Larva	Yes	–	No	US	Kacsoh and Schlenke (2012)
			–	Yes	Yes	AR	Lue et al. (2017)

	Leptopilina heterotoma (Thomson)	Larva	Yes	–	No	CH	Knoll et al. (2017), Girod et al. (2018a, b)
			Yes	–	Yes	IT	Rossi-Stacconi et al. (2015, 2017)
			Yes	–	No	IT	Mazzetto et al. (2016)
			Yes	–	No	IT, US	Miller et al. (2015)
			Yes	–	No	US	Poyet et al. (2013)
	Leptopilina japonica Novkovic & Kimura	Larva	–	Yes	Yes	JP	Novkovic et al. (2011), Kasuya et al. (2013)
			Yes	–	Yes	US	Wang et al. (2018a)
			–	Yes	Yes	CN	Giorgini et al. (2019)
			–	Yes	Yes	KR	Daane et al. (2016)
			Yes	–	Yes	CH	Girod et al. (2018b, c)
Pteromalidae	*Muscidifurax raptorellus* Kogan & Legner	Pupa	Yes	–	Yes	CA	Bonneau et al. (2019)
	Pachycrepoideus vindemmiae Rondani	Pupa	Yes	–	Yes	CH	Knoll et al. (2017)
			–	Yes	Yes	IT	Rossi-Stacconi et al. (2013)
			–	Yes	Yes	US	Miller et al. (2015), Wang et al. (2016b, 2018b)
			Yes	–	Yes	US	Kaçar et al. (2017)
			Yes	–	Yes	CA	Bonneau et al. (2019)
			Yes	–	Yes	FR	Chabert et al. (2012)
			–	Yes	Yes	ES	Gabarra et al. (2015)
			–	Yes	Yes	CN	Giorgini et al. (2019)
			–	Yes	Yes	KR	Daane et al. (2016)
			–	Yes	Yes	MX	García-Cancino et al. (2015)
			Yes	Yes	Yes	BR	Schlesener et al. (2019)
			–	Yes	Yes	AR	Funes et al. (2019)
	Spalangia erythromera Förster	Pupa	Yes	–	Yes	CH	Knoll et al. (2017)
	Spalangia simplex Perkins	Pupa	–	Yes	Yes	MX	García-Cancino et al. (2015)
	Vrestovia fidenas (Walker)	Pupa	Yes	–	Yes	CH	Knoll et al. (2017), Wolf et al. (2019)

[a]ISO Country Codes: AR = Argentina, BR = Brazil, CA = Canada, CH = Switzerland, CN = China, ES = Spain, FR = France, IT = Italy, JP = Japan, MX = Mexico, KR = Republic of Korea (South Korea), US = United States
[b]Dubious association (Nomano et al. 2017)

Interestingly, *G. brasiliensis* would involve four lineages with different geographic distribution and host ranges (Nomano et al. 2017). These morphologically identical lineages could be a complex of cryptic species (Giorgini et al. 2019).

Among the other parasitoid species, 33.4% are native to America, 10% to Europe, and one of them (3.3%), *Asobara citri* Fisher, to Africa (Kacsoh and Schlenke 2012). The remaining 13.3% are present worldwide, such as *Pachycrepoideus vindemmiae* Rondani (Pteromalidae), *Trichopria drosophilae* Perkins (Diapriidae) (Rossi-Stacconi et al. 2013; Wang et al. 2016a, b, c; Daane et al. 2016), *Leptopilina boulardi* Barbotin, Carton & Kelner-Pillault, and *L. clavipes* (Hartig) (Figitidae: Eucoilinae) (Wollmann et al. 2016; Lue et al. 2017). Of the 10 American parasitoid species associated with SWD, six (60%) are undescribed species of eucoilines and diaprines (Table 4.6). The remaining four species are represented by one drosophilid larval parasitoid, *Ganaspis hookeri* Crawford (Lue et al. 2017), and by three generalist parasitoids that attack cycloraphic dipterous pupae, such as the pteromalines *Muscidifurax raptorellus* Kogan & Legner (Bonneau et al. 2019) and *Spalangia simplex* Perkins (García-Cancino et al. 2015) and the diaprine *Trichopria anastrephae* Lima (Krüger et al. 2019). Of the three European parasitoid species, one is drosophilid larval parasitoid, *Leptopilina heterotoma* (Thomson) (Figitidae) (Rossi-Stacconi et al. 2015), and two are generalist pupal parasitoids, *Spalangia erythromera* Förster and *Vrestovia fidenas* (Walker) (Pteromalidae) (Knoll et al. 2017).

Although more larval parasitoids than pupal parasitoids have been associated with the SWD, few species can successfully overcome the defense response of the SWD larva. Parasitoid eggs or larvae are usually encapsulated (Kacsoh and Schlenke 2012; Poyet et al. 2013). Therefore, Asian-native larval parasitoid species have mainly shown high capacity of developing successfully in SWD. Among them, *L. japonica*, *G. brasiliensis*, and *A. japonica* are the most significant parasitoid species, due to their high natural parasitism rate and greater specificity to SWD (Daane et al. 2016; Girod et al. 2018a; Matsuura et al. 2018; Giorgini et al. 2019). Thus, these three Asian parasitoid species were imported for evaluation as classical biological control agents in North America and/or Europe (Daane et al. 2016; Girod et al. 2018b, c; Wang et al. 2018a, 2019a, b; Giorgini et al. 2019).

Among the generalist parasitoid species, pupal parasitoids *P. vindemmiae*, *M. raptorellus*, *V. fidenas*, *T. drosophilae*, and *T. anastrephae* have been mainly evaluated under laboratory conditions, for both their ability to successfully parasitize SWD and their efficiency in reducing the target population (Wang et al. 2016a, b, 2018b; Bonneau et al. 2019; Krüger et al. 2019; Schlesener et al. 2019; Wolf et al. 2019; Yi et al. 2020). Since all of them can attack other dipterous species, their field effectiveness is doubtful. Nevertheless, *T. drosophilae* has a potential as SWD's biological control agent (Mazzetto et al. 2016; Kaçar et al. 2017; Rossi-Stacconi et al. 2017; Pfab et al. 2018; Wang et al. 2018b; Yi et al. 2020). This potentiality was verified through *T. drosophilae* releases in berry fields at Colima and Jalisco, Mexico, where this strategy reduced 50–55% SWD wild populations (Gonzalez-Cabrera et al. 2019). Mass releases of *T. drosophilae* have been recently accomplished in northern Italy (Bioplanet 2020).

Arthropod predators such as ants, spiders, earwigs, and rove beetles have been found attacking SWD larvae and/or pupae inside field-infested fruit (Woltz and Lee 2017). Similarly, several generalist hemipteran species such as *Dicyphus tamaninii* Wagner (Hemiptera: Miridae), *Orius laevigatus* (Fieber), *Cardiastethus nazarenus* Reuter, and *C. fasciventris* Garbiglietti (Hemiptera: Anthocoridae) have also been observed predating on SWD eggs, larvae, and/or pupae, inside fallen fruit in berry growing areas of Northern Spain (Arnó et al. 2012; Gabarra et al. 2015). In addition, *Labidura riparia* Pallas (Dermaptera: Labiduridae) has been recorded as a common and effective SWD larvae and pupae predator, in Spain (Gabarra et al. 2015). Furthermore, several commercially available predator species have been tested under lab conditions for potential use as SWD's biological control agents. The evaluated predator species were *Orius insidiosus* Say (Hemiptera: Anthocoridae), *Dalotia coriaria* Kraatz (Coleoptera: Staphylinidae), *Podisus maculiventris* Say (Hemiptera: Pentatomidae), *Dicyphus hesperus* Knight (Hemiptera: Miridae), and *Chrysoperla carnea* Stephens (Neuroptera: Chrysopidae) (Cuthbertson et al. 2014a; Renkema et al. 2015; Woltz et al. 2015; Renkema and Cuthbertson 2018; Bonneau et al. 2019).

Similarly, several commercial entomopathogenic agents, such as nematodes [*Heterorhabditis bacteriophora* Poinar (Rhabditida: Heterorhabditidae), *Steinernema carpocapsae* (Weiser), *S. feltiae* (Filipjev) and *S. kraussei* (Steiner) Travassos (Rhabditida: Steinernematidae)], and fungi [*Metarhizium brunneum* (= *M. anisopliae*) (Metchnikoff) Sorokin, *Isaria fumosorosea* (= *Paecilomyces fumosoroseus*) (Wize) Brown & Smith, *Lecanicillium (= Verticillium) lecanii* (Zimm.), and *Beauveria bassiana* (Balsamo-Crivelli) Vuillemin (Ascomycota; Hypocreales)], have been tested against SWD under lab conditions (Cuthbertson et al. 2014b; Woltz et al. 2015; Cossentine et al. 2016; Cuthbertson and Audsley 2016; Renkema and Cuthbertson 2018; Yousef et al. 2018; Lee et al. 2019). Bacteria and viruses have been also reported from laboratory trials or natural infestation (Lee et al. 2019). Commercial *Bacillus thuringiensis* var. *kurstakii* or *B. thuringiensis* var. *israeliensis* were tested against SWD in lab trials (Biganski et al. 2018; Cahenzli et al. 2018). More details on the use of SWD's natural enemies as a biological control strategy can be seen in Chap. 8 of this book.

4.9 Concluding Remarks

In brief, egg-laying preference for ripening and ripe healthy small, soft and stone fruits attached to the plant makes SWD a severe and harmful pest for fruit-producing regions of Asia, Europe, and America. Most of the worldwide host fruit species preferred by SWD belong to Rosaceae, a plant family involving several economically important cultivated fruits, such as caneberries, cherries, strawberries, and stone and pome fruits, and a large list of non-crop species. Besides, both non-crop and crop *Vaccinium* L. plants (Ericaceae) have been recorded as recurrent SWD hosts in all continents. In addition to wild species of Rosaceae and Ericaceae families, several Adoxaceae (*Sambucus* spp.), Caprifoliaceae (*Loricera* spp.), and Rhamnaceae (*Rhamnus cathartica* L. and *Frangula* spp.) are crucial host plants for SWD population growth and seasonal dynamics in Europe and the USA. In Mexico and South America, Myrtaceae are fruit species highly susceptible to SWD, playing the important role of facilitating SWD multiplication in wild habitats adjacent to host crop fields.

The high degree of polyphagy shown by SWD either in their native region or in the invaded continents, added to the seasonal phenotypic plasticity that allows adaptations to adverse thermal periods, greatly favors SWD establishment and dispersion in the invaded regions. In turn, complex landscape structures such as forests with alternative non-cultivated host plants, shrub vegetation and flowering field margins, and the distance of these structures to host crops in the different invaded areas, facilitate SWD daily spread between different habitats, which can also determine infestation levels.

In both temperate and subtropical climate regions, the SWD seasonal dynamic is essentially related to temperatures more suitable for its movement between habitats, development, survival, and reproduction. The warmest, as well as the coldest, months of the year reduce, with more or less intensity, SWD populations in summer and winter, respectively. Therefore, SWD populations mostly generate two population peaks, in late spring and mid-autumn. However, as SWD can overwinter locally, adults can be captured throughout the year in the invaded areas.

A better understanding of the relationships between SWD and the components of newly invaded landscapes, in terms of reproductive behavior and interspecific and intergeneric interactions, especially with other frugivorous dipterans, and biological controllers would be very important to achieve. This will allow designing efficacious, efficient, and environmental friendly pest management strategies, maximizing the benefits provided by ecosystem services.

A diverse suite of local predators and parasitoids associated with SWD are found in wild vegetation areas surrounding fruit crop fields in the different invaded regions. These natural enemies should be conserved in their natural habitats. Furthermore, it should be noted that the recently found specialist Asian parasitoids could be significant SWD's biocontrol agents. In addition, several entomopathogens have been laboratory tested, but their suppressive effects on SWD wild populations remains unknown and should be assessed.

References

Agbaba B (2017) A survey of non-crop plants as alternative hosts to raspberries for *Drosophila suzukii* (spotted wing drosophila). Dissertations, South Dakota State University

Alexandre BG (2016) Avaliação das assembleias de Drosophilidae (Insecta, Diptera) em área de restinga no município de Torres-RS. Dissertation, Universidade Federal do Rio Grande do Sul

Aly MFK, Kraus DA, Burrack HJ (2017) Effects of postharvest cold storage on the development and survival of immature *Drosophila suzukii* (Diptera: Drosophilidae) in artificial diet and Fruit. J Econ Entomol 110(1):87–93. https://doi.org/10.1093/jee/tow289

Andreazza F, Bernardi D, Botton M, Nava DE (2015) Índice de infestação natural de *Drosophila suzukii* e *Zaprionus indianus* (Diptera: Drosophilidae) em frutíferas nativas no município de pelotas. In: Abstracts of the XXIV Congresso de Iniciação Científica e XVII Encontro da Pós-Graduação, Universidade Federal de Pelotas, Pelotas, Rio Grande do Sul, 8–9 May 2018

Andreazza F, Baronio CA, Botton M (2016a) Suscetibilidade de bagas de genótipos de videira pela infestação por *Drosophila suzukii* (Diptera: Drosophilidae). Pesq agropec bras 51(5):599–606. https://doi.org/10.1590/S0100-204X2016000500021

Andreazza F, Bernardi D, Marangon RB et al (2016b) Técnica de criação de *Drosophila suzukii* (Matsumura, 1931) (Diptera: Drosophilidae) em dieta artificial. Bol Pesqui Desenvolv 240:1–23

Andreazza F, Haddi K, Oliveira EE (2016c) *Drosophila suzukii* (Diptera: Drosophilidae) arrives at Minas Gerais State, a main strawberry production region in Brazil. Fla Entomol 99(4):796–798. https://doi.org/10.1653/024.099.0439

Andreazza F, Bernardi D, Dos Santos RSS et al (2017) *Drosophila suzukii* in southern neotropical region: current status and futures perspectives. Neotrop Entomol 46(6):591–605. https://doi.org/10.1007/s13744-017-0554-7

Arnó J, Riudavets J, Gabarra R (2012) Survey of host plants and natural enemies of *Drosophila suzukii* in an area of strawberry production in Catalonia (northeast Spain). IOBC/WPRS Bull 80:29–34

Arnó J, Sola M, Riudavets J, Gabarra R (2016) Population dynamics, non-crop hosts, and fruit susceptibility of *Drosophila suzukii* in Northeast Spain. J Pest Sci 89:713–723. https://doi.org/10.1007/s10340-016-0774-3

Asplen MK, Anfora G, Biondi A et al (2015) Invasion biology of spotted wing Drosophila (*Drosophila suzukii*): a global perspective and future priorities. J Pest Sci 88(3):469–494. https://doi.org/10.1007/s10340-015-0681-z

Bal HK, Adams C, Grieshop M (2017) Evaluation of off-season potential breeding sources for spotted wing Drosophila (*Drosophila suzukii* Matsumura) in Michigan. J Econ Entomol 110(6):2466–2470. https://doi.org/10.1093/jee/tox252

Baroffio CA, Richoz P, Fischer S et al (2014) Monitoring *Drosophila suzukii* in Switzerland in 2012. J Berry Res 4:47–52. https://doi.org/10.3233/JBR-140066

Beers EH, Van Steenwyk RA, Shearer PW, Coates WW, Grant JA (2011) Developing *Drosophila suzukii* management programs for sweet cherry in the western United States. Pest Manag Sci 67(11):1386–1395. https://doi.org/10.1002/ps.2279

Bellamy DE, Sisterson MS, Walse SS (2013) Quantifying host potentials: indexing postharvest fresh fruits for spotted wing Drosophila, *Drosophila suzukii*. PLoS One 8:e61227. https://doi.org/10.1371/journal.pone.0061227

Bernardi D, Andreazza F, Botton M et al (2017) Susceptibility and interactions of *Drosophila suzukii* and *Zaprionus indianus* (Diptera: Drosophilidae) in damaging strawberry. Neotrop Entomol 46(1):1–7. https://doi.org/10.1007/s13744-016-0423-9

Bezerra Da Silva CS, Park KR, Blood RA et al (2019) Intraspecific competition affects the pupation behavior of spotted-wing Drosophila (*Drosophila suzukii*). Sci Rep 9:7775. https://doi.org/10.1038/s41598-019-44248-6

Biganski S, Jehle JA, Kleespies RG (2018) *Bacillus thuringiensis* serovar. israelensis has no effect on *Drosophila suzukii* Matsumura. J Appl Entomol 142:33–36. https://doi.org/10.1111/jen.12415

Bioplanet (2020) Veneto has started the launch of antagonist insect against *Drosophila suzukii*. https://www.freshplaza.com/article/9210339/veneto-has-started-the-launch-of-antagonist-insect-against-drosophila-suzukii. Accessed 21 Apr 2020

Bolda M (2008) New fruit fly pest in strawberries and caneberries. https://ucanr.edu/blogs/blogcore/postdetail.cfm?postnum=821. Accessed 14 May 2020

Bonneau P, Renkema J, Fournier V et al (2019) Ability of *Muscidifurax raptorellus* and other parasitoids and predators to control *Drosophila suzukii* populations in raspberries in the laboratory. Insects 10(3):68. https://doi.org/10.3390/insects10030068

Borba RS, Foppa F, Frare JB (2016) Ocorrência de *Drosophila suzukii* (Matsumura, 1931) (Diptera, Drosophilidae) em pêssego. In: Abstracts of the XXVI Congresso Brasileiro de Entomologia, Brazilia, 13–17 March 2016

Briem F, Eben A, Gross J (2016) An invader supported by a parasite: mistletoe berries as a host for food and reproduction of spotted wing drosophila in early spring. J Pest Sci 89(3):749–759. https://doi.org/10.1007/s10340-016-0739-6

Bufington ML, Forshage M (2016) Redescription of *Ganaspis brasiliensis* (Ihering, 1905), new combination, (Hymenoptera: Figitidae) a natural enemy of the invasive *Drosophila suzukii* (Matsumura, 1931) (Diptera: Drosophilidae). Proc Entomol Soc Wash 118(1):1–13. https://doi.org/10.4289/0013-8797.118.1.1

Burrack HJ, Fernandez GE, Spivey T et al (2013) Variation in selection and utilization of host crops in the field and laboratory by *Drosophila suzukii* Matsumura (Diptera: Drosophilidae), an invasive frugivore. Pest Manag Sci 69:1173–1180. https://doi.org/10.1002/ps.3489

CABI EPPO Centre for Agricultural Bioscience International, European and Mediterranean Plant Protection Organization (2016) Distribution maps of plant pests, Map 766 (1st revision). https://www.cabi.org/isc/abstract/20163203811. Accessed 19 Apr 2020

Cahenzli F, Strack T, Daniel C (2018) Screening of 25 different natural crop protection products against *Drosophila suzukii*. J Appl Entomol 142(6):563–577. https://doi.org/10.1111/jen.12510

Calabria G, Máca J, Bachli G et al (2012) First records of the potential pest species *Drosophila suzukii* (Diptera: Drosophilidae) in Europe. J Appl Entomol 136:139–147. https://doi.org/10.1111/j.1439-0418.2010.01583.x

Cichón LI, Lago JD, Garrido SA (2015) Primera detección de *Drosophila suzukii* (Matsumura, 1939) (Diptera: Drosophilidae) en frambuesas del Valle de Río Negro, Argentina. In: Abstracts of the IX Congreso Argentino de Entomología, Universidad Nacional de Misiones, Posadas, 10–12 September 2018

Cini A, Ioratti C, Anfora G (2012) A review of the invasion of *Drosophila suzukii* in Europe and a draft research agenda for integrated pest management. Bull Insectol 65:149–160

Cini A, Anfora G, Escudero-Colomar LA et al (2014) Tracking the invasion of the alien fruit pest *Drosophila suzukii* in Europe. J Pest Sci 87:559–566. https://doi.org/10.1007/s10340-014-0617-z

Chabert S, Allemand R, Poyet M, Eslin P, Gibert P (2012) Ability of European parasitoids (Hymenoptera) to control a new invasive Asiatic pest, Drosophila suzukii. Biol Control 63:40–47. https://doi.org/10.1016/j.biocontrol.2012.05.005

Cossentine J, Robertson M, Buitenhuis R (2016) Impact of acquired entomopathogenic fungi on adult *Drosophila suzukii* survival and fecundity. Biol Control 103:129–137. https://doi.org/10.1016/j.biocontrol.2016.09.002

Cuthbertson AGS, Audsley N (2016) Further screening of entomopathogenic fungi and nematodes as control agents for *Drosophila suzukii*. Insects 7(2):24. https://doi.org/10.3390/insects7020024

Cuthbertson AG, Blackburn LF, Audsley N (2014a) Efficacy of commercially available invertebrate predators against *Drosophila suzukii*. Insects 5(4):952–960. https://doi.org/10.3390/insects5040952

Cuthbertson AGS, Collins DA, Blackburn LF et al (2014b) Preliminary screening of potential control products against *Drosophila suzukii*. Insects 5(2):488–498. https://doi.org/10.3390/insects5020488

Daane KM, Wang X, Biondi A et al (2016) First exploration of parasitoids of *Drosophila suzukii* in South Korea as potential classical biological agents. J Pest Sci 89:823–835. https://doi.org/10.1007/s10340-016-0740-0

Dagatti CV, Marcucci B, Herrera ME et al (2018) Primera detección de *Drosophila suzukii* (Diptera: Drosophilidae) en frutos de zarzamora en Mendoza, Argentina. Rev Soc Entomol Arge 77(3):26–29. https://doi.org/10.25085/rsea.770304

Dalton DT, Walton VM, Shearer PW et al (2011) Laboratory survival of *Drosophila suzukii* under simulated winter conditions of the Pacific Northwest and seasonal field trapping in five primary regions of small and stone fruit production in the United States. Pest Manag Sci 67(11):1368–1374. https://doi.org/10.1002/ps.2280

Dekker T, Revadi S, Mansourian S et al (2015) Loss of *Drosophila* pheromone reverses its role in sexual communication in *Drosophila suzukii*. Proc R Soc B 282:20143018. https://doi.org/10.1098/rspb.2014.3018

Deprá M, Poppe JL, Schmitz HJ et al (2014) The first records of the invasive pest *Drosophila suzukii* in south American continent. J Pest Sci 87:379–383. https://doi.org/10.1007/s10340-014-0591-5

Diepenbrock LM, McPhie DR (2018) Use of a novel host plant family, Basallaceae, by *Drosophila suzukii* (Diptera: Drosophilidae). Proc Entomol Soc Wash 120(2):255–259. https://doi.org/10.4289/0013-8797.120.2.255

Dos Santos LA, Mendes MF, Krüger AP et al (2017) Global potential distribution of *Drosophila suzukii* (Diptera, Drosophilidae). PLoS One 12(3):e0174318. https://doi.org/10.1371/journal.pone.0174318

Dreves AJ (2008) Fruit fly, "Spotted Wing Drosophila," identified in wine grapes. https://today.oregonstate.edu/archives/2009/oct/fruit-fly-"spotted-wing-drosophila"-identified-wine-grapes. Accessed 20 May 2020

Drummond F, Ballman E, Collins J (2019) Population dynamics of spotted wing Drosophila (*Drosophila suzukii* (Matsumura)) in maine wild blueberry (*Vaccinium angustifolium* Aiton). Insects 10(7):205. https://doi.org/10.3390/insects10070205

Efil L (2018) Çanakkale ili çilek alanlarında yeni bir zararlı Drosophila suzukii (Matsumura) (Diptera: Drosophilidae)'nin yayılış alanları ve bulaşıklılığı. Türk Tarım ve Doğa Bilimleri Dergisi 5(3):280–284. https://doi.org/10.30910/turkjans.448354

Elsensohn JE, Loeb GM (2018) Non-crop host sampling yields insights into small-scale population dynamics of *Drosophila suzukii* (Matsumura). Insects 9(1):5. https://doi.org/10.3390/insects9010005

Emiljanowicz LM, Ryan GD, Langille A et al (2014) Development, reproductive output and population growth of the fruit fly pest *Drosophila suzukii* (Diptera: Drosophilidae) on artificial diet. J Econ Entomol 107(4):1392–1398. https://doi.org/10.1603/ec13504

Enriquez T, Colinet H (2017) Basal tolerance to heat and cold exposure of the spotted wing drosophila, *Drosophila suzukii*. PeerJ 5:e3112. https://doi.org/10.7717/peerj.3112

EPPO European and Mediterranean Plant Protection Organization (2014) EPPO gobal database. https://gd.eppo.int/taxon/DROSSU/distribution. Accessed 13 Apr 2020

EPPO European and Mediterranean Plant Protection Organization (2018) EPPO gobal database. https://gd.eppo.int/taxon/DROSSU/hosts. Accessed 13 Apr 2020

EPPO European and Mediterranean Plant Protection Organization (2019) EPPO gobal database. https://gd.eppo.int/taxon/DROSSU/hosts. Accessed 15 Apr 2020

Escobar LI, Ovruski SM, Kirschbaum DS (2018) Foreign invasive pests *Drosophila suzukii* (Matsumura) and *Zaprionus indianus* Gupta (Diptera: Drosophilidae) threaten fruit production in northwestern Argentina. Dros Info Serv 101:9–14

Fraimout A, Loiseau A, Price DK et al (2015) New set of microsatellite markers for the spotted-wing *Drosophila suzukii* (Diptera: Drosophilidae): a promising molecular tool for inferring the invasion history of this major insect pest. Eur J Entomol 112(4):855–859. https://doi.org/10.14411/eje.2015.079

Fraimout A, Jacquemart P, Villarroel B et al (2018) Phenotypic plasticity of *Drosophila suzukii* wing to developmental temperature: implications for flight. J Exp Biol 221(Pt 13):jeb166868. https://doi.org/10.1242/jeb.166868

Funes CF, Escobar LI, Meneguzzi NG et al (2017) Occurrence of *Anastrepha fraterculus* and *Ceratitis capitata* (Diptera: Tephritidae) in organically grown *Rubus* spp. (Rosales: Rosaceae), in two very contrasting environments of northwestern Argentina. Fla Entomol 100:672–674. https://doi.org/10.1653/024.100.0330

Funes CF, Escobar LI, Heredia AM et al (2018a) Primera detección e incidencia de *Drosophila suzukii* (Diptera: Drosophilidae) en cultivos de berries de Tucumán, Argentina. In: Abstracts of the X Congreso Argentino de Entomología, Mendoza, 21–24 May 2018

Funes CF, Escobar LI, Palavecino BE et al (2018b) Detección de *Drosophila suzukii* (Diptera: Drosophilidae) en cultivos de arándano de Tucumán, Argentina. In: Abstracts of the X Congreso Argentino de Entomología, Mendoza, 21–24 May 2018

Funes CF, Gallardo FE, Reche VA et al (2019) Parasitoides de Sudamérica asociados a las plagas invasoras *Drosophila suzukii* y *Zapronius indianus* (Diptera: Drosophilidae) y su potencial como agentes de control biológico. Semiárida Rev Fac Agr UNLPam 29(supl):25–27

Gabarra R, Arnó J, Riudavets J (2012) Primeros resultados sobre *Drosophila suzukii*: huéspedes, susceptibilidad de los frutos y enemigos naturales. Phytoma España 240:46–52

Gabarra R, Riudavets J, Rodríguez GA et al (2015) Prospects for the biological control of *Drosophila suzukii*. Biocontrol 60:331–339. https://doi.org/10.1007/s10526-014-9646-z

Garcia FRM, Wollmann J, Krüger AP et al (2017) Biological control of *Drosophila suzukii* (Matsumura, 1931) (Diptera: Drosophilidae): state of the art and prospects. In: Davenport L (ed) Biological control: methods, applications and challenges. Nova Science Publishers, Inc., New York, pp 1–28

García-Cancino MD, Hernández AG, Cabrera JG et al (2015) Parasitoids of *Drosophila suzukii* (Matsumura) (Diptera: Drosophilidae) in Colima, Mexico. Southwest Entomol 40(4):855–858. https://doi.org/10.3958/059.040.0418

Gargani E, Bagnoli B, Simoni S (2015) *Drosophila suzukii* in Tuscany (Italy), from cherry crops to vineyards and beyond. International Centre for Advanced Mediterranean Agronomic Studies (CIHEAM). Watch Letter n°33. https://www.iamm.ciheam.org/publications/177/019_Gargani.pdf. Accessed 3 Jun 2020

Garrido SA, Cichón LI, Lago JD, Gallardo FE, Navarro MD (2018) Primer registro de Leptopilina boulardi (Hymenoptera: Figitidae) asociada a Drosophila suzukii (Diptera: Drosophilidae) en el Alto Valle de Rio Negro y Neuquén, Patagonia Argentina. Rev Soc Entomol Arge 77:22–27

Gao HH, Zhai YF, Chen H, Wang YM, Liu Q, Hu QL, Ren FS, Yu Y (2018) Ecological niche difference associated with varied ethanol tolerance between Drosophila suzukii and Drosophila melanogaster (Diptera: Drosophilidae). Fla Entomol 101:498–504. https://doi.org/10.1653/024.101.0308

Geisler FCS, Santos J, Holdefer DR et al (2015) Primeiro registro de *Drosophila suzukii* (Matsumura, 1931) (Diptera: Drosophilidae) para o estado do Paraná, Brasil e de novos hospedeiros. Rev Css Amb 9:125–129. https://doi.org/10.18316/1981-8858.15

Giorgini M, Wang X, Wang Y et al (2019) Exploration for native parasitoids of *Drosophila suzukii* in China reveals a diversity of parasitoid species and narrow host range of the dominant parasitoid. J Pest Sci 92:509–522. https://doi.org/10.1007/s10340-018-01068-3

Girod P, Borowiec N, Buffington M et al (2018a) The parasitoid complex of *D. suzukii* and other fruit feeding Drosophila species in Asia. Sci Rep 8:11839. https://doi.org/10.1038/s41598-018-29555-8

Girod P, Lierhmann O, Urvois T et al (2018b) Host specifcity of Asian parasitoids for potential classical biological control of *Drosophila suzukii*. J Pest Sci 91:1241–1250. https://doi.org/10.1007/s10340-018-10033-z

Girod P, Rossignaud L, Haye T et al (2018c) Development of Asian parasitoids in larvae of *Drosophila suzukii* feeding on blueberry and artifcial diet. J Appl Entomol 142(5):483–494. https://doi.org/10.1111/jen.12496

Gong X, Bräcker L, Bölke N et al (2016) Strawberry accessions with reduced *Drosophila suzukii* emergence from fruits. Front Plant Sci 7:1880. https://doi.org/10.3389/fpls.2016.01880

Gonsebatt G, San Pedro P, Fanara JJ et al (2017) *Drosophila suzukii* (Matsumura) (Diptera: Drosophilidae) en durazneros (*Prunus persica*) en el sur de la provincia de Santa Fé: primer registro. In: Abstracts of the XIX Congreso y XXXVII Reunión Anual de Biología, Santa Fé, 28 November 2017

Gonzales G, Mary A, Goñi B (2015) *Drosophila suzukii* (Matsumura) found in Uruguay. Dros Info Serv 98:103–107

Gonzalez-Cabrera J, Moreno-Carrillo G, Sanchez-Gonzalez JA (2019) Single and combined release of *Trichopria drosophilae* (Hymenoptera: Diapriidae) to control *Drosophila suzukii* (Diptera: Drosophilidae). Neotrop Entomol 48:949–956. https://doi.org/10.1007/s13744-019-00707-3

Grassi A, Palmieri L, Giongo L (2009) Nuovo fitofago per i piccolo frutti in Trentino. Terra Trent. 55:19–23

Grassi A, Giongo L, Palmieri L (2011) *Drosophila (Sophophora) suzukii* (Matsumura), new pest of soft fruits in Trentino (North Italy) and in Europe. IOBC/WPRS Bull 70:121–128

Grassi A, Gottardello A, Dalton DT et al (2018) Seasonal reproductive biology of *Drosophila suzukii* (Diptera: Drosophilidae) in temperate climates. Environ Entomol 47(1):166–174. https://doi.org/10.1093/ee/nvx195

Guerrieri E, Giorgini M, Cascone P et al (2016) Species diversity in the parasitoid genus *Asobara* (Hymenoptera: Braconidae) from the native area of the fruit fly pest *Drosophila suzukii* (Diptera: Drosophilidae). PLoS One 11(2):e0147382. https://doi.org/10.1371/journal.pone.0147382

Hamby KA, Bolda MP, Sheehan ME et al (2014) Seasonal monitoring for *Drosophila suzukii* (Diptera: Drosophilidae) in California commercial raspberries. Environ Entomol 43:1008–1018. https://doi.org/10.1603/EN13245

Hamby KA, Bellamy DE, Chiu JC et al (2016) Biotic and abiotic factors impacting development, behavior, phenology, and reproductive biology of *Drosophila suzukii*. J Pest Sci 89(3):605–619. https://doi.org/10.1007/s10340-016-0756-5

Harris DW, Hamby KA, Wilson HE et al (2014) Seasonal monitoring of *Drosophila suzukii* (Diptera: Drosophilidae) in a mixed fruit production system. J Asia Pac Entomol 17(4):857–864. https://doi.org/10.1016/j.aspen.2014.08.006

Hauser M (2011) A historic account of the invasion of *Drosophila suzukii* (Matsumura) (Diptera: Drosophilidae) in the continental United States, with remarks on their identification. Pest Manag Sci 67(11):1352–1357. https://doi.org/10.1002/ps.2265

Haye T, Girod P, Cuthbertson AGS et al (2016) Current SWD IPM tactics and their practical implementation in fruit crops across different regions around the world. J Pest Sci 89(3):643–651. https://doi.org/10.1007/s10340-016-0737-8

Ideo S, Watada M, Mitsui H, Kimura MT (2008) Host range of *Asobara japonica* (Hym.: Braconidae), a larval parasitoid of drosophilid flies. Entomol Sci 11(1):1–6. https://doi.org/10.1111/j.1479-8298.2007.00244.x

Jakobs R, Gariepy TD, Sinclair BJ (2015) Adult plasticity of cold tolerance in a continental-temperate population of *Drosophila suzukii*. J Insect Physiol 79:1–9. https://doi.org/10.1016/j.jinsphys.2015.05.003

Junior Z, Fornazier MJ, Andreazza F et al (2018) Spread of two invasive flies (Diptera: Drosophilidae) infesting commercial fruits in Southeastern Brazil. Fla Entomol 101(3):522–525. https://doi.org/10.1653/024.101.0328

Kaçar G, Wang XG, Biondi A, Daane KM (2017) Linear functional response by two pupal *Drosophila* parasitoids foraging within single or multiple patch environments. PLoS One 12(8):e0183525. https://doi.org/10.1371/journal.pone.0183525

Kacsoh BZ, Schlenke TA (2012) High hemocyte load is associated with increased resistance against parasitoids in *Drosophila suzukii*, a relative of *D. melanogaster*. PLoS One 7(4):e34721. https://doi.org/10.1371/journal.pone.0034721

Kaneshiro KY (1983) *Drosophila (Sophophora) suzukii* (Matsumura). Proc Hawaii Entomol Soc 24:179

Kasuya N, Mitsui H, Ideo S et al (2013) Ecological, morphological and molecular studies on *Ganaspis* individuals (Hymenoptera: Figitidae) attacking *Drosophila suzukii* (Diptera: Drosophilidae). Appl Entomol Zool 48:87–92. https://doi.org/10.1007/s13355-012-0156-0

Kellermann V, Loeschcke V, Hoffmann AA et al (2012) Phylogenetic constraints in key functional traits behind species' climate niches: patterns of desiccation and cold resistance across 95 *Drosophila* species. Evol 66(11):3377–3389. https://doi.org/10.1111/j.1558-5646.2012.01685.x

Kenis M, Tonina L, Eschen R et al (2016) Non-crop plants used as hosts by *Drosophila suzukii* in Europe. J Pest Sci 89:735–748. https://doi.org/10.1007/s10340-016-0755-6

Kidera H, Takahashi KH (2020) Chemical cues from competitors change the oviposition preference of *Drosophila suzukii*. Entomol Exp Appl 168(4):304–310. https://doi.org/10.1111/eea.12889

Kienzle R, Groß LB, Caughman S et al (2020) Resource use by individual *Drosophila suzukii* reveals a fexible preference for oviposition into healthy fruits. Sci Rep 10:3132. https://doi.org/10.1038/s41598-020-59595-y

Kimura MT (1988) Adaptations to temperate climates and evolution of overwintering strategies in the *Drosophila melanogaster* species group. Evol 42:1288–1297. https://doi.org/10.1111/j.1558-5646.1988.tb04188.x

Kimura MT (2004) Cold and heat tolerance of drosophilid flies with reference to their latitudinal distributions. Oecologia 140:442–449. https://doi.org/10.1007/s00442-004-1605-4

Kinjo H, Kunimi Y, Ban T et al (2013) Oviposition efficacy of *Drosophila suzukii* (Diptera: Drosophilidae) on different cultivars of blueberry. J Econ Ent 106:1767–1771. https://doi.org/10.1603/EC12505

Kinjo H, Kunimi Y, Nakai M (2014) Effects of temperature on the reproduction and development of *Drosophila suzukii* (Diptera: Drosophilidae). Appl Entomol Zool 49(2):297–304. https://doi.org/10.1007/s13355-014-0249-z

Knoll V, Ellenbroek T, Romeis J et al (2017) Seasonal and regional presence of hymenopteran parasitoids of Drosophilain Switzerland and their ability to parasitize the invasive *Drosophila suzukii*. Sci Rep 7:40697. https://doi.org/10.1038/srep40697

Koch JB, Dupuis JR, Jardeleza MK et al (2020) Population genomic and phenotype diversity of invasive *Drosophila suzukii* in Hawai'i. Biol Invasions 22:1753–1770. https://doi.org/10.1007/s10530-020-02217-5

Krüger AP, Schlesener DCH, Martins LN et al (2018a) Effects of irradiation dose on sterility induction and quality parameters of *Drosophila suzukii* (Diptera: Drosophilidae). J Econ Entomol 111(2):741–746. https://doi.org/10.1093/jee/tox349

Krüger AP, Schlesener DC, Martins LN et al (2018b) Radiation effects on *Drosophila suzukii* (Diptera: Drosophilidae) reproductive behaviour. J Appl Entomol 143:88–94. https://doi.org/10.1111/jen.12563

Krüger AP, Scheunemann T, Vieira JGA et al (2019) Effects of extrinsic, intraspecific competition and host deprivation on the biology of *Trichopria anastrephae* (Hymenoptera: Diapriidae) reared on *Drosophila suzukii* (Diptera: Drosophilidae). Neotrop Entomol 48(6):957–965. https://doi.org/10.1007/s13744-019-00705-5

Lasa R, Tadeo E, Dinor LA et al (2017) Fruit firmness, superficial damage and location modulate infestation by *Drosophila suzukii* and *Zaprionus indianus* (Diptera: Drosophilidae): The case of guava *Psidium guajava* L., in Veracruz, Mexico. Entomol Exp Appl 162:4–12. https://doi.org/10.1111/eea.12519

Lauyé MA (2017) Diversidad de Drosophilidae (Diptera) en frutos de interés productivo y agronómico, en localidades de la región sur de Uruguay, con especial atención a la especie invasora *Drosophila suzukii* (Matsumura, 1931). Dissertation, Universidad de la República (Uruguay)

Lavagnino NJ, Diaz BM, Cichón LI et al (2018) New records of the invasive pest *Drosophila suzukii* (Matsumura) (Diptera: Drosophilidae) in the South American continent. Rev Soc Entomol Argent 77(1):27–31. https://doi.org/10.25085/rsea.770105

Lee RE (1991) Principles of insect low temperature tolerance. In: Insects at low temperature. Springer, Boston, pp 17–46

Lee JC, Bruck DJ, Curry H et al (2011a) The susceptibility of small fruits and cherries to the spotted-wing drosophila, *Drosophila suzukii*. Pest Manag Sci 67(11):1358–1367. https://doi.org/10.1002/ps.2225

Lee JC, Bruck DJ, Dreves AJ et al (2011b) In focus: spotted wing drosophila, *Drosophila suzukii*, across perspectives. Pest Manag Sci 67(11):1349–1351. https://doi.org/10.1002/ps.2271

Lee JC, Burrack HJ, Barrantes LD et al (2012) Evaluation of monitoring traps for *Drosophila suzukii* (Diptera: Drosophilidae) in North America. J Econ Entomol 105(4):1350–1357. https://doi.org/10.1603/EC12132

Lee JC, Dreves AJ, Cave AM et al (2015) Infestation of wild and ornamental noncrop fruits by *Drosophila suzukii* (Diptera: Drosophilidae). Ann Entomol Soc Am 108(2):117–129. https://doi.org/10.1093/aesa/sau014

Lee JC, Wang XG, Daane KM et al (2019) Biological control of spotted-wing Drosophila (Diptera: Drosophilidae)-current and pending tactics. J Integr Pest Manag 10(1):22. https://doi.org/10.1093/jipm/pmz021

Little CM, Chapman TW, Moreau DL et al (2017) Susceptibility of selected boreal fruits and berries to the invasive pest *Drosophila suzukii* (Diptera: Drosophilidae). Pest Manag Sci 73(1):160–166. https://doi.org/10.1002/ps.4366

Little CM, Chapman TW, Hillier NK (2020) Plasticity is key to success of *Drosophila suzukii* (Diptera: Drosophilidae) invasion. J Insect Sci 20(3):5. https://doi.org/10.1093/jisesa/ieaa034

Lochbaum T (2017) *Drosophila suzukii* (Diptera: Drosophilidae): Distribución estacional, incidencia y estrategias preliminares de manejo en el departamento Confluencia, Neuquén. Primer Encuentro Nacional Técnico-Científico sobre la plaga *Drosophila suzukii*. SENASA, Buenos Aires

Lue CH, Mottern JL, Walsh GC et al (2017) New record for the invasive spotted wing Drosophila, *Drosophila suzukii* (Matsumura, 1931) (Diptera: Drosophilidae) in Anillaco, Western Argentina. Proc Entomol Soc Wash 119(1):146–150. https://doi.org/10.4289/0013-8797.119.1.146

Magnacca KN, Foote D, O'Grady PM (2008) A review of the endemic Hawaiian Drosophilidae and their host plants. Zootaxa 1728:1–58

Matsuura A, Mitsui H, Kimura MT (2018) A preliminary study on distributions and oviposition sites of *Drosophila suzukii* (Diptera: Drosophilidae) and its parasitoids on wild cherry tree in Tokyo, central Japan. Appl Entomol Zool 53(1):47–53. https://doi.org/10.1007/s13355-017-0527-7

Mazzetto F, Pansa MG, Ingegno BL et al (2015) Monitoring of the exotic fly *Drosophila suzukii* in stone, pome and soft fruit orchards in NW Italy. J Asia Pac Entomol 18(2):321–329. https://doi.org/10.1016/j.aspen.2015.04.001

Mazzetto F, Marchetti E, Amiresmaeili N et al (2016) Drosophila parasitoids in northern Italy and their potential to attack the exotic pest *Drosophila suzukii*. J Pest Sci 89(1):837–850. https://doi.org/10.1007/s10340-016-0746-7

Mazzetto F, Lessio F, Giacosa S et al (2020) Relationships between *Drosophila suzukii* and grapevine in North-western Italy: seasonal presence and cultivar susceptibility. Bull Insectol 73(1):29–38

Mazzi D, Bravin E, Meraner M et al (2017) Economic impact of the introduction and establishment of *Drosophila suzukii* on Sweet cherry production in Switzerland. Insects 8:18. https://doi.org/10.3390/insects8010018

Mazzoni V, Anfora G, Virant-Doberlet M (2013) Substrate vibrations during courtship in three *Drosophila* species. PLOS One 8(11):e80708. https://doi.org/10.1371/journal.pone.0080708

Miller B, Anfora G, Bufngton M et al (2015) Seasonal occurrence of resident parasitoids associated with *Drosophila suzukii* in two small fruit production regions of Italy and the USA. B Insectol 68(2):255–263

Mitsui H, Kimura MT (2010) Distribution, abundance and host association of two parasitoid species attacking frugivorous drosophilid larvae in central Japan. Eur J Entomol 107(4):535–540. https://doi.org/10.14411/eje.2010.061

Mitsui H, Takahashi KH, Kimura MT (2006) Spatial distributions and clutch sizes of *Drosophila* species ovipositing on cherry fruits of different stages. Popul Ecol 48(3):233–237. https://doi.org/10.1007/s10144-006-0260-5

Mitsui H, Van Achterberg K, Nordlander G et al (2007) Geographical distributions and host associations of larval parasitoids of frugivorous Drosophilidae in Japan. J Nat Hist 41:1731–1738. https://doi.org/10.1080/00222930701504797

Mitsui H, Beppu K, Kimura MT (2010) Seasonal life cycles and resource uses of flower- and fruit-feeding Drosophilid flies (Diptera: Drosophilidae) in central Japan. Entomol Sci 13:60–67. https://doi.org/10.1111/j.1479-8298.2010.00372.x

Morales K (2020) The spotted wing Drosophila in the south of the world: Chilean case and its first productive impacts. In: Invasive species—introduction pathways, economic impact, and possible management options. https://www.intechopen.com/online-first/the-spotted-wing-drosophila-in-the-south-of-the-world-chilean-case-and-its-first-productive-impacts. Accessed 15 Apr 2020

Müller FA, Nava DE (2014) Primeiro relato de *Drosophila suzukii* (Diptera: Drosophilidae) em frutos de araçá e pitanga em municípios da região sul do Rio Grande do Sul, Brasil. In: Abstracts of the XXV Congresso Brasileiro de Entomologia, Universidade Federal de Goiás, 4–18 September 2014

Nava DE, Botton M, Bernardi D et al (2015) Bioecologia, monitoramento e controle de *Drosophila suzukii* na cultura do morangueiro. Série documentos 398:1–28

Nomano FY, Mitsui H, Kimura MT (2015) Capacity of Japanese *Asobara* species (Hymenoptera: Braconidae) to parasitize a fruit pest *Drosophila suzukii* (Diptera: Drosophilidae). J Appl Entomol 139:105–113. https://doi.org/10.1111/jen.1214

Nomano FY, Kasuya N, Matsuura A et al (2017) Genetic differentiation of *Ganaspis brasiliensis* (Hymenoptera: Figitidae) from East and Southeast Asia. Appl Entomol Zool 52:429–437. https://doi.org/10.1111/jen.12141

Novkovic B, Mitsui H, Suwito A, Kimura MT (2011) Taxonomy and phylogeny of *Leptopilina* species (Hymenoptera: Cynipoidea: Figitidae) attacking frugivorous drosophilid flies in Japan, with description of three new species. Entomol Sci 14:333–346. https://doi.org/10.1111/j.1479-8298.2011.00459.x

Nunes AM, Schlesener DCH, Souza DS et al (2014) Primeiros registros de *Drosophila suzukii* (Diptera: Drosophilidae) em agroecossistemas na metade sul do Rio Grande do Sul. In: Abstracts of the XXV Congresso Brasileiro de Entomologia, Universidade Federal de Goiás, 4–18 September 2014

O'Grady PM, Beardsley JW, Perreira WD (2002) New records for introduced Drosophilidae (Diptera) in Hawaii. Bishop Mus Occ Pap 69:34–35

Oliveira AS, Amaral Neto JAB, Santos RSS (2015) Primeiro registro de *Drosophila suzukii* (Diptera: Drosophilidae) em pomar de macieira em Vacaria- Rio Grande do Sul. In: Abstracts of the XIII Encontro de Iniciação Científica e IX Encontro de Pós-Graduandos da Embrapa Uva e Vinho, Bento Gonçalves, 16–17 July 2015

Orhan A, Aslantaş R, Önder BŞ et al (2016) First record of the invasive vinegar fly *Drosophila suzukii* (Matsumura) (Diptera: Drosophilidae) from eastern Turkey. Turk J Zool 40:290–293. https://doi.org/10.3906/zoo-1412-25

Panel ADC, Zeeman L, van der Sluis BJ et al (2018) Overwintered *Drosophila suzukii* are the main source for infestations of the first fruit crops of the season. Insects 9:145. https://doi.org/10.3390/insects9040145

Pfab F, Stacconi MVR, Anfora G et al (2018) Optimized timing of parasitoid release: a mathematical model for biological control of *Drosophila suzukii*. Theor Ecol 11:489–450. https://doi.org/10.1007/s12080-018-0382-3

Poyet M, Havard S, Prévost G et al (2013) Resistance of *Drosophila suzukii* to the larval parasitoids *Leptopilina heterotoma* and *Asobara japonica* is related to haemocyte load. Physiol Entomol 38:45–53. https://doi.org/10.1111/phen.12002

Poyet M, Le Roux V, Gibert P et al (2015) The wide potential trophic niche of the Asiatic fruit fly *Drosophila suzukii*: The key of Its invasion success in temperate Europe? PLoS One 10(11):e0142785. https://doi.org/10.1371/journal.pone.0142785

Rendon D, Walton V, Tait G et al (2019) Interactions among morphotype, nutritions, and temperature impact fitness of an invasive fly. Ecol Evol 9(5):2615–2252. https://doi.org/10.1002/ece3.4928

Renkema JM, Cuthbertson AGS (2018) Impact of multiple natural enemies on immature *Drosophila suzukii* in strawberries and blueberries. Biocontrol 63:719–728. https://doi.org/10.1007/s10526-018-9874-8

Renkema JM, Telfer Z, Gariepy T et al (2015) *Dalotia coriaria* as a predator of *Drosophila suzukii*: functional responses, reduced fruit infestation and molecular diagnostics. Biol Control 89:1–10. https://doi.org/10.1016/j.biocontrol.2015.04.024

Revadi S, Lebreton S, Witzgall P et al (2015) Sexual behavior of *Drosophila suzukii*. Insects 6(1):183–196. https://doi.org/10.3390/insects6010183

Rossi-Stacconi M, Grassi A, Dalton D et al (2013) First field records of *Pachycrepoideus vindemiae* as a parasitoid of *Drosophila suzukii* in European and Oregon small fruit production areas. Entomol 1(1):11–16. https://doi.org/10.4081/entomologia.2013.e3

Rossi-Stacconi MV, Bufington M, Daane KM et al (2015) Host stage preference, efficacy and fecundity of parasitoids attacking *Drosophila suzukii* in newly invaded areas. Biol Control 84:28–35. https://doi.org/10.1016/j.biocontrol.2015.02.003

Rossi-Stacconi MV, Kaur R, Mazzoni V et al (2016) Multiple lines of evidence for reproductive winter diapause in the invasive pest *Drosophila suzukii*. J Pest Sci 89(3):689–700. https://doi.org/10.1007/s10340-016-0753-8

Rossi-Stacconi MV, Panel A, Baser N et al (2017) Comparative life history traits of indigenous Italian parasitoids of *Drosophila suzukii* and their effectiveness at different temperatures. Biol Control 112:20–27. https://doi.org/10.1016/j.biocontrol.2017.06.003

SAG Servicio Agrícola y Ganadero (2019) Agrícola/Plagas y Enfermedades/Plagas Relevantes Presentes/*Drosophila suzukii*. https://www.sag.gob.cl/ambitos-de-accion/drosophila-suzukii. Accessed 16 May 2020

Sampson BJ, Mallette T, Addesso KM et al (2016) Novel aspects of *Drosophila suzukii* (Diptera: Drosophilidae) biology and an improved method for culturing this invasive species with a modified *D. melanogaster* diet. Fla Entomol 99(4):774–780. https://doi.org/10.1653/024.099.0433

Santadino MV, Riquelme Virgala MB, Ansa MA et al (2015) First record of *Drosophila suzukii* (Diptera: Drosophilidae) in blueberry (*Vaccinium* spp.) from Argentina. Rev Soc Entomol Argen 74(3-4):183–185

Santos RSS (2014) Ocorrência de *Drosophila suzukii* (Matsumura, 1931) (Diptera: Drosophilidae) atacando frutos de morango no Brasil. http://ainfo.cnptia.embrapa.br/digital/bitstream/item/106368/1/Comunicado-Tecnico-159.pdf. Accessed 20 Apr 2020

Sapir Y, Mazer SJ, Holzapfel C (2008) Sex ratio. In: Erik JS, Brian F (eds) Encyclopedia of ecology. Academic, Oxford, pp 3243–3248

Sarto V, Sorribas R (2011) *Drosophila suzukii* (Matsumura 1931) nueva amenaza para las producciones agrícolas. Phytoma España 234:1–6

Sasaki M, Sato R (1995) Bionomics of the cherry Drosophila, *Drosophila suzukii* Matsumura (Diptera: Drosophilidae) in Fukushima Prefecture. Ann Rep Plant Prot North Jpn 46:170–172

Schlesener DCH, Wollmann J, Pazini JB et al (2017) Effects of insecticides on adults and eggs of *Drosophila suzukii* (Diptera, Drosophilidae). Rev Colomb Entomol 43(2):208–214. https://doi.org/10.25100/socolen.v43i2.5945

Schlesener DCH, Wollmann J, Kruger AP et al (2018) Biology and fertility life table of *Drosophila suzukii* on artificial diets. Entomol Exp Appl 166(11–12):932–936. https://doi.org/10.1111/eea.12736

Schlesener DCH, Wollmann J, Pazini JB et al (2019) Insecticide toxicity to *Drosophila suzukii* (Diptera: Drosophilidae) parasitoids: *Trichopria anastrephae* (Hymenoptera: Diapriidae) and *Pachycrepoideus vindemmiae* (Hymenoptera: Pteromalidae). J Econ Entomol 112(3):1197–1206. https://doi.org/10.1093/jee/toz033

Schlesener DCH, Wollmann J, Kruger AP et al (2020) Effect of temperature on reproduction, development, and phenotypic plasticity of *Drosophila suzukii* in Brazil. Entomol Exp Appl. https://doi.org/10.1111/eea.12954

Shaw B, Brain P, Wijnen H et al (2018) Reducing *Drosophila suzukii* emergence through inter-species competition. Pest Manag Sci 74(6):1466–1471. https://doi.org/10.1002/ps.4836

Shearer PW, West JD, Walton VM et al (2016) Seasonal cues induce phenotypic plasticity of *Drosophila suzukii* to enhance winter survival. BMC Ecol 16:11. https://doi.org/10.1186/s12898-016-0070-3

Shrader ME (2017) *Drosophila suzukii* (Matsumura) (Diptera: Drosphilidae): risk assessment for an invasive vinegar fly in Virginia vineyards. Dissertation, Virginia Polytechnic Institute and State University

Shrader ME, Burrack HJ, Pfeiffer DG (2020) Effects of interspecific larval competition on developmental parameters in nutrient sources between *Drosophila suzukii* (Diptera: Drosophilidae) and *Zaprionus indianus*. J Econ Entomol 113(1):230–238. https://doi.org/10.1093/jee/toz297

Skalski JR, Ryding KE, Millspaugh JJ (2005) Wildlife demography: analysis of sex, age, and count data. Academic, San Diego

Snellings Y, Herrera B, Wildemann B et al (2018) The role of cuticular hydrocarbons in mate recognition in *Drosophila suzukii*. Sci Rep 8:4996. https://doi.org/10.1038/s41598-018-23189-6

Steffan SA, Lee JC, Singleton ME et al (2013) Susceptibility of cranberries to *Drosophila suzukii* (Diptera: Drosophilidae). J Econ Entomol 106(6):2424–2427. https://doi.org/10.1603/EC13331

Stephens AR, Asplen MK, Hutchison WD et al (2015) Cold hardiness of winter-acclimated *Drosophila suzukii* (Diptera: Drosophilidae) adults. Environ Entomol 44(6):1619–1626. https://doi.org/10.1093/ee/nvv134

Stewart TJ (2015) Agricultural landscape ecology of spotted wing drosophila in the central valley of California. Dissertation, Jordan College of Agricultural Sciences and Technology California State University

Stockton D, Wallingford A, Loeb G (2018) Phenotypic plasticity promotes overwintering survival in a globally invasive crop pest, *Drosophila suzukii*. Insects 9:105. https://doi.org/10.3390/insects9030105

Stockton D, Wallingford A, Rendon D et al (2019) Interactions between biotic and abiotic factors affect survival in overwintering *Drosophila suzukii* (Diptera: Drosophilidae). Environ Entomol 48(2):454–464. https://doi.org/10.1093/ee/nvy192

Sward GFH (2017) Evaluating host plant use by spotted wing Drosophila, *Drosophila suzukii*, in Minnesota. Dissertation, University of Minnesota

Swoboda-Bhattarai KA, Burrack HJ (2016) *Drosophila suzukii* infestation in ripe and ripening caneberries. Acta Hortic 1133(65):419–430. https://doi.org/10.17660/ActaHortic.2016.1133.65

Tait G, Cabianca A, Grassi A et al (2020) *Drosophila suzukii* daily dispersal between distinctly different habitats. Entomol Gen 40(1):25–37. https://doi.org/10.1127/entomologia/2019/0876

Thistlewood HMA, Gill P, Beers EH et al (2018) Spatial analysis of seasonal dynamics and overwintering of *Drosophila suzukii* (Diptera: Drosophilidae) in the Okanagan-Columbia Basin, 2010-2014. Environ Entomol 47(2):221–232. https://doi.org/10.1093/ee/nvx178

Tochen S, Dalton DT, Wiman N et al (2014) Temperature related development and population parameters for *Drosophila suzukii* (Diptera: Drosophilidae) on cherry and blueberry. Environ Entomol 43(2):501–510. https://doi.org/10.1603/EN13200

Tonina L, Mori N, Giomi F, Battisti A (2016) Development of *Drosophila suzukii* at low temperatures in mountain areas. J Pest Sci 89:667–678. https://doi.org/10.1007/s10340-016-0730-2

Toxopeus J, Jakobs R, Ferguson LV et al (2016) Reproductive arrest and stress resistance in winter-acclimated *Drosophila suzukii*. J Insect Physiol 89:37–51. https://doi.org/10.1016/j.jinsphys.2016.03.006

Vilaire A, Lavine LC, Walsh DB (2011) The interspecific competition between invasive *Drosophila suzukii* and endemic *Drosophila melanogaster* on fruit hosts. In: Abstracts of the Conference: Entomological Society of America Annual Meeting, Reno, NV, 13–16 November 2011

Vilela CR, Mori L (2014) The invasive spotted-wing Drosophila (Diptera, Drosophilidae) has been found in the city of São Paulo (Brazil). Rev Bras Entomol 58(4):371–375. https://doi.org/10.1590/S0085-56262014000400004

Wallingford AK, Lee JC, Loeb GM (2016) The influence of temperature and photoperiod on the reproductive diapause and cold tolerance of spotted-wing drosophila, *Drosophila suzukii*. Entomol Exp Appl 159(3):327–337. https://doi.org/10.1111/eea.12443

Walsh DB, Bolda MP, Goodhue RE et al (2011) *Drosophila suzukii* (Diptera: Drosophilidae): invasive pest of ripening soft fruit expanding its geographic range and damage potential. J Integr Pest Manag 2(1):G1–G7. https://doi.org/10.1603/IPM10010

Wang XG, Kaçar G, Biondi A et al (2016a) Life-history and host preference of *Trichopria drosophilae*, a pupal parasitoid of spotted wing drosophila. Biocontrol 61:387–397. https://doi.org/10.1007/s10526-016-9720-9

Wang XG, Kaçar G, Biondi A et al (2016b) Foraging efficiency and outcomes of interactions of two pupal parasitoids attacking the invasive spotted wing Drosophila. Biol Control 96:64–71. https://doi.org/10.1016/j.biocontrol.2016.02.004

Wang XG, Stewart TJ, Biondi A et al (2016c) Population dynamics and ecology of *Drosophila suzukii* in Central California. J Pest Sci 89(3):701–712. https://doi.org/10.1007/s10340-016-0747-6

Wang XG, Nance AH, Jones JML et al (2018a) Aspects of the biology and reproductive strategy of two Asian larval parasitoids evaluated for classical biological control of *Drosophila suzukii*. Biol Control 121:58–65. https://doi.org/10.1016/j.biocontrol.2018.02.010

Wang XG, Serrato MA, Son Y et al (2018b) Thermal performance of two indigenous pupal parasitoids attacking the invasive *Drosophila suzukii* (Diptera: Drosophilidae). Environ Entomol 47(3):764–772. https://doi.org/10.1093/ee/nvy053

Wang XG, Kaçar G, Daane KM (2019a) Temporal dynamics of host use by *Drosophila suzukii* in California's San Joaquin Valley: implications for area-wide pest management. Insects 10(7):206. https://doi.org/10.3390/insects10070206

Wang XG, Hogg BN, Hougardy E et al (2019b) Potential competitive outcomes among three solitary larval endoparasitoids as candidate agents for classical biological control of *Drosophila suzukii*. Biol Control 130:18–26. https://doi.org/10.1016/j.biocontrol.2018.12.003

Wang XG, Biondi A, Daane KM (2020) Functional responses of three candidate Asian larval parasitoids evaluated for classical biological control of Drosophila suzukii (Diptera: Drosophilidae). J Econ Entomol 113(1):73–80. https://doi.org/10.1093/jee/toz265

Weydert C, Mandrin JF (2013) Le ravageur émergent *Drosophila suzukii*: situation en France et connaissances acquises en verger (2eme partie). Infos CTIFL 292:32–40

Weydert C, Trottin Y, Mandrin JF et al (2016) *Drosophila suzukii* connaissance du ravageur, moyens de protection. Infos CTIFL:1–16

Wilson HE, Hamby KA, Zalom FG (2013) Host susceptibility of "French Prune" Prunus domestica to *Drosophila suzukii* (Diptera: Drosophilidae). Acta Hortic (985):249–254. https://doi.org/10.17660/ActaHortic.2013.985.32

Wiman NG, Walton VM, Dalton DT et al (2014) Integrating temperature-dependent life table data into a matrix projection model for *Drosophila suzukii* population estimation. PLoS One 9(9):e106909. https://doi.org/10.1371/journal.pone.0106909

Wolf S, Baur H, Collatz J (2019) Life history of *Vrestovia fidenas*, a potential control agent of *Drosophila suzukii*. Biocontrol 64(3):263–275. https://doi.org/10.1007/s10526-019-09933-5

Wollmann J, Schlesener DCH, Ferreira MS et al (2016) Parasitoids of Drosophilidae with potential for parasitism on *Drosophila suzukii* in Brazil. Dros Info Serv 99:38–42

Wollmann J, Schlesener DCH, Ferreira MS et al (2019) Population dynamics of *Drosophila suzukii* (Diptera: Drosophilidae) in berry crops in southern Brazil. Neotrop Entomol 48(4):699–705. https://doi.org/10.1007/s13744-019-00686

Woltering SB, Romeis J, Collatz J (2019) Influence of the rearing host on biological parameters of *Trichopria drosophilae*, a potential biological control agent of Drosophila suzukii. Insects 10(6):183. https://doi.org/10.3390/insects10060183

Woltz JM, Lee JC (2017) Pupation behavior and larval and pupal biocontrol of *Drosophila suzukii* in the field. Biol Control 110:62–69. https://doi.org/10.1016/j.biocontrol.2017.04.007

Woltz JM, Donahue KM, Bruck DJ et al (2015) Efficacy of commercially available predators, nematodes and fungal entomopathogens for augmentative control of *Drosophila suzukii*. J Appl Entomol 139:759–770. https://doi.org/10.1111/jen.12200

Yi C, Cai P, Lin J et al (2020) Life history and host preference of Trichopria *drosophilae* from Southern China, one of the effective pupal parasitoids on the Drosophila Species. Insects 11(2):103. https://doi.org/10.3390/insects11020103

Yousef M, Aranda-Valera E, Quesada-Moraga E (2018) Lure-and-infect and lure-and-kill devices based on *Metarhizium brunneum* for spotted wing drosophila control. J Pest Sci 91:227–235. https://doi.org/10.1007/s10340-017-0874-8

Yu D, Zalom FG, Hamby KA (2013) Host status and fruit odor response of *Drosophila suzukii* (Diptera: Drosophilidae) to figs and mulberries. J Econ Entomol 106(4):1932–1937. https://doi.org/10.1603/EC12480

Zengin E, Karaca I (2019) Dynamics of trapped adult populations of *Drosophila suzukii* Matsumura (Diptera: Drosophilidae) and its parasitoids in Uşak Province, Turkey. Egypt J Biol Pest Co 29:43. https://doi.org/10.1186/s41938-019-0147-3

Zerulla FN, Schmidt S, Streitberger M et al (2015) On the overwintering ability of *Drosophila suzukii* in South Tyrol. J Berry Res 5:41–48. https://doi.org/10.3233/JBR-150089

Chapter 5
Basis for Area-Wide Management of *Drosophila suzukii* in Latin America

Flávio Roberto Mello Garcia

Abstract *Drosophila suzukii* (Matsumura) is a species native to Western Asia and have a rapid expansion worldwide; occurrences were recorded in North America and Europe in 2008, South America in 2013, and Africa in 2020. In Latin America (LA), *D. suzukii* occurs in Argentina, Brazil, Chile, Mexico, and Uruguay. *D. suzukii* is a polyphagous species with 52 host plants from 23 families in LA, most of these hosts are exotic. In LA, there are 12 species of parasitoids of *D. suzukii* belonging to Diapriidae, Figitidae, and Pteromalidae, promising native parasitoids in the control of this pest, such as *Trichopria anastrephae* Lima (Hymenoptera, Diapriidae). This chapter presents results from monitoring, biological control, chemical control, cultural control, Sterile Insect Technique that served as the basis for the area-wide management of *D. suzukii* under LA conditions.

Keywords Monitoring · Biological control · Chemical control · Sterile insect technique · Neotropical region

5.1 Introduction

Latin America (LA) includes most of the Neotropical Region. The Neotropical region comprises South America, Central America, and reaches as far north as central Mexico (Morrone 2014). LA is the largest tropical fruit exporting region in the world and accounts for about 25% of the total worldwide production of these fruits, with an annual production of about 54 million tons. The region generates an estimated total export value of around US $ 11 billion (FAO 2019).

F. R. M. Garcia (✉)
Departamento de Ecologia, Zoologia e Genética, Instituto de Biologia, Universidade Federal de Pelotas, Pelotas, RS, Brazil
e-mail: flavio.garcia@ufpel.edu.br

F. R. M. Garcia (ed.), *Drosophila suzukii Management*,
https://doi.org/10.1007/978-3-030-62692-1_5

One of the threats to this production is the recent introduction of *Drosophila suzukii* (Matsumra) or Spotetted Wing Drosophila (SWD). Currently, SWD is distributed in five Latin American countries: Argentina, Brazil, Chile, Mexico, and Uruguay. The first detection in LA was in Mexico (Trujillo-Arriaga 2011), 2 years later arrived in Brazil (Deprá et al. 2014), years after being detected in Brazil, it was detected in the other LA countries, in Argentina (Díaz et al. 2015; Santadino et al. 2015), Chile (Rojas et al. 2019), and in Uruguay (González et al. 2015).

The first research project in South America on SWD was the project entitled "Study of the biology, ecology, and genetics of Brazilian populations of the invasive species *D. suzukii* approved in 2014, a partner of two Brazilian universities (Universidade Federal de Pelotas and Universidade Federal do Rio Grande do Sul)" and the project entitled "Interactions between *Drosophila suzukii* (Diptera: Drosophilidae), their hosts and their parasitoids (Hymenoptera) in the Brazilian Pampa Biome" coordinated by the Universidade Federal de Pelotas, both financed by the Brazilian National Council for Scientific and Technological Development (CNPq). These projects generated the first publications on biological, ecological, and population control aspects of *D. suzukii* in Latin America.

The distribution of *D. suzukii* in LA may be higher than the one currently recorded. This is due to the lack of entomologists trained in the identification of the species in some countries, the lack of monitoring and the low density at the beginning of the invasion of *D. suzukii*; in addition, the species often inhabits areas of natural forests, which are often rarely monitored. Although countries in Asia, Europe, and the United States have advanced in the management of SWD, it is important that LA create and adapt the management techniques of this species to the conditions of the region. Thus, the objective of this chapter is to present the current situation of integrated management of *D. suzukii* in Latin America, and to present the most appropriate forms in the management of this species for the region. This chapter was organized on the basis of an integrated area-wide pest management approach (AW-IPM). AW-IPM is a coordinated, sustainable, and preventive approach that targets pest populations in all areas, including non-commercial urban areas (Hendrichs et al. 2007). Thus, the objective of this chapter is to present the current situation of integrated management of *D. suzukii* in Latin America, and to present the most appropriate forms in the management of this species for the region.

5.2 Status of the Distribution of *Drosophila suzukii* in Latin America

The first detection of *D. suzukii* in Latin America occurred in Mexico in 2011 (Trujillo-Arriaga 2011) and was detected in 2013 in Brazil (Deprá et al. 2014), in 2014 in Uruguay, in 2015 in Argentina (Cichón et al. 2016) and Chile (Rojas et al. 2019). However, there are previous records of SWD occurrence in Costa Rica in 1997 and in Ecuador in 1998 (Hauser 2011), such specimens were not found in any insect collection, and this species had not been collected elsewhere in these countries (Andreazza et al. 2017a).

The central region of southern Brazil, the southern half of Paraguay, all of Uruguay and the regions to the east and south of Argentina as potential distribution areas. On the Pacific coast, the entire coastline of Chile is indicated as a potential distribution area of the SWD. The areas of greatest environmental suitability for *D. suzukii* are in southern Chile, Uruguay, on the south coast and in southern Brazil, and along small range on the northern coast of Argentina (dos Santos et al. 2017).

In Mexico, *D. suzukii* was first detected in Tijuana and Ensenada, Baja California, Ziracuaretiro, Michoacán, and several municipalities in the states of Aguas Calientes, Guanajuato, and State of Mexico. In 2014, the species was detected in the municipalities of San Juan del Rio, in Querétaro; Ixtlán, Uruapan, Taretan, Tocumbo, Periban, Chavinda, Jacona, Zamora and Pajacuaran, and Michoacánem (Santo-Ramos et al. 2014; Lasa and Tadeo 2015). The state of Jalisco in Mexico presents the most favorable environmental conditions for *D. suzukii* populations in the country, followed by the states of Michoacán and Oaxaca (Castro-Sosa et al. 2017).

The population of *D. suzukii* from Brazil originates in North America, with a genetic mixture between individuals of *D. suzukii* from the southwest and Eastern regions of the USA (Fraimout et al. 2017). Currently the species is found in eight states of the country, the first records were in Rio Grande do Sul and Santa Catarina states (Deprá et al. 2014), it was then registered in Paraná (Geisler et al. 2015), Federal District (Paula et al. 2014), Rio de Janeiro (Bitner-Mathé et al. 2014), São Paulo (Vilela and Mori 2014), Minas Gerais (Andreazza et al. 2016), and Espirito Santo (Zanúncio Junior et al. 2018). SWD is distributed in 26 municipalities in Brazil (Table 5.1), the state of Rio Grande do Sul presents the highest number of municipalities with occurrence of SWD (10 municipalities), followed by the state of Santa Catarina with six municipalities (Table 5.1). In the Federal District, Minas Gerais and Rio de Janeiro, the species was registered in only one municipality (Table 5.1). In the case of the Federal District and Rio de Janeiro, the species was detected in natural forest areas (Bitner-Mathé et al. 2014; Paula et al. 2014). Southern Brazil is the most climatically favorable area for SWD development and where potential economic losses are expected to be the greatest (Benito et al. 2016; dos Santos et al. 2017). The SWD expanded from southern to southeastern Brazil, aided by human-mediated transport of fruits from region to region (Ferronato et al. 2019). Benito et al. (2016) estimated monetary economic losses of US$ 21.4 million for peaches and US$ 7.8 million for figs in Brazil.

The invasion of SWD in Argentina must have occurred in at least two events, originating from previously invaded areas, i.e., North America and Brazil (de la Vega et al. 2020). There are SWD larvae in cherries in this country which can cause problems exporting these fruits due to quarantine issues with Australia and New Zealand (Cichón et al. 2019). The species was first detected in Argentina in Rio Negro in 2015 attacking raspberries (Cichón et al. 2016); in the same year, it was collected in traps in orchards of Entre Ríos (Díaz et al. 2015; Lavagnino et al. 2018) and in Buenos Aires in blueberries (Santadino et al. 2015). In La Rioja, *D. suzukii* was collected in traps (Lue et al. 2017), in Santa Fé, was collected in peach trees

Table 5.1 States and municipalities in Brazil where *Drosophila suzukii* were detected in Brazil

State	Municipality	References
Federal District	Brasilia	Paula et al. (2014)
Espirito Santo	Domingos Martins	Zanúncio Júnior et al. (2018)
	Santa Maria de Jetiba	Zanúncio Júnior et al.(2018)
	Serra	Zanúncio Júnior et al. (2018)
Minas Gerais	Ervália	Andreazza et al. (2016)
Paraná	Porto Vitória	Geisler et al. (2015)
	União da Vitória	Geisler et al. (2015)
Rio de Janeiro	Petrópolis	Bitner-Mathé et al. (2014)
Rio Grande do Sul	Capão do Leão	Andreazza et al. (2017a)
	Cerrito	Andreazza et al. (2017a)
	Erechim	Deprá et al. (2014)
	Farroupilha	Foppa et al. (2018)
	Morro Redondo	Wollmann et al. (2019a, b)
	Osório	Deprá et al. (2014)
	Pelotas	Wollmann et al. (2019a, b)
	Torres	Andreazza et al. (2017a)
	Vacaria	Santos (2014)
	Vila Maria	Deprá et al. (2014)
Santa Catarina	Botuverá	Deprá et al. (2014)
	Frei Rogério	Zazycki et al. (2019)
	Lages	Souza et al. (2017)
	Nova Veneza	Deprá et al. (2014)
	São Joaquim	Andreazza et al. (2017a)
	Santo Amaro da Imperatriz	Fischer et al. (2017)
São Paulo	Campinas	Louzeiro et al. (2019)
	São Paulo	Vilela and Mori (2014)

(Gonsebatt et al. 2017), in Tucumán on *Opuntia ficus-indica* (Cacataceae) (Lavagnino et al. 2018), and in Mendoza in the cultivation of blackberry (Dagatti et al. 2018). Approximately 65% of adults of *D. suzukii* were recovered from guavas collected in the canopy of trees in northwestern Argentina (Escobar et al. 2018).

In Uruguay, the SWD was the most abundant species (96%) of the total *Drosophila* flies emerged from ripened, decayed, or damaged blueberries collected at rural Canelones Department and collected from banana-baited traps in urban Montevideo city (González et al. 2015).

In Chile, the SWD was reported in Central Valley (32°3′47′S and 36°49′37′S) in 2015 using traps (Medina-Muñoz et al. 2015), according to SAG (2020), samples were analyzed by the Laboratory of Entomology and Biotechnology SAG, and it was determined that species was not *D. suzukii* and was actually *Drosophila amplitnis* Malloch. The confirmation of the presence of the species in the country occurred

2 years later through specimens collected in traps in the Los Lagos Region (Rojas et al. 2019). However, in 2014, fruits infested of cherries and blueberries were found in Brazil imported from Chile. Currently in the country, *D. suzukii* occurs in seven regions: Araucanía, Aysén, Biobío, Los Lagos, Los Ríos, Maule, and Ñuble e O'Higgins (SAG 2020). SWD has dispersed to other regions of the central-south region of the country, currently in the pest category present with restricted distribution. In cherries the losses are 1, 2.16, and 2.7 ton/hectare equivalent to 5.000–17.550 USD/hectare, and in blueberies the losses are 1–1.5 ton/hectare or a loss equivalent to 4000 USD/hectare (Buzzetti 2020).

5.3 Monitoring of *Drosophila suzukii* in Latin America

Monitoring is an important step for detection, dispersion studies, population dynamics, and decision-making in an area-wide approach. This makes it important to know which is the best attractive and trap for capturing SWD.

In a study carried out with laboratory cages, it was observed that SWD is more attracted to raspberry pulp alone or with sucrose than to apple vinegar (ACV) or Suzukii Trap® and similar to bakery yeast + sucrose. The red-black stripe trap with a hemispherical dome-shaped lid with which apple cider vinegar + 10% ethanol and the additional tube device was baited with a fermentation mixture of sugar and yeast proved to be very efficient in capturing SWD in the laboratory, in a guava orchard, and in a commercial blackberry orchard (Lasa et al. 2017a).

In another field study, the myxture purchased Merlot™ wine (60%) + apple vinegar (40%) + sugarcane molasses (20 g/L) (WVM) showed a higher attraction (61.97% of the captured insects) to the WVM mixture than to Ceratrap®, Torula®, Biofruit®, SuzikTrap®, and apple vinegar; however, Suzukii Trap® and apple vinegar were the most selective to non-target insects. Immature females showed a preference for Biofruit®, apple vinegar, and WVM (Wollmann et al. 2019a).

In Argentina, in Norpatagonia the weather conditions of spring and end of summer (beginning of the year) are very favorable for the development of SWD. Obviously, the most important attacks in the region occur during the months of March/April. In the spring, the attacks begin with their low population density (Cichón et al. 2016).

The highest SWD population peaks in southern Brazil occur during late spring to mid-fall. Temperature is the factor that most influenced the seasonality of the SWD population in the field, promoting low species catches during winter (Wollmann et al. 2019b). The control level in Mexico is 2–3 flies/trap/week in two consecutive weeks (personal information of Dr. Rodrigo Lasa), which can be adopted in all LA countries.

5.4 Cultural and Physical Control of *Drosophila suzukii* in Latin America

Cultural control consists of the use of cultural treatment and physical control of physical processes to reduce the population density of a pest population (Garcia 2014). One of the bases of cultural control is the knowledge of the host plants of the pest under control. The knowledge of possible alternative hosts allows temporal prediction of how SDW migrates and returns to natural or agricultural habitats around the fields (Andreazza et al. 2017a). *D. suzukii* is more abundant in fruits still attached to the tree compared to fallen fruit (Lasa et al. 2015). In Brazil, *D. suzukii* presents a higher infestation in blackberry (40–65% infestation) and strawberry (approximately 30% infestation) (Wollmann et al. 2020). In contrast, blueberry is less infested (<7% infestation). The cattley guava (*Psidium cattleianum*) and Surinam cherry (*Eugenia uniflora*) are hosts that serve to multiply SWD in off-season periods (Wollmann et al. 2020). This species is the most common Drosophilidae species that emerges from guava fruits in Brazil (Mendes et al. 2019). *D. suzukii* is a polyphagous species with 52 host plants from 23 families in Latin America, most of these hosts are exotic (66.8%) (Table 5.2). Most of these host plants are common to most LA countries.

Some measures of cultural control should be taken to prevent the populations of *D. suzukii* from increasing rapidly and thus minimize the use of chemical control. Servicio Agricola y Ganedero (SAG 2020) from Chile proposed some important measures that can be used in the cultural management of *D. suzukii* to be carried out in the country, but which can be used in other Latin American countries. They are as follows:

1. Periodic collection and destruction of fallen fruits and ripe fruits that remain in the plant after harvest as these fruits can house SWD larvae and pupae that continue to reproduce. In addition, it reduces fly oviposition sites and reduces food supply for adults.
2. SWD has the ability to stay in crops from year to year, surviving on alternative hosts, such as *Eriobotrya japonica* (Wollmann et al. 2019b, 2020). Thus, it is necessary to collect fruits, clean up, and eventually eliminate alternative host plants present in the areas surrounding the cultivation. Drastic elimination of native host plants in the area adjacent to the orchard should be avoided since females may be forced to ovipositar in the fruits of the orchard.
3. These fruits can be buried in a ditch on the ground and deposited after the fruits must be covered with at least 30 cm of soil or placed in a hermetically sealed clear plastic bag and leave it exposed for at least 5 days.

It is very important to take control measures of SWD also in urban areas, in grocery stores cross contamination can occur, i.e., field fruits can be infested by SWD in grocery stores when preventive measures are not taken. It is necessary that the owners of grocery stores eliminate residues of host fruits using solarization or freezing before being discarded (Santos et al. 2017).

Table 5.2 List of hosts of *Drosophila suzukii* in Latin America, E, exotic, N, Native

Host plant	Common name	Origin	References
Actinidiaceae			
Actinidia chinensis Planch	Kiwi	E	Andreazza et al. (2017a)
Aquifoliaceae			
Ilex aquifolium L.	Holly	E	SAG (2020)
Araliaceae			
Fatsia japonica (Thunb.)	Glossy-leaf paper plant	E	SAG (2020)
Hedera helix L	Common ivy	E	SAG (2020)
Hedera rhombea Siebold and Zucc	Japanese ivy	E	SAG (2020)
Arecaceae			
Butia capitata (Mart.)	*Butia*	N	Beatriz Goñi—personal information
Butia yatay (Mart.)	*Butia*	N	Beatriz Goñi—personal information
Cacatcaeae			
Opuntia ficus-indica (L.)	Prickly pear	E	Lavagnino et al. (2018)
Caricaceae			
Carica papaya L.	Papaya	N	Zanúncio Junior et al. (2018)
Cornaceae			
Comus sp.	Dogwood	E	SAG (2020)
Ebenaceae			
Dispyro kaki Thunberg	Kaki	E	Andreazza et al. (2017a), Lauyé (2017)
Elaeocarpaceae			
Aristotelia chilensis (Molina)	Chilean wineberry	N	Buzzetti (2020)
Ericaceae			
Vaccinium ashei Reade	Rabbiteye Blueberry	E	González et al. (2015)
Vaccinium corymbosum L.	Northern Highbush Blueberry	E	Vilela and Mori (2014), Santadino et al. (2015), Lauyé (2017), Funes et al. (2019), Buzzetti (2020)
Ginkgoaceae			
Ginkgo biloba L.	Ginkgo	E	Beatriz Goñi—personal information
Loranthaceae			
Notanthera heterophylla Ruiz and Pav.	Quintral del boldo	N	SAG (2020)
Malpighiaceae			
Malpighia emarginata DC	Barbados cherry	N	Louzeiro et al. (2019, 2020)
Monimiaceae			

(continued)

Table 5.2 (continued)

Host plant	Common name	Origin	References
Peumus boldus Molina	Boldo	N	SAG (2020)
Moraceae			
Morus nigra L.	Berry	E	Beatriz Goñi—personal information
Ficus carica L.	Figs	E	Lauyé (2017)
Myrtaceae			
Acca sellowiana Burret	Feijoa	N	Andreazza et al. (2017a), Lauyé (2017), Souza et al. (2017)
Eugenia involucrata DC.	Cherry of Rio Grande	N	Geisler et al. (2015)
Eugenia uniflora L.	Surinam cherry	N	Geisler et al. (2015)
Luma apiculata DC	Chilean myrtle	N	SAG (2020)
Myrceugenia planipes (Hook. et Harn.)	Valdivia's patagua	N	SAG (2020)
Psidium cattleianum Sabine	Cattley guava	N	Andreazza et al. (2017a), Lauyé (2017), Zanúncio Junior et al. (2018), Wollmann et al. (2020)
Psidium guajava L.	Common guava	N	Andreazza et al. (2017a), Lasa et al. (2017b), Escobar et al. (2018), Mendes et al. (2019)
Syzygium cumini (L.) Skeels	Jambul	E	Rampasso and Vilela (2017)
Ugni molinae Turcz.	Chilean guava	N	SAG (2020)
Lythraceae			
Punica granatum L.	Pomegranate	E	Beatriz Goñi—personal information
Passifloraceae			
Passiflora caerulea L.	Passionfruit	N	Beatriz Goñi—personal information
Rosaceae			
Eriobotrya japonica (Thunb.)	Loquat	E	Geisler et al. (2015)
Fragaria x *ananassa* Duch.	Strawberry	E	Andreazza et al. (2017a), Schlesener et al. (2017), Zanúncio Junior et al. (2018), Wollmann et al. (2020)
Malus domestica L.	Apple	E	Andreazza et al. (2017a)
Prunus armeniaca L.	Siberian apricot	E	
Prunus avium L.	Sweet cherry	E	Andreazza et al. (2017a), Buzzetti (2020)
Prunus cerasifera Ehrh	Cherry Plum	E	Buzzetti (2020)
Prunus cerasus	Sour Cherry L.	E	Buzzetti (2020)
Prunus domestica L	Common plum	E	Andreazza et al. (2017a)
Prunus persica L.	Peach	E	Geisler et al. (2015), Foppa et al. (2018)
Prunus serotina Ehrhart	Wild black cherry	E	Andreazza et al. (2017a)
Pyrus communis L.	European pear	E	Andreazza et al. (2017a), Lauyé (2017)

(continued)

Table 5.2 (continued)

Host plant	Common name	Origin	References
Pyrus pyrifolia (Burm.) Nak	Asian pear	E	Andreazza et al. (2017a)
Rosa moschata Herm.	Musk Rose	E	SAG (2020)
Rubus fruticosus L.	Blackberry	E	Andreazza et al. (2017a), Lasa et al. (2017b), Funes et al. (2019)
Rubus idaeus L.	Raspberry	E	Alexandre (2016), Cichón et al. (2016), Andreazza et al. (2017a), Escobar et al. (2018), Funes et al. (2019), Lauyé (2017), Lasa et al. (2019)
Rubus ulmifolius Schott	Elmleaf Blackberry	E	Lauyé (2017), Buzzetti (2020)
Rubiaceae			
Psychotria suterella Muell.	–	N	Conde et al. (2019)
Rutaceae			
Citrus sinensis L.	Orange	E	Lauyé (2017)
Sapindaceae			
Allophylus edulis (ASt.Hil)	*Chal-chal*	N	Beatriz Goñi -personal information
Vitaceae			
Vitis labrusca L	Fox Grape	E	Andreazza et al. (2017a)
Vitis vinifera L.	Grape vine	E	Andreazza et al. (2017a), Lauyé (2017)

Regarding physical control, it is suggested in Argentina the use of nets with a permeability of 0.98 to 1 mm × 0.6 mm (80 g) in cherry crops and other small fruit crops (Cichón et al. 2016).

5.5 Biological Control of *Drosophila suzukii* in Latin America

Among the SWD biological control agents, parasitoids are the most commonly studied and the ones presenting the upmost probability for success (Garcia et al. 2017). In Latin America, 12 species of parasitoids of *D. suzukii* belonging to Diapriidae, Figitidae, and Pteromalidae are recorded (Table 5.3). The highest wealth of parasitoid species of *D. suzukii* in the Region is found in the Figitidae family with seven species of three genera, followed by Diapriidae with three species of one genus and Pteromalidae with two species of two genera. Parasitoids were recorded in Argentina, Brazil, and Mexico, and there were no records for Chile and Uruguay so far (Table 5.3).

Two important species of parasitoids of SWD are found in the genus *Trichopria*, *Trichopria drosophilae* Perkins (introduced), and *Trichopria anastrephae* Lima (native). *T. drosophiliae* seems to be cosmopolitan with a wide distribution, and has been reported to attack pupae of *D. suzukii* in Europe, North America, and Asia (Yi et al. 2020). *T. anastrephae* (Fig. 5.1) is located only in South America, distributed

Table 5.3 Parasitoids of *Drosophila suzukii* recorded in Latin America. Host stage attached, HAS, Country, ARG, Argentina, BRA, Brazil, MEX, Mexico

Parasitoids	HAS	Country	References
Diapriidae			
Trichopria anastrephae Lima	pupa	BRA	Wollmann et al. (2016), Krüger et al. (2019a), Schlesener et al. (2019)
Trichopria drosophilae Perkins	pupa	MEX	Garcia-Cancino et al. (2015), González-Cabrera et al. (2019)
Trichopria sp.	pupa	ARG	Funes et al. (2019)
Figitidae			
Dieucoila sp.	larva	ARG	Funes et al. (2019)
Ganaspis brasiliensis Ihering	larva	MEX	González-Cabrera et al. (2020)
Ganaspis hookeri Crawford	larva	ARG	Lue et al. (2017)
Ganaspis sp.	larva	ARG	Funes et al. (2019)
Leptopilina boulardi Barbotin, Carton and Kelner-Pillault	larva	ARG BRA MEX	Garcia-Cancino et al. (2015) Wollmann et al. (2016), Garrido et al. (2018) González-Cabrera et al. (2019)
Leptopilina clavipes (Hartig)	larva	ARG	Lue et al. (2017)
Hexacola sp.	larva	ARG	Funes et al. (2019)
Pteromalidae			
Pachycrepoideus vindemmiae Rondani	pupa	ARG BRA MEX	Garcia-Cancino et al. (2015), Moreno-Carrillo et al. (2015), Funes et al. (2019), Schlesener et al. (2019)
Spalangia simplex Perkins	pupa	MEX	Garcia-Cancino et al. (2015)

Fig. 5.1 *Trichopria anastrephae* (Hymenoptera: Diapriidae), one of the most promising parasitoids to be used in biological control programs of *Drosohila suzukii* in Latin America

in Argentina, Brazil (Goiás, Minas Gerais, Rio de Janeiro, Rio Grande do Sul, and Santa Catarina), and Venezuela, parasitizing pupae of the genus *Anastrepha* (Diptera: Tephritidae) (Cruz et al. 2011; Garcia and Corseuil 2004; Garcia and Ricalde 2013); however, this species was recently observed parasitizing *D. suzukii* only in the state of Rio Grande do Sul (Brazil) (Garcia et al. 2017). The Insect Ecology Laboratory of the Department of Ecology, Zoology and Genetics of the Biology Institute of the Universidade Federal de Pelotas established the colony of *T. anastrephae* with finality of studies aiming at the use of this species in future biological control programs. *T. anastrephae* should be reared in 24-h-old *D. suzukii* pupae, which should be exposed to parasitism for 24 h at a density of 15 pupae per female, and the parasitoids should be fed pure honey (Vieira et al. 2020). The extrinsic intraspecific competition among females of *T. anastrephae* result in a higher number of oviposition scars on host and an extension of egg-to-adult period, suggesting the occurrence of superparasitism (Krüger et al. 2019a, b).

Two species of the genus *Leptopilina* were reported attacking *D. suzukii* in LA. *Leptopilina boulardi* Barbotin, Carton and Kelner-Pillault is a generalist koinobiont endoparasitoid of Drosophilidae larvae of African origin, cosmopolitan distribution and very abundant in the holartic region (Lue et al. 2016), and *Leptopilina clavipes* (Hartig) is found in Argentina (Funes et al. 2019).

In a study of the release of *T. drosophilidae* and *L. boulardi*, simple or combined in Mexico, it was observed that parasitism by *T. drosophilae* increased about four times in places of release compared to places where there was no release, and the parasitism of *L. boulardi* increased 2.8 times. In addition, a reduction of 50 and 55% was recorded in the population of *D. suzukii* in the sites with simple and combined releases, respectively. In this case, the results indicate that the unique release of *T. drosophilae* is a more economical option to reduce populations of *D. suzukii* (González-Cabrera et al. 2019).

The parasitoid *Pachycrepoideus vindemmiae* (Rondani) is an ectopasitic pupal parasitoid and was registered in pupae of SWD in Argentina, Brazil, and Mexico (Table 5.3). Its occurrence associated to *D. suzukii* has been verified in Asia, Europe, and Americas. This generalist parasitoid is able to attack species from different insect orders, such as Diptera, Hemiptera, Hymenoptera, and Lepidoptera. Furthermore, it is considered a facultative hyperparasitoid; in other words, it is able to parasite other primary parasitoids that integrate biological control programs for pests in the Tephritidae family [*Diachasmimorpha fullawayi* (Silvestri), *D. kraussii* (Fullaway), *Diachasmimorpha longicaudata* (Ashmead), *D. tryoni* (Cameron), *Psyttalia concolor* (Szépligeti), *Psyttalia humilis* (Silvestri) (Hymenoptera: Braconidae), *Coptera silvestrii* (Kieffer) (Hymenoptera: Diapriidae), and *Tetrastichus giffardianus* Silvestri (Hymenoptera: Eulophidae)]. Due to its low specificity and hyperparasitism (Garcia et al. 2017) and because it is a very sensitive species to insecticides (Schlesener et al. 2019), this parasitoid is not suitable to integrate a *D. suzukii* management program (Garcia et al. 2017).

The genus *Ganaspis* has three species associated with SWD, *Ganaspis brasiliensis* Ihering in Mexico, *Ganaspis hookeri* Crawford, and *Ganaspis* sp. in Argentina. Recently, the parasitoid *Ganaspis brasiliensis* Ihering was detected as a parasite of SWD in Mexico, although not parasitizing *D. suzukii*, this species has already been detected in Brazil, Guadeloupe, and Panama (González-Cabrera et al. 2020).

Other biological control agents that have been promising in the biological control of SWD in Brazil are entomopathogenic nematodes. The nematodes

Heterorhabditis indica IBCBn 05 and *Heterorhabditis amazonensis* IBCBn 24 present the highest mortality in pupae with 33.0 and 43.0% in laboratory conditions (Brida et al. 2019).

The mites *Stratiolaelaps scimitus* Berlese (Mesostigmata: Laelapidae) and *Macrocheles muscaedomestauae* (Scopoli) (Mesostigmata: Macrochhelidae) have been shown to be efficient predators of SWD eggs. The number of preys consumed ranged from 5 to 14 for *S. scimitus* and from 3 to 17 for *M. muscaedomesae* under laboratory conditions (Silva et al. 2018).

5.6 Chemical Control of *Drosophila suzukii* in Latin America

The most commonly used control method for SWD control is chemical control. However, most LA countries do not have products registered for the control of populations of this species. Organophosphorates, pyretroids, and spinosine showed good results in contact control and residual power over *D. suzukii*, in cherry, raspberry, blueberry, strawberry, and grapevine. The systemic neonicotinoids and organophosphorates present ovidal action and the ability to control larvae within the fruits (Schlesener et al. 2015). The insecticides deltamethrin, dimethoate, spinosad, fenitrothion, phosmet, malathion, methidathion, and zeta-cypermethrin resulted in mortality to 100% of *D. suzukii* 3 days after the treatment (DAT) and insecticides fenitrothion, malathion, and methidathion deemed 100% of the eggs not viable (Schlesener et al. 2017). Organophosphates (dimethoate and malathion), spinosyns (spinosad and spinetoram), pyrethroid (lambda-cyhalothrin), and diamide (cyantraniliprole) insecticides exhibited high toxicity to larvae of SWD (Andreazza et al. 2017b). Regarding the selectivity of insecticides to the parasitoids of SWD occurring in Brazil, it was observed that spinosyns (spinosade and spinetoram) and abamectin caused high mortality rates in *P. vindemmiae*, but were harmless to *T. anastrephae*. Neonicotinoids, organophosphates, and pyrethroids caused high mortality rates, regardless of species (Schlesener et al. 2019).

In Brazil, a single insecticide authorized for the control of SWD is spinetoram in blackberry, raspberry, blueberry, and grape (AGROFIT 2020) at concentrations of 96–160 g of commercial product per hectare in blackberry, raspberry, and blueberry and 96–120 g of commercial product for 100 liters of water in the grape. Although SWD causes about 30% of loss in strawberry production (Santos 2014), there is no insecticide registered for the control of SWD; this is probably due to the risk of contamination from chemical residues in fruit. This risk is highest during the preharvest or ripening periods when the likelihood of infestations by SWD (Andreazza et al. 2017b).

Currently, there are no insecticides registered for SWD control in Argentina and Uruguay. Nevertheless, the insecticides abamectin, lambda-cyhalothrin, spinetoram, spirotetramat, emamectin benzoate, and methomyl are candidates for registration in Argentina (Cichón et al. 2019). In Mexico, the insecticides zeta cypermethrin, spinosad, spinotoram, ciantrilipol, malathion, and novaluron are apllied when are collected 2 or 3 SWD flies per trap per week in two consecutive weeks (personal information of Dr. Rodrigo Lasa). Chile is the country with the highest number of insecticides for the control of SWD, the country presents a list of 21 insecticides authorized for the control of the pest in 21 crops (SAG 2020).

5.7 Sterile Insect Technique in *Drosophila suzukii* Management in Latin America

The Sterile Insect Technique (SIT) is considered a type of autocotide control, where the pest is used for its own control, this technique meets the current requirements of the fruit importing countries and the consumer internal market (Dias and Garcia 2014). Given its great efficiency in fruit fly control, this has become an important tool in the control of SWD. The first step toward success in the use of SIT is to obtain a suitable breeding technique for this purpose, or to present a high number of pupae and maximum bioconversion. Adults of SWD developed in the coconut fiber + Brewer's yeast diet provide the highest number of pupae per gram of diet and the maximum bioconversion (6%). The use of 30 × 40 × 30 cm Plexiglas cages, each loaded with 5000 adults and stocked with 500 g of coconut fiber and Brewer's yeast diet distributed in 15 × 5 × 10 cm plastic trays with a diet layer of 3 cm thick, allows a minimum production of 84,000 pupae of *D. suzukii* per day (Aceituno-Medina et al. 2020).

The first research project with SIT in South America is the project entitled "Evaluating the use of Sterile Insects and pupal parasitoids to manage *Drosophila suzukii* in greenhouse" started in 2017 in Brazil. The project is coordinated by the Insect Ecology Laboratory of Universidade Federal de Pelotas and financed by the International Atomic Energy Agency (IAEA). Thus, the aim of this proposal is to obtain essential information that turns possible the use of SIT and biological control for *D. suzukii* in greenhouse. From this project, our team found that female sterility of SWD female is achieved at 75 Gy of gamma radiation, while an adequate level of male sterility (99.67%) is obtained at 200 Gy (Krüger et al. 2018). Male sterility of SWD does not influence mating and remating likelihood; however, copula duration of sterile males is shorter compared to fertile males. On the other hand, sterile females are less likely to mate (Krüger et al. 2019b).

5.8 Conclusions

L.A. has contributed to the SWD management studies, while the governments of the countries should invest in the creation of national programs for the management of *D. suzukii*; in addition, joint cooperation between countries where the species has been detected becomes important for the optimization of financial and human resources. It is important to increase the monitoring of the species on a large scale, aiming to know the dynamics of the dispersion of the species in the region. Similarly, studies of trilotrophic interactions are fundamental to the knowledge of hosts, parasitoids, and predators and their interactions with each other and with *D. suzukii* in the region. The parasitoids *T. anastrephae* (native) and *T. drosophilidae* (exotic) are potential agents for future biological control programs in L.A. However, many studies should be carried out to verify whether there is competition, intraguild predation, or synergism in the use of these species in regions where *T. anastrephae* inhabits. SIT is an important technique for the management of SWD in the countries of the region, mainly in Chile, where the species could even be eradicated due to its geographical conditions.

Acknowledgments I would like to thank the Brazilian National Council for Scientific and Technological Development (CNPq) research productivity grant and the International Atomic Energy Agency (IAEA) for the project funding entitled evaluating the use of Sterile Insects and pupal parasitoids to managing *Drosophila suzukii* in greenhouse, which I am the coordinator. To Dr. Rodrigo Lasa (Instituto de Ecología AC) for information on insecticides authorized in Mexico for the control of SWD and Dr. Beatriz Goñi (Universidad de la República) for the information of SDW hosts in Uruguay.

References

Aceituno-Medina M, Ordoñez A, Carrasco M et al (2020) Mass rearing, quality parameters, and bioconversion in *Drosophila suzukii* (Diptera: Drosophilidae) for sterile insect technique purposes. J Econ Entomol 113:1097–1104. https://doi.org/10.1093/jee/toaa022

AGROFIT (2020) Sistema de agrotóxicos fitossanitários. https://agrofitagricultura. Accessed 22 Jun 2020

Alexandre BG (2016) Avaliação das assembleias de Drosophilidae (Insecta, Diptera) em área de restinga no município de Torres-RS. Dissertation, Universidade Federal do Rio Grande do Sul

Andreazza F, Haddi K, Oliveira EE et al (2016) *Drosophila suzukii* (Diptera: Drosophilidae) arrives at Minas Gerais State, a main strawberry production region in Brazil. Fla Entomol 99:796–798. https://doi.org/10.1653/024.099.0439

Andreazza F, Bernardi D, Dos Santos RSS et al (2017a) *Drosophila suzukii* in southern neotropical region: current status and futures perspectives. Neotrop Entomol 46:591–605. https://doi.org/10.1007/s13744-017-0554-7

Andreazza F, Bernardi D, Baronio CA et al (2017b) Toxicities and effects of insecticidal toxic baits to control *Drosophila suzukii* and *Zaprionus indianus* (Diptera: Drosophilidae). Pest Manag Sci 73:146–152. https://doi.org/10.1002/ps.4348

Benito NP, Lopes-da-Silva M, Santos RSS (2016) Potential spread and economic impact of invasive *Drosophila suzukii* in Brazil. Pesq Agropec Bras 51:571–578. https://doi.org/10.1590/S0100-204X2016000500018

Bitner-Mathé BC, Victorino J, Faria FS (2014) *Drosophila suzukii* has been found in tropical Atlantic rainforest in southeastern Brazil. Dros Inf Serv 97:136–113

Brida AL, Wilcken SRS, Garrigós Leite L et al (2019) Virulence of entomopathogenic nematode to pupae and adults of *Drosophila suzukii* in laboratory. Revista Científica Rural 21:126–138. https://doi.org/10.30945/rcr-v21i2.2725

Buzzetti K (2020) The spotted wing Drosophila in the south of the world: Chilean case and its first productive impacts. In Invasive species—Introduction pathways, economic impact, and

possible management options. https://www.intechopen.com/online-first/the-spotted-wing-drosophila-in-the-south-of-the-world-chilean-case-and-its-first-productive-impacts. Accessed 10 Mar 2020
Castro-Sosa R, Castillo-Peralta MR, Monterroso-Rivas AI et al (2017) Potential distribution of *Drosophila suzukii* (Diptera: Drosophilidae) in relation to alternate hosts in Mexico. Fla Entomol 100:787–794. https://doi.org/10.1653/024.100.0403
Cichón L, Garrido S, Lago J (2016) *Drosophila suzukii*. Uma nueva plaga presente em la Norpatagonia. Ediciones INTA, Buenos Aires
Cichón L, Araque L, Garrido S et al (2019) Insecticidas con posibilidad de registro en cereza en Argentina y sus implicâncias para la exportación de frutas frescas. RIA 45:285–291
Conde N. et al. (2019) Emergence record of Drosophilidae from fruits of *Psychotria suterella* Muell. Arg. (Rubiaceae). In: Abstracts of the XI Symposium of Ecology, Genetics, and Evolution of Drosophila, Universidade Federal de Pelotas, Pelotas, p 17
Cruz PP, Neutzling AS, Garcia FRM (2011) Primeiro registro de Trichopria anastrephae, parasitoide de moscas-das-frutas, no Rio Grande do Sul. Ciência Rural 41:1297–1299. https://doi.org/10.1590/S0103-84782011000800001
Dagatti CV, Marcucci B, Herrera ME et al (2018) Primera detección de *Drosophila suzukii* (Diptera: Drosophilidae) en frutos de zarzamora en Mendoza, Argentina. Rev Soc Entomol Arg 77:26–29. https://doi.org/10.25085/rsea.770304
de la Vega GJ, Corley JC, Soliani C (2020) Genetic assessment of the invasion history of *Drosophila suzukii* in Argentina. J Pest Sci 93:63–75. https://doi.org/10.1007/s10340-019-01149-x
Deprá M, Poppe JL, Schmitz HJ et al (2014) The first records of the invasive pest *Drosophila uzukii* in the South American continent. J Pest Sci 87:379–383. https://doi.org/10.1007/s10340-014-0591-5
Dias NP, Garcia FRM (2014) Fundamentos da técnica do inseto estéril (TIE) para o controle de moscas-das-frutas (Diptera, Tephritidae). O Biológico 76:57–62
Díaz BM et al (2015) Deteccíon de la mosca plaga *Drosophila suzukii* (Matsumura) en la regíon de Concordia (Entre Ríos). In Abstracts of XV Jornadas Fitosanitarias Argentinas, Universidad Nacional Noroeste, Santa Fe, 10-07 September 2015
dos Santos LA, Mendes MF, Krüger AP et al (2017) Global potential distribution of *Drosophila suzukii* (Diptera, Drosophilidae). PLoS One 12:e0174318. https://doi.org/10.1371/journal.pone.0174318
Escobar LI, Ovruski SM, Kirschbaum DS (2018) Foreign invasive pests *Drosophila suzukii* (Matsamura) and *Zaprionus indianus* Gupta (Diptera: Drosophilidae) threaten fruit production in northwestern Argentina. Dros Inf Serv 101:9–14
FAO (2019). http://www.fao.org/brasil/noticias/detail-events/pt/c/1193684/. Accessed 22 Jun 2020
Ferronato P, Woch AL, Soares PL et al (2019) A Phylogeographic approach to the *Drosophila suzukii* (Diptera: Drosophilidae) invasion in Brazil. J Econ Entomol 112:425–433. https://doi.org/10.1093/jee/toy321
Fischer L et al (2017) Drosophilids as indicators of environmental quality in remnant of Atlantic Forest in Santa Catarina. In: Abstracts of the X Simpósio de Ecologia, Genética e Evolução de Drosophila, Ouro Preto, p 70
Foppa F, Borba RS, Secretti T et al (2018) Ocorrência de *Drosophila suzukii* (Matsumura) (Diptera, Drosophilidae) na cultura do pessegueiro, em Farroupilha, na Serra Gaúcha, RS. EntomoBrasilis 11:178–184. https://doi.org/10.12741/ebrasilis.v11i3.763
Fraimout AD, Fellous SV, Hufbauer S et al (2017) Deciphering the routes of invasion of *Drosophila suzukii* by means of ABC random forest. Mol Biol Evol 34:980–996. https://doi.org/10.1093/molbev/msx050
Funes CF, Gallardo FE, Reche VA et al (2019) Parasitoides de Sudamérica asociados a las plagas invasoras *Drosophila suzukii* y *Zaprionus indianus* (Diptera: Drosophilidae) y su potencial como agentes de control biológico. Semiárida 29:25–29
Garcia FRM (2014) Zoologia Agrícola: manejo ecológico de pragas. Editora Rígel, Porto Alegre

Garcia FRM, Corseuil E (2004) Native hymenopteran parasitoids associated with fruit flies (Diptera: Tephritoidea) in Santa Catarina State, Brazil. Fla Entomol 87:517–521. https://doi.org/10.1653/0015-4040(2004)087[0517:NHPAWF]2.0.CO;2

Garcia FRM, Ricalde MP (2013) Augmentative biological control using parasitoids for fruit fly management in Brazil. Insecta 4:55–70. https://doi.org/10.3390/insects4010055

Garcia FRM, Wollmann J, Krüger AP et al (2017) Biological control of *Drosophila suzukii* (Diptera, Drosophilidae): state of the art and prospects. In: Davenport L (ed) Biological control: methods, applications and challenges. Nova Science Publishers, New York, pp 1–27

Garcia-Cancino MDG, Hernández AG, Cabrera JG et al (2015) Parasitoides de *Drosophila suzukii* (Matsumura) (Diptera: Drosophilidae) en Colima, México. Southwestern Entomol 40:855–858. https://doi.org/10.3958/059.040.0418

Garrido SA, Cichón L, Lago JD et al (2018) First record of *Leptopilina boulardi* (Hymenoptera: Figitidae) asociated to *Drosophila suzukii* (Diptera: Drosophilidae) in Alto Valle of Río Negro and Neuquén, Patagonia, Argentina. Rev Soc Entomol Arg 77:22–27. https://doi.org/10.25085/rsea.770202

Geisler FCS, Santos J, Holdefer DR et al (2015) First record of *Drosophila suzukii* (Matsumura, 1931) (Diptera: Drosophilidae) for the State of Paraná, Brazil and new hosts. RCA 9:125–129. https://doi.org/10.18316/1981-8858.15

Gonsebatt G et al (2017) *Drosophila suzukii* (Matsumura) (Diptera: Drosophilidae) en durazneros (*Prunus persica*) en el sur de la provincia de Santa Fé: primer registro. In: Abstracts of the XIX Congreso y de la XXXVII Reunión Anual de la Sociedad de Biología de Rosario, p 73

González G, Mary AL, Goñi B (2015) *Drosophila suzukii* (Matsumura) found in Uruguay. Dros Inf Serv 98:103–107

González-Cabrera J, Moreno-Carrillo G, Sanchez-Gonzalez JA et al (2019) Single and combined release of *Trichopria drosophilae* (Hymenoptera: Diapriidae) to control *Drosophila suzukii* (Diptera: Drosophilidae). Neotrop Entomol 48:949–956. https://doi.org/10.1007/s13744-019-00707-3

González-Cabrera J, Cordoba-Urtiz EG, Moreno-Carrillo G et al (2020) First report of the parasitoid *Ganaspis brasiliensis* Ihering (Hymenoptera: Figitidae) in Mexico. Entomol News 129:67–70. https://doi.org/10.3157/021.129.0110

Hauser M (2011) Historic account of the invasion of *Drosophila suzukii* (Matsumura) (Diptera: Drosophilidae) in the continental United States, with remarks on their identification. Pest Manag Sci 67:1352–1357. https://doi.org/10.1002/ps.2265

Hendrichs J, Kenmore P, Robinson A, Vreysen MJ (2007) Area-wide integrated pest management (AW-IPM): principles, practice and prospects. In: Vreysen MJ, Robinson A, Hendrichs J (eds) Area-wide control of insect pests: from research to field implementation. Springer, Netherlands, pp 3–33

Krüger AP, Schlesener DCH, Martins LN et al (2018) Effects of irradiation dose on sterility induction and quality parameters of *Drosophila suzukii* (Diptera: Drosophilidae). J Econ Entomol 111:741–746. https://doi.org/10.1093/jee/tox349

Krüger AP, Scheunemann T, Vieira JGA et al (2019a) Effects of extrinsic, intraspecific competition and host deprivation on the biology of *Trichopria anastrephae* (Hymenoptera: Diapriidae) reared on *Drosophila suzukii* (Diptera: Drosophilidae). Neotrop Entomol 48:957–965. https://doi.org/10.1007/s13744-019-00705-5

Krüger AP, Schlesener DCH, Martins LN et al (2019b) Radiation effects on *Drosophila suzukii* (Diptera: Drosophilidae) reproductive behaviour. J Appl Entomol 143:88–94. https://doi.org/10.1111/jen.12563

Lasa R, Tadeo E (2015) Invasive drosophilid pests *Drosophila suzukii* and *Zaprionus indianus* (Diptera: Drosophilidae) in Veracruz, Mexico. Fla Entomol 98:987–988. https://doi.org/10.1653/024.098.0332

Lasa R, Tadeo E, Toledo-Hérnandez RA et al (2017a) Improved capture of *Drosophila suzukii* by a trap baited with two attractants in the same device. PLoS One 12(11):e0188350. https://doi.org/10.1371/journal.pone.0188350

Lasa R, Tadeo E, Dinorín L et al (2017b) Fruit firmness, superficial damage, and location modulate infestation by *Drosophila suzukii* and *Zaprionus indianus*: the case of guava in Veracruz, Mexico. Entomol Exp Appl 162:4–12. https://doi.org/10.1111/eea.12519

Lasa R, Toledo-Hernández RA, Rodríguez D et al (2019) Raspberry as a source for the development of *Drosophila suzukii* attractants: Laboratory and commercial polytunnel trials. Insects 10:137. https://doi.org/10.3390/insects10050137

Lauyé MA (2017) Diversidad de Drosophilidae (Diptera) en frutos de interés productivo y agronómico, en localidades de la región sur de Uruguay, con especial atención a la especie invasora *Drosophila suzukii* (Matsumura, 1931). Dissertation, Universidad de la República (Uruguay)

Lavagnino NJ1, Diaz, Beatriz M et al. (2018) New records of the invasive pest *Drosophila suzukii* (Matsumura) (Diptera: Drosophilidae) in the South American continent. Rev Soc Entomol Arg 77(1):27–31. https://doi.org/10.25085/rsea.770105

Louzeiro LR, Souza-Filho MF, Bulgarelli CA (2019) Infestation of *Malpighia emarginata* (Malpighiaceae) by *Drosophila suzukii* (Diptera: Drosophilidae) in São Paulo State, Brazil. Fla Entomol 102:645–648. https://doi.org/10.1653/024.102.0336

Louzeiro LR, Souza-Filho MF, Raga A et al (2020) Infestation indices and characterization of injuries of *Drosophila suzukii* Matsumura (Diptera: Drosophilidae) in Barbados cherry. Entomobrasilis 13:e898. https://doi.org/10.12741/ebrasilis.v13.e898

Lue CH, Driskell AC, Leips J et al (2016) Review of the genus *Leptopilina* (Hymenoptera, Cynipoidea, Figitidae, Eucoilinae) from the Eastern United States, including three newly described species. J Hymenop Res 53:35–76. https://doi.org/10.3897/jhr.53.10369

Lue CH, Mottern JL, Guillermo C, Walsh GC et al (2017) New record for the invasive Spotted Wing Drosophila, *Drosophila suzukii* (Matsumura, 1931) (Diptera: Drosophilidae) in Anillaco, western Argentina. Proc Entomol Soc Wash 119:146–150. https://doi.org/10.4289/0013-8797.119.1.146

Medina-Muñoz MC, Lucero X, Severino C et al (2015) Drosophila suzukii arrived in Chile. Dros Inf Serv 98:75

Mendes MF, Machado DR, Garcia FRM et al (2019) The use of the trophic resource by exotic and native species Drosophilidae: fruit colonization on the plant. Dros Inf Serv 102:18–20

Moreno-Carrillo GM, Vélez BRV, González JAS et al (2015) Trampeo y registro del parasitoide *Pachycrepoideus vindemmiae* (Rondani) (Hymenoptera: Pteromalidae) sobre *Drosophila suzukii* (Matsumura) (Diptera: Drosophilidae) en México. Southwestern Entomol 40:199–203. https://doi.org/10.3958/059.040.0118

Morrone JJ (2014) Biogeographical regionalisation of the neotropical region. Zootaxa 3782(1):1–110

Paula MA, Lopes PHS, Tidon R (2014) First record of *Drosophila suzukii* in the Brazilian Savanna. Dros Inf Serv 97:113–115

Rampasso AS, Vilela CR (2017) Fruits from three ornamental plants as breeding sites of invasive and native Drosophilidae (Diptera) in the capital of the State of São Paulo, Brazil. In: Abstracts of the X Simpósio de Ecologia, Genética e Evolução de Drosophila, Ouro Preto, p 1

Rojas E, Andrade J, Concha C et al (2019) Manual de reconocimiento. Estados de desarrollo de *Drosophila suzukii* (Diptera: Drosophilidae) y otras especies del género, comunes en el sur de Chile. Ministerio de Agricultura, Servicio Agrícola y Ganadero, Santiago

SAG (2020) Drosophila suzukii. https://www.sag.gob.cl/ambitos-de-accion/drosophila-suzukii. Accessed 22 Jun 2020

Santadino MV, Riquelme Virgala MB, Ansa MA et al (2015) First record of *Drosophila suzukii* (Diptera: Drosophilidae) in blueberry (*Vaccinium* spp.) from Argentina. Rev Soc Entomol Arg 74:183–185

Santo-Ramos M, Bello-Rivera A, Gómez-Paula RJ et al (2014) Monitoring of Spotted Wing Drosophila (*Drosophila suzukii* Mats.) and assessment of the new attractant SuzukiiTrap® in Tijuana, Baja California, Mexico. JFAE 12:349–355

Santos RSS (2014) *Drosophila suzukii* (Matsumura, 1931) (Diptera: Drosophilidae) atacando frutos de morangueiro no Brasil. Enciclopédia Biosfera 10:4005–4010

Santos RSS, Goulart NFB, Bizotto LA et al (2017) Occurrence and quantification of *Drosophila suzukii* in the urban area of Vacaria, RS. Rev Elet Cient UERGS 3:588–599

Schlesener DCH, Wollmann J, Nunes AM et al (2015) *Drosophila suzukii*: nova praga para a fruticultura brasileira. O Biológico 77:45–51

Schlesener DCH, Wollmann J, Pazini JB et al (2017) Effects of insecticides on adults and eggs of *Drosophila suzukii* (Diptera, Drosophilidae). Rev Colomb Entomol 43:208–214. https://doi.org/10.25100/socolen.v43i2.5945

Schlesener DCH, Wollmann J, Pazini JB et al (2019) Insecticide toxicity to *Drosophila suzukii* (Diptera: Drosophilidae) parasitoids: *Trichopria anastrephae* (Hymenoptera: Diapriidae) and *Pachycrepoideus vindemmiae* (Hymenoptera: Pteromalidae). J Econ Entomol 112:1197–1206. https://doi.org/10.1093/jee/toz033

Silva IB et al (2018) Eficiência de ácaros na predação de *Drosophila suzukii* (Matsumura, 1931) (Diptera: Drosophilidae). In: Abstracts of the XXVII Congresso Brasileiro de Entomologia e X Congresso Latino-Americano de Entomologia. p 614

Souza GK, Pikart TG, Oliveira VL et al. (2017) *Acca sellowiana* (Myrtaceae): A New Alternative Host for *Drosophila suzukii* (Diptera: Drosophilidae) in Brazil. Fla Entomol 100:190–191. https://doi.org/10.1653/024.100.0131

Trujillo-Arriaga J (2011) Detection of spotted wing Drosophila (*Drosophila suzukii* Matsumura) in the municipality of Los Reyes, State of Michoacan, Mexico. https://www.pestalerts.org/official-pest-report/detection-spotted-winged-drosophila-drosophila-suzukii-matsumura-municipality. Accessed 4 June 2020

Vieira JGA, Krüger AP, Scheuneumann T et al (2020) Some aspects of the biology of *Trichopria anastrephae* (Hymenoptera: Diapriidae), a resident parasitoid attacking *Drosophila suzukii* (Diptera: Drosophilidae) in Brazil. J Econ Entomol 113:81–87. https://doi.org/10.1093/jee/toz270

Vilela R, Mori L (2014) The invasive spotted-wing Drosophila (Diptera, Drosophilidae) has been found in the city of São Paulo (Brazil). Rev Bras Entomol 58:371–375. https://doi.org/10.1590/S0085-56262014000400004

Wollmann J, Schlesener DCH, Ferreira MS et al (2016) Parasitoids of Drosophilidae with potential for parasitism on *Drosophila suzukii* in Brazil. Dros Inf Serv 99:38–42

Wollmann J, Schlesener DCH, Vieira JGA et al (2019a) Evaluation of food baits to capture *Drosophila suzukii* in the southern of Brazil. An Acad Bras Cienc 91:e20180375. https://doi.org/10.1590/0001-3765201920180375

Wollmann J, Schlesener DCH, Garcia MS et al (2019b) Population dynamics of *Drosophila suzukii* (Diptera: Drosophilidae) in berry crops in Southern Brazil. Neotrop Entomol 48:699–705. https://doi.org/10.1007/s13744-019-00686-5

Wollmann J, Schlesener DCH, Mendes SR et al (2020) Infestation index of *Drosophila suzukii* (Diptera: Drosophilidae) in small fruit in Southern Brazil. Arq Inst Biol 87:e0432018. https://doi.org/10.1590/1808-1657000432018

Yi C, Cai P, Lin J et al (2020) Life history and host preference of *Trichopria drosophilae* from Southern China, one of the effective pupal parasitoids on the *Drosophila* species. Insects 11:103. https://doi.org/10.3390/insects11020103

Zanúncio Junior JS, Fornazier JS, Andreazza MJ et al (2018) Spread of two invasive flies (Diptera: Drosophilidae) infesting commercial fruits in Southeastern Brazil. Fla Entomol 101:522–525. https://doi.org/10.1653/024.101.0328

Zazycki LF, Vargas JP, Bonfim VS et al (2019) Ampliação da área ocorrência de *Drosophila suzukii* (Diptera, Drosophilidae) em Santa Catarina, Brasil. Revista de Ciências Agroveterinárias 18:526–529. https://doi.org/10.5965/223811711832019526

Chapter 6
Progress and Challenges in Building Monitoring Systems for *Drosophila suzukii*

Hannah Burrack, Jana C. Lee, Cesar Rodriguez-Saona, and Greg Loeb

Abstract After the introduction of an invasive species is confirmed in a new area, the next step in mitigation and management step is surveying to determine the range and extent of spread. Determining which monitoring methods are most effective and efficient is essential to determining range and spread, and research efforts at the onset of a biological invasive often focus on developing these technologies. This was certainly the case for *Drosophila suzukii*. A substantial amount of research in the early stages of the *D. suzukii* invasion into North America, and Europe focused on the design of monitoring traps and the identification of attractants. Later efforts shifting to interpreting monitoring data in the context of population estimation and crop risk. In this chapter, we review the current state of knowledge surrounding adult and larval *D. suzukii* monitoring methods, and how data generated using these methods can be applied to research and management questions.

Keywords Insect trapping · Post-harvest sampling · Trap design · Host-plant volatiles · Chemical ecology · Invasive species

H. Burrack (✉)
North Carolina State University, Raleigh, NC, USA
e-mail: hjburrac@ncsu.edu

J. C. Lee
USDA ARS, Corvallis, OR, USA
e-mail: Jana.Lee@ars.usda.gov

C. Rodriguez-Saona
Rutgers University, Chatsworth, NJ, USA
e-mail: crodriguez@njaes.rutgers.edu

G. Loeb
Cornell University, Geneva, NY, USA
e-mail: gme1@cornell.edu

F. R. M. Garcia (ed.), *Drosophila suzukii Management*,
https://doi.org/10.1007/978-3-030-62692-1_6

6.1 Introduction

Effective insect monitoring systems rely on three essential processes; they must be able to first capture, retain, and preserve the target organism for identification. In the context of invasive species, monitoring systems are utilized first to determine its range and spread (Wittengber and Cock 2001), and following establishment, monitoring tools are important in determining the need for management action and evaluating the effect of management programs. Monitoring systems used in detecting and delineating spread of invasive species are ideally highly attractive in order to capture small numbers of introduced or dispersing insects. In the case of *Drosophila suzukii* Matsumura, initial monitoring methods were developed based on those used for other drosophilid flies and were non-species-specific (Lee et al. 2012).

In order to be most useful in integrated pest management programs, monitoring systems should be as specific to the target pest as possible to reduce by-catch, and therefore reduce the labor and expertise needed to service traps, and produce data correlated with crop damage. For example, traps containing pheromone lures can be used to time pesticide applications (Witzgall et al. 2010). Research efforts following the global range expansion of *D. suzukii* have focused on refining monitoring tools in order to improve specificity, efficiency, and correlation with crop risk.

6.2 Trap Designs and *D. suzukii* Behavior

Traps for *D. suzukii* are provisioned with either a liquid bait or an attractant within a separate container or sachet which can either be used in a trap with a liquid drowning solution or a dry sticky trap. Many traps optimization experiments process flies after collection and also rely on trap liquid as a preservative. Dry sticky traps are not commonly used alone for *D. suzukii.*

Physical traps work in three phases: (1) attraction, typically via visual or volatile cues, (2) entrance (for container traps), and (3) retention.

The first phase of drawing in *D. suzukii* is done with an attractive odor combined with visual cues. Since many lures are liquid baits, the trap itself can also be designed to encourage bait volatilization. Entrance area and orientation influence entry. Quick killing agents and entry points that are difficult to exit retain flies. An ideal trap design is also easy to use. Physical traps that enable more bait volatilization such as increasing the surface area or wideness of the trap are effective but can make traps more physically cumbersome and increase nontarget captures (Lee et al. 2013). Smaller traps are more efficient when adjusting for bait volume and trap area (Drummond et al. 2018).

Early comparative studies comparing a range of attributes identified particular trends that have been further investigated (Fig. 6.1). In an early study replicated across seven US states, plastic traps with hole on the side or mesh screening, and red, white, and clear coloration was compared using apple cider vinegar (Lee et al.

Fig. 6.1 Trap designs evaluated for *Drosophila. suzukii*. (**a**) Trap colors compared in field experiments, photo: J. Lee. (**b**) Zorro trap, photo: A. Knight. Typical deli trap with side mesh (**c**) or holes (**d**), photo: J. Lee. (**e**) Scentry trap, photo: J. Lee. (**f**) Yellow dome trap, photo: D. Cha

2012). Though traps differed, those with greater entry area caught more flies. The greater entry area likely allowed more bait to volatilize out of the trap, and increase the probability of *D. suzukii* to enter. Another study compared a standard clear cup with four holes to a dome yellow trap both baited with either apple cider vinegar or a four-component synthetic lure, and the dome trap caught more than the cup in two locations (Cha et al. 2013). Here, coloration, larger entry area, and possibly orientation may have led to higher catches in the dome trap which has a large central opening where flies enter through the bottom.

6.2.1 Visual Attraction

Studies testing *D. suzukii* response to colors in closed arenas have identified red, black, and yellow as attractive. *D. suzukii* in outdoor cages were trapped more in soapy water cups that had yellow or red paint cards affixed to the clear bottom when 23 yellow hues out of 52 and 12 red hues out of 52 trapped more than the control (Supplemental in Lee et al. 2013). *D. suzukii* landed more on red and black cards (Basoalto et al. 2013), and red and yellow disks than other colors in the laboratory (Kirkpatrick et al. 2016). *D. suzukii* were trapped more on red or black sticky spheres in lab and semi-field trials (Rice et al. 2016).

Further field tests have confirmed the attractiveness of these colors. More *D. suzukii* were trapped on sticky spheres of red or black color than other colors including yellow in a 2-year peach trial (Rice et al. 2016) and in a raspberry tunnel trial (Kirkpatrick et al. 2017). Also, more flies were trapped in black or red cup traps than yellow or clear traps in a raspberry field (Renkema et al. 2014). The addition of yellow sticky cards in baited clear cups did not improve capture (Iglesias et al. 2014). In contrast, other field studies have found yellow traps to perform well. More *D. suzukii* were trapped in red, yellow, and black cup traps than clear traps in a 7 state/province study in North America across 7 crops (Lee et al. 2013). Yellow and red cup traps caught similar amounts in a study done at 17 nursery production sites (Addesso et al. 2015). The inconsistent results with yellow may depend on the hue, several of the studies have reported L*a*b* values, and less than half of the yellow hues out of 52 were attractive to *D. suzukii* in outdoor cages (Lee et al. 2013). To date, commercial traps for SWD usually incorporate red and black, or are clear. The commercial Scentry trap which has a red exterior caught more SWD in blueberry fields than a clear plastic commercial trap or homemade clear deli traps when traps used the same bait (Harmon et al. 2019).

Visual contrast is also attractive to *D. suzukii*. Basoalto et al. (2013) was the first to show that a black and red contrast container trap caught more than all red or all black coloration in lab tests, and have hence used the "Zorro" trap. Likewise, Kirkpatrick et al. (2016) found that *D. suzukii* landed on all non-fluorescent colors at similar rates when placed on a black background, showing how contrast can affect attractiveness. Further studies with black and red have revealed better capture when the black is directly contrasted to red, rather than have both colors spaced on a clear container trap (Lasa et al. 2017).

6.2.2 Entry

In traps which rely on a drowning solution, *D. suzukii* sometimes land on the exterior but do not enter. Not surprisingly, when similar traps with greater entry areas were tested in the field, more flies were trapped (Renkema et al. 2014). The entry position differs on traps from the side or top on cup traps, to center bottom for dome traps. Whether a particular entry point facilitates more *D. suzukii* to enter, or retains them better is not known. Traps with a mesh entry on the side caught more than those with entry on the top in a multicrop experiment (Lee et al. 2013). In contrast, traps with holes on lids under a close-fitting lid improved captures compared to holes on the side in a single crop experiment in strawberries (Renkema et al. 2014). Given these differences, the effect of entry point may be dependent on field conditions. If rain or irrigation occurs, a cover is needed for traps that have entry points on the lid to prevent traps from overflowing. There is also evidence that suggests that *D. suzukii* entering a trap are qualitatively different that those remaining on the exterior and that male *D. suzukii* may use the exterior of traps to intercept mates rather than entering (Swoboda-Bhattarai et al. 2017). These differences may result in variable trap entry based on crop and insect phenology.

6.2.3 Retention

For liquid baited traps, *D. suzukii* walk inside before they drown, and it is possible that some escape. To minimize escape, bait solutions include a killing agent, or soap to drown flies. Research conducted early in the *D. suzukii* invasion in North American included yellow sticky cards hung inside the middle or wrapped around the sides of traps but found no trapping advantage (Burrack et al. 2012; Iglesias et al. 2014). There is similarly no advantage to adding yellow sticky cards inside red cup traps (Drummond et al. 2018). A dome cover can lower the likelihood of escape. The number of flies that escaped a clear cup trap with a dome lid was half of those escaping a similar trap with a flat lid (Lasa et al. 2017). The greater area in the dome may reduce the likelihood of flies finding exit holes.

Recently, insecticide coating and use of tunnel entries to retain *D. suzukii* have been compared (Van Kerckvoorde et al. 2020). When a variety of commercial *D. suzukii* traps were tested, coating the inner surface with cypermethrin or deltamethrin resulted in 4.9–7.4 higher retention in the lab. Trapping efficiency was improved in the field by 1.2–4.5 times. Traps with conical tunnel entries as opposed to holes also increase retention in lab and field study by 1.5 times.

6.2.4 Sticky Traps

One advantage of sticky traps is that they only require *D. suzukii* to alight on it, and therefore entering and retention are no longer a design concern. Additionally, these traps do not require handling liquids or straining samples, and males can be visually counted in the field. Female *D. suzukii*, on the other hand, may be difficult to confirm on sticky traps even under a microscope. This may be particularly important in early season when females are often captured more frequently.

Of sticky traps compared, red sticky spheres baited with Scentry lure have captured more *D. suzukii* than clear cup traps with the same lure or yeast bait in raspberry tunnels (Kirkpatrick et al. 2017). Red sphere traps are often used in attract and kill programs where there is no need to count insects on the surface. If used for monitoring, sphere traps are more cumbersome to service and to count *D. suzukii*. Thus, additional studies were done with red panel traps or combination yellow panel-red sphere trap (Kirkpatrick et al. 2018a). The panel trap caught a similar number or more *D. suzukii* than clear deli traps using the same bait in cherry orchards and raspberry tunnels. The authors concluded that a trap with both visual and olfactory cues is better than a trap with only olfactory cues and that a dry sticky trap can perform similar or better than the common clear liquid trap.

While numerous effective traps have been developed, a more difficult challenge has been identifying attractants that are both efficient and relatively specific for *D. suzukii*. This remains a key challenge for adult *D. suzukii* monitoring programs.

6.3 Trap Attractants and Relationship to SWD Chemical Ecology

Since its first detection, there has been interest among the scientific community in the discovery of effective baits and lures to attract *D. suzukii* adults. Here we define a "bait" as any substance (usually a liquid) used to capture *D. suzukii* flies, while a "lure" is defined as a synthetic blend made of attractive volatile compounds. The main goal of this work has been to identify attractive baits or lures that are: (a) effective at capturing flies, (b) selective, i.e., minimize attraction of nontargets, (c) provide early fly detection, and (d) estimate the risk of fruit infestation.

Early trapping methods used apple cider vinegar as a bait. However, it became soon clear that this bait does not provide early enough warning to prevent fruit infestation. As a result, research was initiated to improve the bait or develop a lure to trap *D. suzukii* flies. These studies have focused on attractive compounds from fermenting and ripening fruit, yeast, and leaves as well as *D. suzukii*-derived pheromones. Although our understanding of the roles these chemical cues at play during *D. suzukii* foraging has increased in recent years, it is still unclear how flies respond to some of these cues under natural conditions and how the cues interact. *D. suzukii* flies utilize chemical cues for three distinct foraging behaviors: habitat (host-plant and food) location, oviposition site location, and mate location (Cloonan et al. 2018) (Fig. 6.2). Odors from fermenting fruit likely guide unmated and mated flies to the host-plant habitat and location of food sources. These odors, together with

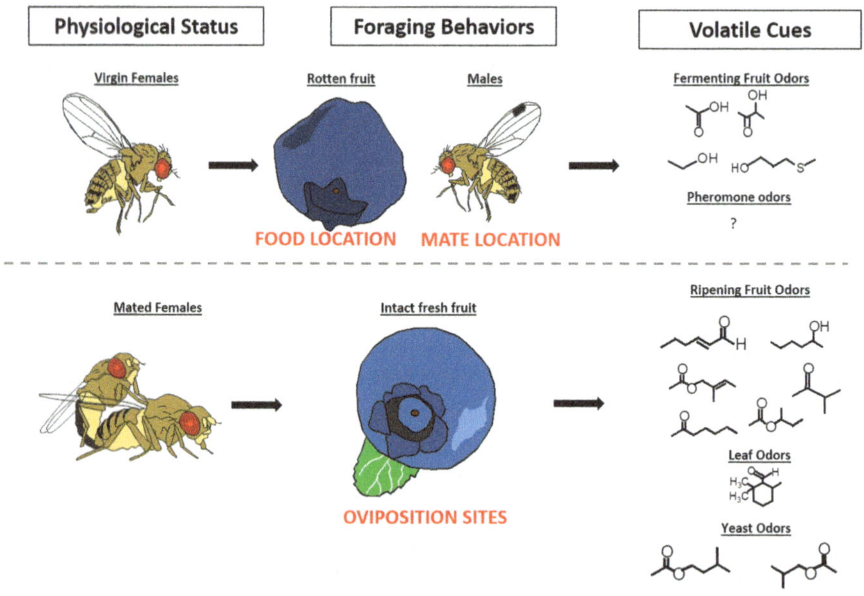

Fig. 6.2 Possible volatile cues used by *Drosophila suzukii* virgin and mated females during foraging for food, mates, and oviposition sites. (Adapted from Cloonan et al. 2018)

pheromones, are also likely involved in mate finding and recognition. Once mated, females then utilize odors from ripening fruits to find oviposition sites. Although more research is needed, yeast and leaf odors may interact with odors from ripening fruit to enhance mated female attraction to oviposition sites. Below we summarize studies on the chemical cues involved in these behaviors.

6.3.1 Fermenting Fruit Odors

Fermenting fruit odors play a crucial role in *D. suzukii* host-plant habitat and food location. After emergence, adults need carbohydrate-rich foods as energy sources to increase their longevity and reproduction (Tochen et al. 2016; Young et al. 2018). Thus, odors from fermenting fruit can guide flies to these foods. In early studies, Landolt et al. (2012a) placed different combinations of wine, apple cider vinegar, ethanol, and acetic acid (the primary components of wine and vinegar, respectively) in blackberry fields in Oregon, USA. These authors hypothesized that much or all of the flies' response to vinegar and wine is due to acetic acid and ethanol, respectively. They found that mixtures of merlot wine and apple cider vinegar were more attractive to female *D. suzukii* than mixtures of ethanol and acetic acid alone (Landolt et al. 2012a).

Follow-up experiments examined mixtures of wine, vinegar, ethanol, and acetic acid for field attraction and found that mixtures of wine/vinegar, ethanol/vinegar, and acetic acid/wine capture similar numbers of *D. suzukii* flies, indicating that these mixtures would result in synergistic fly attraction (Landolt et al. 2012b). Building upon these findings, Cha et al. (2012) investigated the headspace of rice vinegar and merlot wine for antennally active compounds to *D. suzukii* using gas chromatography coupled with electroantennographic detection (GC-EAD). Rice vinegar contained seven EAD-active compounds, namely ethyl acetate, 3-hydroxybutanone (acetoin), ethyl lactate, isoamyl acetate, 2-methylbutyl acetate, ethyl-3-hydroxybutyrate (grape butyrate), and 2-phenylethanol, while merlot wine contained 13 EAD-active compounds, the following six in addition to the previous ones: ethyl butyrate, 1-hexanol, methionol, isoamyl lactate, ethyl sorbate, and diethyl succinate. In laboratory assays and field experiments, an eight-component wine blend [acetic acid + ethanol + acetoin + grape butyrate + methionol + isoamyl lactate + 2-phenylethanol + diethyl succinate] and a five-component vinegar blend [acetic acid + ethanol + acetoin + grape butyrate + 2-phenylethanol] were more attractive than the acetic acid plus ethanol mixture, and as attractive as the wine plus vinegar mixture (Cha et al. 2012).

Subsequent studies in wild blackberry and blueberry field trials with several lure formulations found that mixtures of acetic acid, ethanol, acetoin, ethyl lactate, and methionol caught more flies than apple cider vinegar, but caught similar numbers of flies compared to wine and vinegar baits (Cha et al. 2013). Dropout studies revealed that only acetic acid, ethanol, acetoin, and methionol are necessary to elicit *D. suzukii* attraction in the field (Cha et al. 2014). This four-component blend is now used in two commercial lures: the Pherocon® SWD Dual-Lure™ (proprietary blend, Trécé Inc., USA) and Scentry® (proprietary blend, Scentry Biologicals Inc., USA) lures (Fig. 6.3). In an effort to increase the efficiency of this four-component blend, Cha et al. (2017) found that increasing the release rates of acetoin and acetic acid increases *D. suzukii* attraction; however, an increase in the release rate of methionol did not improve the trap capture (Cha et al. 2014).

Besides being more user-friendly than baits, a major benefit of using the commercially available lures is that they capture *D. suzukii* flies earlier than apple cider vinegar. In a large, multistate comparison of different baits and lures in the USA, the Pherocon® SWD lure suspended over apple cider vinegar captured *D. suzukii* flies between 1 and 2 weeks earlier than apple cider vinegar and, most importantly, detected their presence prior to fruit infestation (Burrack et al. 2015). In northern latitude blueberries, where winters are able to knock down *D. suzukii* populations, the Scentry® lure can detect adult activity 1–5 weeks before fruit infestation, but this varies between regions and crops. In raspberries, for example, the Scentry® lure detected fly activity the same week of fruit infestation (Cha et al. 2018).

Trap selectivity is, however, a challenging problem with these commercial lures since they capture large numbers of nontarget drosophila flies in the field. Adding apple cider vinegar to the Pherocon® SWD lure increases *D. suzukii* attraction in the field but does not increase its specificity (Frewin et al. 2017). Less than half of all flies caught by the Pherocon® SWD lure suspended over apple cider vinegar were *D. suzukii* (Burrack et al. 2015). In an attempt to find a more selective lure, Feng et al. (2018) analyzed headspace volatiles from fresh and fermented apple juices. They identified a quinary blend consisting of acetoin, ethyl octanoate, ethyl acetate, phenethyl alcohol, and acetic acid that was more efficient and selective for *D. suzukii*

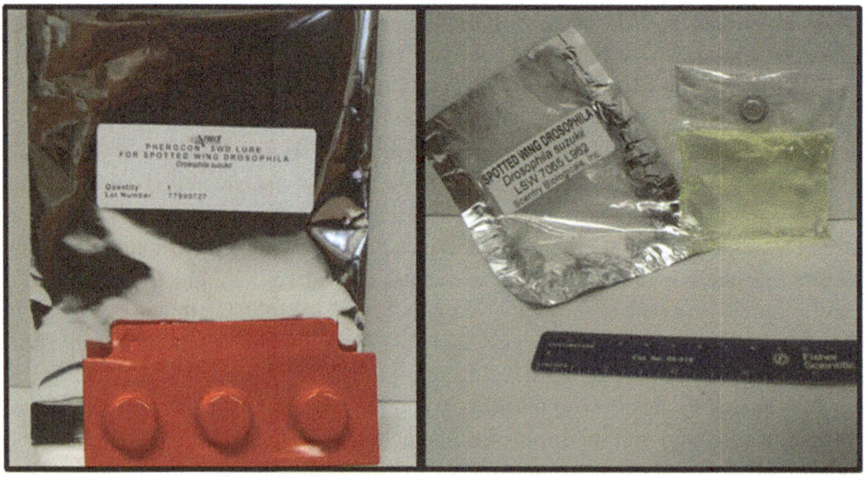

Fig. 6.3 Currently available commercial lures for *Drosophila suzukii*: the Pherocon® SWD Dual-Lure™ (proprietary blend, Trécé Inc., USA), and Scentry® (proprietary blend, Scentry Biologicals Inc., USA)

than apple cider vinegar and Scentry® lure under field conditions. There also appear to qualitative differences in female flies attracted to fruit and trap attractants with reproductively immature females distinguishing between bait types while reproductively mature females may not (Swoboda-Bhattarai et al. 2017; Burrack et al. 2015; Wollman et al. 2019).

6.3.2 Ripening Fruit Odors

Odors from ripening fruit should be important for mated *D. suzukii* females searching for oviposition sites, given their relatively unique host niche among Drosophilidae. *Drosophila suzukii* flies show differential attraction to several crop host fruits. In dual-choice laboratory assays, Revadi et al. (2015) found that odors from whole raspberry, strawberry, blueberry, and cherry fruits were all more attractive to adult *D. suzukii* than clean air. In similar choice studies, Abraham et al. (2015) found that adult flies are more attracted to odors from raspberry fruit extracts than to fruit odor extracts from blueberries and cherries. GC-EAD assays from the raspberry headspace extracts revealed that *D. suzukii* antennae can detect the following 11 compounds: butyl acetate, hexanol, 2-heptanone, 3-methyl-1-butanone, *trans*-2-hexanal, 3-methyl-2-butenyl acetate, 2-heptanol, hexanol, *cis*-3-hexanol, 6-methyl-5-hepten-2-ol, and linalool (Abraham et al. 2015). This 11-component blend was attractive to both male and female adult flies versus a blank control; however, it was not more attractive than the raspberry extract (Abraham et al. 2015). *Drosophila suzukii* antennae can also detect the fresh strawberry volatiles methyl isovalerate, butyl acetate, isopentyl acetate, and hexyl acetate (Keesey et al. 2015).

6.3.3 Yeast Odors

Studies on the mutualistic associations between *D. suzukii* and microbial (fungi and bacteria) species may help researchers improve the attraction of synthetic lures in the field (Hamby and Becher 2016). Yeasts are important for the survival and development of *Drosophila* species and even increase adult fecundity (Simmons and Bradley 1997). Baker's yeast, *Saccharomyces cerevisiae*, is incorporated into *D. suzukii* diets because it is necessary for larval growth and development (Starmer 1981; Starmer and Aberdeen 1990; Becher et al. 2012). Because of these fitness benefits, it is reasonable to assume that *D. suzukii* would be attracted to yeast odors present in their environment.

Hamby et al. (2012) identified several yeast species from *D. suzukii* alimentary canals and larval frass in cherry and raspberry fields, including *Hanseniaspora uvarum*, *Pichia kluyveri*, and *Pichia terricola*. *Hanseniaspora uvarum* was the most abundant yeast species present in the fly's alimentary canals and larval frass, and this yeast's odors were the most attractive of six yeast species tested in binary or multi-choice assays (Hamby et al. 2012). Subsequent GC-EAD analysis revealed isobutyl acetate and isoamyl acetate as the volatiles from *H. uvarum* eliciting the greatest antennal responses in *D. suzukii* (Scheidler et al. 2015). Interestingly, isoamyl acetate is also found in the headspace of several attractive fruits such as strawberry, blueberry, cherry, blackberry, and raspberry (Dekker et al. 2015). In wind tunnel bioassay, Mori et al. (2017) found that mated *D. suzukii* females are more attracted to odors from both blueberry fruit and *H. uvarum* than unmated females, and mated females consume more *H. uvarum* yeast than unmated ones; however, blueberry fruit treated with *H. uvarum* did not elicit more oviposition versus blueberries without the yeast. In the laboratory, mated females laid fewer eggs on blueberries when given access to *H. uvarum* in the arena for feeding versus arenas that did not contain the yeast for feeding (Mori et al. 2017), suggesting a possible trade-off between yeast consumption and oviposition in *D. suzukii*. Conversely, *D. suzukii* females laid more eggs on cherry fruits infested with *Candida* sp. and *S. cerevisiae* than un-infested cherries (Bellutti et al. 2018). The role of yeast odors on *D. suzukii* female attraction and oviposition and its interactions with fruit odors needs further investigation. In fact, *D. suzukii* attraction to the *H. uvarum* volatiles isobutyl acetate and isoamyl acetate in the field increased post-harvest compared to during the fruiting season, indicating that background fruit odors may influence the fly response to yeast odors (Cloonan et al. 2019).

Since vinegar is the by-product of acetic acid bacterial metabolism, Mazzetto et al. (2016) investigated the attraction of adult *D. suzukii* to pure cultures of several acetic acid bacterial species under laboratory conditions. In two-choice attraction assays, adult flies were more attracted to pure cultures of *Gluconobacter oxydans*, *Gluconobacter kanchanaburiensis*, and *Gluconobacter saccharivorans* versus blank media controls. Ethanol, acetic acid, 2-propanol, 2-propanone, benzaldehyde, 2-methylpropanoic acid, 3-methylbutanoic acid, 2-methylbutanoic acid, and acetaldehyde all elicited EAD responses from the headspace of these bacteria (Mazzetto et al. 2016).

6.3.4 Leaf Odors

Previous studies suggest that *D. suzukii* flies cue in on fermenting and ripening fruit odors as well as on yeast odors to find suitable hosts for egg-laying and feeding purposes; however, less is known about their attraction to odors from other plant tissues. In laboratory two-choice assays, Keesey et al. (2015) found that *D. suzukii* are attracted to the strawberry leaf odor β-cyclocitral, and the authors concluded that this chemical is used as a possible long-range cue in attracting flies to the vicinity of a fruiting plant. In contrast, antennae of the close-related species, *Drosophila melanogaster*, did not respond to this compound, or there was any behavioral response (Keesey et al. 2015).

6.3.5 Pheromone Odors

Drosophila melanogaster and *Drosophila sechellia* females are known to produce the male attractant dienes, 7,11-heptacosadiene and 7,11-nonacosadiene (Antony et al. 1985), while monoenes, such as 7-tricosene, are mostly found on males (Everaerts et al. 2010). However, there is no clear role of these cuticular hydrocarbons on *D. suzukii* mating behavior. Snellings et al. (2018) reported four major cuticular hydrocarbons, 9-tricosene, 7-tricosene, 5-tricosene, and tricosane, in *D. suzukii* sexually mature flies that negatively regulated courtship and mating.

In *D. melanogaster*, males produce the pheromone 11-*cis*-vaccenyl acetate (cVA) that influences many behaviors including aggregation of both females and males (Kurtovic et al. 2007; Weng et al. 2013), increasing the acceptance of males to unmated females (Weng et al. 2013), reducing courtship among mated males (Zawistowski and Richmond 1986; Ejima et al. 2007; Kurtovic et al. 2007), and reducing the courtship behaviors toward mated females (Scott 1986; Ejima et al. 2007; Ziegler 2013). An oxygenated hydrocarbon 3-O-acetyl-1,3-dihydroxy-octacosa-11,19-diene (CH_5O_3) is also found in ejaculatory bulbs of *D. melanogaster* (Yew et al. 2009). Like cVA, CH_5O_3 is transferred to females during mating and decreases courtship of males after mating. The ejaculatory bulb of *D. suzukii* is smaller relative to *D. melanogaster*, and it does not produce cVA (Dekker et al. 2015) although the cVA-sensitive odor receptor, Or67d, is functional on *D. suzukii* antennae. The *D. suzukii* sensilla expressing Or67d, and the corresponding antennal lobe glomerulus (DA1), are, however, reduced compared to *D. melanogaster* (Dekker et al. 2015).

6.3.6 Interactions Among Different Odors

Odors from different sources (i.e., fermenting and ripening fruit, yeast, leaves, and pheromones) may interact synergistically to increase *D. suzukii* attraction and improve lure selectivity. For example, the addition of the Scentry® lure to a yeast (Baker's yeast, *S. cerevisiae*) and sugar bait increased *D. suzukii* captures, but it had low selectivity for *D. suzukii* (Jaffe et al. 2018); this increased attraction is likely due to the fact that both the bait and the lure emit similar volatile compounds. The addition of volatiles to an attractive blend can, however, result in lower *D. suzukii* attraction if these additional volatiles are perceived by flies as out of the "right" context. For example, the addition of the propanol, phenethyl acetate, formic acid, acetic acid, and valeric acid all reduced *D. suzukii* attraction when mixed with apple cider vinegar (Kleiber et al. 2014). In laboratory and field studies, the addition of the leaf volatile β-cyclocitral or volatiles from the yeast *H. uvarum* (isobutyl acetate and isoamyl acetate) reduced the response of *D. suzukii* to an attractive blend based on fermenting odors (acetic acid, ethanol, methionol, and acetoin) without improving its selectivity (Cloonan et al. 2019). However, combinations of volatiles in the "right" context can increase attraction. For example, a synergistic interaction was found between β-cyclocitral and cherry juice in the attraction of *D. suzukii* females, and between β-cyclocitral and isoamyl acetate on males (Piñero et al. 2019), indicating that combinations of foliage- and fruit- or yeast-based odors may be needed to increase *D. suzukii* attraction. The addition of β-cyclocitral also increased attraction of *D. suzukii* males and females to individual fruit-based volatiles in caged behavioral assays (Bolton et al. 2019). Altogether, these studies suggest that odors from ripening fruit may synergize with odors from yeast and leaves to attract mated *D. suzukii* females seeking for oviposition sites ("right" context) (Fig. 6.2). On the other hand, attraction of *D. suzukii* flies to fermenting fruit odors may be inhibited by yeast and leaf odors because these flies are seeking food and mates and, thus, these odor combinations might not be in the "right" context. Hence, finding the "right" combination of odors may help improve the selectivity of lures for capturing *D. suzukii* and should be the subject of future research.

6.4 Influences of Seasonal Biology on Trap Captures

Part of the reason for *D. suzukii*'s success as an invasive species is its ability to survive and thrive in a broad range of climate conditions from sub-tropical locations such as Florida or Hawaii to cold-temperate regions such as northern USA into southern Canada and northern Europe (Asplen et al. 2015; Little et al. 2020). *D. suzukii* is considered a cool-temperate species of *Drosophila* (Kimura 1988) with its distribution in northern latitudes constrained by winter cold. In addition to winter cold, hot, and dry conditions in the summer likely also limit its distribution and population dynamics (Winkler et al. 2020).

Time of year, crop phenology, non-crop host availability, and geographic region all influence seasonal abundance of *D. suzukii*. In general, adult trap captures are routinely higher in late fall, after susceptible crop hosts are no longer available (Arnó et al. 2016). This is likely due in part to population increases in overlapping generations, and differential attraction to baits and lures in absence of fruit may also contribute to this observation.

6.4.1 Seasonal Activity Patterns

Trapping conducted in many parts of its current distribution reflect *D. suzukii* seasonality. There are at least three patterns observed based on year-long trapping studies. In more northern latitudes in Europe and North America *D. suzukii* or southern latitudes in South America, there is a period during the mid and late winter where no adult flies are captured, a gradual increase over the summer, large peak in the fall, and finally a dramatic decline after temperatures consistently drop below freezing (Zerulla et al. 2015; Thistlewood et al. 2018; Guédot et al. 2018). A second pattern found in more moderate Mediterranean climates such as California is characterized by reduced but continual trap captures through the winter, peak captures in the spring and early summer, a decline over the hot, dry summer, and a second increase in the fall (Harris et al. 2014; Hamby et al. 2014; Wang et al. 2016; Arnó et al. 2016). Intermediate patterns are observed in more moderate but seasonal environments such as the Pacific Northwest, southeastern USA, northern Italy, Germany, and southern Brazil where *D. suzukii* is captured throughout the year but with fewest captures occurring during winter and a summer decline due to hot temperatures depending on region (Rossi-Stacconi et al. 2016; Panel et al. 2018; Renkema et al. 2018; Briem et al. 2018; Klesener et al. 2018; Thistlewood et al. 2018).

The lack of captures, or significantly reduced captures, in winter in colder regions or higher elevations raises questions about *D. suzukii*'s ability to survive winter, with some speculating that they die out with recolonization from more temperate regions each growing season. Although evidence does indicate *D. suzukii* displays seasonal migration at least between low to high altitude locations (Mitsui et al. 2010; Tait et al. 2018), it also appears clear that *D. suzukii* is adapted to survive prolong periods of cold, including limited below freezing conditions (Stockton et al. 2018, 2019). Related, *D. suzukii* has the capacity to enter a form of reproductive diapause induced by a combination of short day-length and cold temperatures (Wallingford et al. 2016; Toxopeus et al. 2016). Cessation of reproduction, however, is readily broken with increasing temperatures in the spring (Panel et al. 2018; Grassi et al. 2018).

D. suzukii shows morphological and physiological changes in response to cold temperatures during development in late summer and fall that contribute to its ability to survive winter conditions (Jakobs et al. 2015; Shearer et al. 2016; Wallingford and Loeb 2016; Stockton et al. 2018, 2019, 2020; Leach et al. 2019a; Rendon et al. 2019) and also likely influences trapping, especially in the spring. Adult winter morph flies are characterized by darker color and increased tolerance to cold temperatures (Shearer et al. 2016; Leach et al. 2019a; Stockton et al. 2019, 2020). Winter morph flies also appear to differ in their behavior relative to summer morphology flies which influences their capacity to move and adjust their locations to potentially more favorable habitats (Wallingford et al. 2018) and are overall less responsive to odor cues compared to summer morph flies (Kirkpatrick et al. 2018b). The lack of captures in the winter in more cold-temperate regions, therefore, is probably due to a combination of low populations, low or no flight activity due to low temperatures, and changes in behavior that reduce the efficiency of adult traps including trap location and potentially type of lures being used. Developing more effective trapping methods for winter morph flies in the spring may provide insights into forecasting infestation risk and potentially reducing their abundance by targeting habitats where *D. suzukii* overwinters.

Trapping studies conducted over multiple years in different regions have allowed investigators to explore how previous year variables, including overwintering abiotic conditions, relate to current season *D. suzukii* population dynamics and infestation risk. For moderate and cold-temperate regions, there is often significant year-to-year variation in when first trap captures occur and overall *D. suzukii* abundance. For example, in some years, adult *D. suzukii* does not present a significant risk for mid-season crops such as sweet cherries, summer raspberries, and early-maturing blueberries since they only reach significant numbers after harvest while in other years, these crops are at risk. In analyses of several independent trapping data sets, cold winters with prolonged temperatures below freezing are significantly correlated with later trap captures and later and smaller peak populations of *D. suzukii* thereby potentially providing a means to forecast growing seasons with high risk of infestation (Rossi-Stacconi et al. 2016; Thistlewood et al. 2018; Leach et al. 2019b).

6.4.2 Daily Activity Patterns

In addition to seasonal activity patterns, there is also variation in *D. suzukii* attraction over the course of the day. In general, these patterns are similar to those of other Drosophilidae (Swoboda-Bhattarai and Burrack 2020). During hot summer months, *D. suzukii* appear crepuscular with more flies captured in traps baited with fermentation baits or lures during the first 4 hours after sunrise and the last 4 hours before sunset (Swoboda-Bhattarai and Burrack 2020; Evans et al. 2017). These time periods also coincide with when most egg-laying appears to occur. *D. suzukii* is likely partitioning its activity throughout the day to take advantage of lower temperatures during dawn and dusk. These same observations would be difficult to repeat in periods of the year with lower fly densities, but it would be interesting to determine if daily activity patterns shift in cooler weather.

6.5 Relationship Between Trap Captures and Fruit Infestation and Applications for Management

The majority of *D. suzukii* monitoring research has focused on capturing and tracking adult flies, but in commercial fruit production, growers are primarily concerned with preventing fruit infestation. Fruit containing eggs and small larvae may appear otherwise undamaged; therefore, infestation cannot be reliably assessed by external observation alone. Fruit growers are encouraged to sample ripening and ripe fruit regularly, at least weekly or at each harvest, whichever is more frequent, in order to monitor control and assess marketability.

Adult trap captures have been demonstrated to poorly correlate with both fruit infestation timing and rate within the same fields as traps, regardless of bait or lure type. Flies may be captured in higher densities in areas or times where fruit are not available (Burrack et al. 2015; Harris et al. 2014; Joshi et al. 2017). For example, fruit infestation in spring harvested organic raspberries in California were high, but trap captures during the same period were very low (Hamby et al. 2014). Therefore, fly traps may provide only limited actionable management information for a given crop. In early fruiting crops in northern North American or southern South American regions, first adult trap capture may be useful for timing the start of insecticide treatments for *D. suzukii*. However, in later fruiting crops that are harvested during the time of the year when *D. suzukii* are abundant, trap captures do not currently provide guidance for management recommendations. Instead, fruit ripening is a better indicator of risk and need for management.

Fruit growers need to assess fruit infestation in order to determine the efficacy of their management tactics and to determine if they can market their fruit, and fruit marketers need to determine if fruit is infested prior to purchase or sale. Effective fruit sampling methods are also essential for assessing the effectiveness of post-harvest control tactics such as cold storage (Saeed et al. 2020; Aly et al. 2017), irradiation (Follett et al. 2014; Kim et al. 2016a, 2018), modified atmosphere, or combinations thereof (Follett et al. 2018). Research studies on these methods typically hold treated *D. suzukii* until pupation or adult emergence, but that is impractical for fruit marketers or growers who may be utilizing post-harvest controls.

In addition to assessing the success of control tactics and determining marketability, the ability to track fruit infestation can reveal patterns that can be exploited to enhance management tactics. *D. suzukii* infestation rates are higher in blackberries in dense in the center of the plant than in the outer portions (Diepenbrock and Burrack 2017), and this observation led to development of pruning strategies to reduce infestation (Schöneberg et al. 2020).

A range of different extraction methods have been used to observe other internally feeding dipteran pests, primarily tephritid fruit flies, including immersion in hot water, brown sugar solution, or salt solution (Yee 2014). Several of these methods have been adapted for use in quantifying *D. suzukii* within fruit. Sugar (Shaw et al. 2019) or salt (Van Timmeren et al. 2017) extraction is the most widely recommended methods and generally perform similar to each other, although work directly comparing fruit sample methods has only been done in artificially infested fruit (Shaw et al. 2019). In both methods, fruit are gently crushed, but not shredded or pulverized, and covered in solution, either 18% sucrose or 7.5–8% NaCl (Shaw et al. 2019). Larvae exit the fruit into the solution after as little as 15 minutes, but it is recommended to hold fruit for an hour to ensure as many larvae exit as possible. Larvae can then be observed and enumerated. Large second instar and third instar larvae can be observed with the naked eye within the same container as the fruit, but magnification is necessary to observe first instar and small second instar larvae. Filtering the liquid allows for enumeration of all larval stages (Van Timmeren et al. 2017), and reusable fine mesh filters work well for this purpose. Salt or sugar extraction is not a reliable means of observing eggs in fruit; direct microscopic observation of individual fruit is necessary to identify and enumerate eggs.

In most crops, *D. suzukii* can readily be distinguished from other internally feeding pests such as fruitworm caterpillars (*Acrobasis vaccinii*, *Grapholita packardi*) and tephritid fruit flies (*Anastrepha fraterculus*, *Rhagoletis cerasi*, *R. cingulata*, *R. mendax*, *R. indifferens*). Eggs, pupae, and adult *D. suzukii* can all be visually distinguished from other co-occurring Drosophilidae, at least within North America

(Hauser 2011), but larvae cannot. The inability to differentiate *D. suzukii* larvae from non-pest drosophilids complicates fruit infestation monitoring. Because non-pest *Drosophila* spp. are incapable of infesting sound, undamaged fruit, sampling fruit which does not appear externally damaged will reduce the likelihood of non-target detection. In instances where the exact identity of larvae within fruit is important to determine and rearing pupae or adults is impractical, molecular diagnostic tools have been developed to differentiate *D. suzukii* genetic material from other drosophilids in fruit (Dhami and Kumarasinghe 2014; Kim et al. 2014, 2016b) and, interestingly, in predator guts (Wolf et al. 2018).

6.6 Conclusions

While significant progress has been made in developing monitoring systems for *D. suzukii* that capture large numbers of flies and track relative abundance, much work remains to improve trap selectively, increase trap capture correlation with crop risk, and integrate monitoring into management systems. Understanding the highly variable impact of *D. suzukii* across host crops and regions as well as differing trap performance among crops (Burrack et al. 2015) has revealed additional challenges in improving attractants, traps, and fruit monitoring strategies. Future goals include further exploiting differences in the biological differences between *D. suzukii* and other drosophilid, refining crops-specific monitoring strategies and management recommendations, determining fine-scale effects of weather and the environment on trap performance, and integrating information on seasonal variability into monitoring programs.

References

Abraham J, Zhang A, Abubeker SU et al (2015) Behavioral and antennal responses of spotted wing drosophila, *Drosophila suzukii*, to volatiles from fruit extracts. Environ Entomol 44:356–367. https://doi.org/10.1093/ee/nvv013

Addesso KM, Oliver JB, O'Neal PA (2015) Survey for spotted-wing drosophila (Diptera: Drosophilidae) in the five-county nursery production region of middle Tennessee, USA. Fla Entomol 98:1050–1055. https://doi.org/10.1653/024.098.0407

Aly MFK, Kraus DA, Burrack HJ (2017) Effects of postharvest cold storage on the development and survival of immature *Drosophila suzukii* (Diptera: Drosophilidae) in artificial diet and fruit. J Econ Entomol 110:87–93. https://doi.org/10.1093/jee/tow289

Antony C, Davis TL, Carlson DA et al (1985) Compared behavioral responses of male *Drosophila melanogaster* (Canton S) to natural and synthetic aphrodisiacs. J Chem Ecol 11:1617–1629. https://doi.org/10.1007/BF01012116

Arnó J, Solà M, Riudavets J et al (2016) Population dynamics, non-crop hosts, and fruit susceptibility of *Drosophila suzukii* in Northeast Spain. J Insect Sci 89:713–723. https://doi.org/10.1007/s10340-016-0774-3

Asplen MK, Anfora G, Biondi A et al (2015) Invasion biology of spotted wing drosophila (*Drosophila suzukii*): a global perspective and future priorities. J Pest Sci 88:469–494. https://doi.org/10.1007/s10340-015-0681-z

Basoalto E, Hilton R, Knight A (2013) Factors affecting the efficacy of a vinegar trap for *Drosophila suzukii* (Diptera; Drosophilidae). J Appl Entomol 137:562–570. https://doi.org/10.1111/jen.12053

Becher PG, Flick G, Rozpędowska E et al (2012) Yeast, not fruit volatiles mediate *Drosophila melanogaster* attraction, oviposition and development. Funct Ecol 2:822–828. https://doi.org/10.1111/j.1365-2435.2012.02006.x

Bellutti N, Gallmetzer A, Innerebner G et al (2018) Dietary yeast affects preference and performance in *Drosophila suzukii*. J Pest Sci 91:651–660. https://doi.org/10.1007/s10340-017-0932-2

Bolton LG, Piñero JC, Barrett BA (2019) Electrophysiological and behavioral responses of *Drosophila suzukii* (Diptera: Drosophilidae) towards the leaf volatile β-cyclocitral and selected fruit-ripening volatiles. Environ Entomol 48:1049–1055. https://doi.org/10.1093/ee/nvz092

Briem F, Golla B, Hoffmann C et al (2018) Explorative data analysis of *Drosophila suzukii* trap catches from a seven-year monitoring program in southwest Germany. Insects 9:25. https://doi.org/10.3390/insects9040125

Burrack HJ, Smith JP, Pfeiffer DG et al (2012) Using volunteer-based networks to track *Drosophila suzukii* (Diptera: Drosophilidae) an invasive pest of fruit crops. J Integr Pest Manag 3(4):B1–B5. https://doi.org/10.1603/IPM12012

Burrack HJ, Asplen M, Bahder L et al (2015) Multistate comparison of attractants for monitoring *Drosophila suzukii* (Diptera: Drosophilidae) in blueberries and caneberries. Environ Entomol 44:7004–7712. https://doi.org/10.1093/ee/nvv022

Cha D, Adams T, Rogg H, Landolt PJ (2012) Identification and field evaluation of fermentation volatiles from wine and vinegar that mediate attraction of spotted wing drosophila, *Drosophila suzukii*. J Chem Ecol 38:1419–1431. https://doi.org/10.1007/s10886-012-0196-5

Cha DH, Hesler SP, Cowles RS et al (2013) Comparison of a synthetic chemical lure and standard fermented baits for trapping *Drosophila suzukii* (Diptera: Drosophilidae). Environ Entomol 42:1052–1060. https://doi.org/10.1603/EN13154

Cha DH, Adams T, Werle CT et al (2014) A four-component synthetic attractant for *Drosophila suzukii* (Diptera: Drosophilidae) isolated from fermented bait headspace. Pest Manag Sci 70(2):324–331. https://doi.org/10.1002/ps.3568

Cha DH, Landolt PL, Adams TB (2017) Effect of chemical ratios of a microbial-based feeding attractant on trap catch of *Drosophila suzukii* (Diptera: Drosophilidae). Environ Entomol 46(4):907–915. https://doi.org/10.1093/ee/nvx079

Cha DH, Hesler SP, Wallingford AK et al (2018) Comparison of commercial lures and food baits for early detection of fruit infestation risk by *Drosophila suzukii* (Diptera: Drosophilidae). J Econ Entomol 111(2):645–652. https://doi.org/10.1093/jee/tox369

Cloonan KR, Abraham J, Angeli S et al (2018) Advances in the chemical ecology of the spotted wing drosophila (*Drosophila suzukii*) and its applications. J Chem Ecol 44:922–939. https://doi.org/10.1007/s10886-018-1000-y

Cloonan KR, Hernández-Cumplido J, De Sousa ALV et al (2019) Laboratory and field evaluation of host-related foraging odor-cue combinations to attract *Drosophila suzukii* (Diptera: Drosophilidae). J Econ Entomol 112:2850–2860. https://doi.org/10.1093/jee/toz224

Dekker T, Revadi S, Mansourian S et al (2015) Loss of *Drosophila* pheromone reverses its role in sexual communication in *Drosophila suzukii*. Proc R Soc B 282:20143018. https://doi.org/10.1098/rspb.2014.3018

Dhami MK, Kumarasinghe L (2014) A HRM real-time PCR assay for rapid and specific identification of the emerging pest spotted-wing drosophila (*Drosophila suzukii*). PLoS One 9:e98934. https://doi.org/10.1371/journal.pone.0098934

Diepenbrock LM, Burrack HJ (2017) Variation of within-crop microhabitat use by *Drosophila suzukii* (Diptera: Drosophilidae) in blackberry. J Appl Entomol 141:1–7. https://doi.org/10.1111/jen.12335

Drummond FA, Collins JA, Al-Najjar G, Chrstensen J (2018) Itty-bitty traps for monitoring spotted wing Drosophila (Drosophila suzukii Matsumura), does size matter? North American Blueberry Research and Extension Workers Conference 4. https://digitalcommons.libratry.umanine.edu/nabrew2018/proceeedingpapers/4

Ejima A, Smith BPC, Lucas C, van der Goes van Naters W, Miller CJ, Levine JD, Griffith LC (2007) Generalization of courtship learning in *Drosophila* is mediated by *cis*-Vaccenyl acetate. Curr Biol 17:599–605. https://doi.org/10.1016/j.cub.2007.01.053

Evans RK, Toews MD, Sial AA (2017) Diel periodicity of *Drosophila suzukii* (Diptera: Drosophilidae) under field conditions. PLoS One 12:e0171718. https://doi.org/10.1371/journal.pone.0171718

Everaerts C, Farine J-P, Cobb M et al (2010) *Drosophila* cuticular hydrocarbons revisited: mating status alters cuticular profiles. PLoS One 5:e9607. https://doi.org/10.1371/journal.pone.0009607

Feng Y, Bruton R, Park A et al (2018) Identification of attractive blend for spotted wing drosophila, *Drosophila suzukii*, from apple juice. J Pest Sci 91:1251–1267. https://doi.org/10.1007/s10340-018-1006-9

Follett PA, Swedman A, Price DK (2014) Postharvest irradiation treatment for quarantine control of *Drosophila suzukii* (Diptera: Drosophilidae) in fresh commodities. J Econ Entomol 107:964–969. https://doi.org/10.1603/EC14006

Follett PA, Swedman A, Mackey B (2018) Effect of low-oxygen conditions created by modified atmosphere packaging on radiation tolerance in *Drosophila suzukii* (Diptera: Drosophilidae) in sweet cherries. J Econ Entomol 111:141–145. https://doi.org/10.1093/jee/tox337

Frewin AJ, Renkema J, Fraser H et al (2017) Evaluation of attractants for monitoring *Drosophila suzukii* (Diptera: Drosophilidae). J Econ Entomol 110:1156–1163. https://doi.org/10.1093/jee/tox081

Grassi A, Gottardello A, Dalton DT et al (2018) Seasonal reproductive biology of *Drosophila suzukii* (Diptera: Drosophilidae) in temperate climates. Environ Entomol 47:166–174. https://doi.org/10.1093/ee/nvx195

Guédot C, Avanesyan A, Hietala-Henschell K (2018) Effect of temperature and humidity on the seasonal phenology of *Drosophila suzukii* (Diptera: Drosophilidae) in Wisconsin. Environ Entomol 47:1365–1375. https://doi.org/10.1093/ee/nvy159

Hamby KA, Becher PG (2016) Current knowledge of interactions between *Drosophila suzukii* and microbes, and their potential utility for pest management. J Pest Sci 89:621–630. https://doi.org/10.1007/s10340-016-0768-1

Hamby KA, Hernández A, Boundy-Mills K et al (2012) Associations of yeasts with spotted-wing Drosophila (*Drosophila suzukii*; Diptera: Drosophilidae) in cherries and raspberries. Appl Environ Microbiol 78:4869–4873. https://doi.org/10.1128/AEM.00841-12

Hamby KA, Bolda MP, Sheehan ME et al (2014) Seasonal monitoring for *Drosophila suzukii* (Diptera: Drosophilidae) in California commercial raspberries. Environ Entomol 43:1008–1018. https://doi-org.prox.lib.ncsu.edu/10.1603/EN13245

Harmon DS, Haseeb M, Kanga LHB et al (2019) Evaluation of monitoring traps and lures for *Drosophila suzukii* (Diptera: Drosophilidae) in berry plantings in Florida. Insects 10:313. https://doi.org/10.3390/insects10100313

Harris DW, Hamby KA, Wilson HE et al (2014) Seasonal monitoring of *Drosophila suzukii* (Diptera: Drosophilidae) in a mixed fruit production system. J Asia Pac Entomol 17:857–864 https://doi.org/10.1016/j.aspen.2014.08.006

Hauser M (2011) A historic account of the invasion of *Drosophila suzukii* (Matsumura) (Diptera: Drosophilidae) in the continental United States, with remarks on their identification. Pest Manag Sci 67:1352–1357. https://doi.org/10.1002/ps.2265

Iglesias LE, Nyoike TW, Liburd OE (2014) Effect of trap design, bait type, and age on captures of *Drosophila suzukii* (Diptera: Drosophilidae) in berry crops. J Econ Entomol 107:1508–1518. https://doi.org/10.1603/EC13538

Jaffe BD, Avanesyan A, Bal HK et al (2018) Multistate comparison of attractants and the impact of fruit development stage on trapping *Drosophila suzukii* (Diptera: Drosophilidae) in raspberry and blueberry. Environ Entomol 47:935–945. https://doi.org/10.1093/ee/nvy052

Jakobs R, Gariepy TD, Sinclair BJ (2015) Adult plasticity of cold tolerance in a continental-temperate population of *Drosophila suzukii*. J Insect Physiol 79:1–9. https://doi.org/10.1016/J.JINSPHYS.2015.05.003

Joshi NK, Butler B, Demchak K et al (2017) Seasonal occurrence of spotted wing drosophila in various small fruits and berries in Pennsylvania and Maryland. J Appl Entomol 141:156–160. https://doi.org/10.1111/jen.12325

Keesey IW, Knaden M, Hansson BS (2015) Olfactory specialization in *Drosophila suzukii* supports an ecological shift in host preference from rotten to fresh fruit. J Chem Ecol 41:121–128. https://doi.org/10.1007/s10886-015-0544-3

Kim SS, Tripodi AD, Johnson DT et al (2014) Molecular diagnostics of *Drosophila suzukii* (Diptera: Drosophilidae) using PCR-RFLP. J Econ Entomol 107:1292–1294. https://doi.org/10.1603/EC13389

Kim J, Kim J, Park CG (2016a) X-ray radiation and developmental inhibition of *Drosophila suzukii* (Matsumura) (Diptera: Drosophilidae). Int J Radiat Biol 92:849–854. https://doi.org/10.1080/09553002.2016.1230236

Kim YA, Hur JH, Lee GS et al (2016b) Rapid and highly accurate detection of *Drosophila suzukii*, spotted wing Drosophila (Diptera: Drosophilidae) by loop-mediated isothermal amplification assays. J Asia Pac Entomol 19:1211–1216. https://doi.org/10.1016/j.aspen.2016.10.015

Kim J, Kim J, Lee YJ, Park CG (2018) Developmental inhibition of *Drosophila suzukii* by ionizing radiation. Entomol Res 48:331–338. https://doi.org/10.1111/1748-5967.12283

Kimura MT (1988) Adaptations to temperature climates and evolution of overwintering strategies in the *Drosophila melanogaster* species group. Evolution 42:1288–1297

Kirkpatrick DM, McGhee PS, Hermann SL et al (2016) Alightment of spotted wing drosophila (Diptera: Drosophilidae) on odorless disks varying in color. Environ Entomol 45:185–191. https://doi.org/10.1093/ee/nvv155

Kirkpatrick DM, McGhee PS, Gut LJ et al (2017) Improving monitoring tools for spotted wing drosophila, *Drosophila suzukii*. Entomol Exp Appl 164:87–93. https://doi.org/10.1111/eea.12602

Kirkpatrick DM, Gut LJ, Miller JR (2018a) Development of a novel dry, sticky trap design incorporating visual cues for *Drosophila suzukii* (Diptera: Drosophilidae). J Econ Entomol 111:1775–1779. https://doi.org/10.1093/jee/toy097

Kirkpatrick DM, Leach HL, Xu P et al (2018b) Comparative antennal and behavioral responses of summer and winter morph *Drosophila suzukii* (Diptera: Drosophilidae) to ecologically relevant volatiles. Environ Entomol 47:700–706. https://doi.org/10.1093/ee/nvy046

Kleiber JR, Unelius CR, Lee JC et al (2014) Attractiveness of fermentation and related products to spotted wing drosophila (Diptera: Drosophilidae). Environ Entomol 43:439–447. https://doi.org/10.1603/EN13224

Klesener DF, Silva dos Santos RS, Gebler L et al (2018) Population fluctuation and infestation of *Drosophila suzukii* in berry crops in southern Brazil. Afr J Agric Res 13:499–511. https://doi.org/10.5897/AJAR2018.12999

Kurtovic A, Widmer A, Dickson B (2007) A single class of olfactory neurons mediates behavioural responses to a *Drosophila* sex pheromone. Nature 446:542–546. https://doi.org/10.1038/nature05672

Landolt PL, Adams T, Davis TS et al (2012a) Spotted wing drosophila, *Drosophila suzukii* (Diptera: Drosophilidae), trapped with combinations of wines and vinegars. Fla Entomol 95(2):326–332. https://doi.org/10.1653/024.095.0213

Landolt PL, Adams T, Rogg H (2012b) Trapping spotted wing drosophila, *Drosophila suzukii* (Matsumura) (Diptera: Drosophilidae), with combinations of vinegar and wine, and acetic acid and ethanol. J App Entomol 136:148–154. https://doi.org/10.1111/j.1439-0418.2011.01646.x

Lasa R, Tadeo E, Toledo-Hernandez RA et al (2017) Improved capture of Drosophila suzukii by a trap baited with two attractants in the same device. PLoS One 12:e0188350. https://doi.org/10.1371/journal.pone.0188350

Leach H, Stone J, Van Timmeren S et al (2019a) Stage-specific and seasonal induction of the overwintering morph of spotted wing drosophila (Diptera: Drosophilidae). J Insect Sci 19(4):5. https://doi.org/10.1093/jisesa/iez067

Leach H, Van Timmeren S, Wetzel W et al (2019b) Predicting within- and between-year variation in activity of the invasive spotted wing drosophila (Diptera: Drosophilidae) in a temperate region. Environ Entomol 48:1223–1233. https://doi.org/10.1093/ee/nvz101

Lee JC, Burrack HJ, Barrantes LD et al (2012) Evaluation of monitoring traps for *Drosophila suzukii* (Diptera: Drosophilidae) in North America. J Econ Entomol 105:1350–1357. https://doi.org/10.1603/Ec12132
Lee JC, Shearer PW, Barrantes LD et al (2013) Trap designs for monitoring *Drosophila suzukii* (Diptera: Drosophilidae). Environ Entomol 42:1348–1355. https://doi.org/10.1603/EN13148
Little CM, Chapman TW, Hillier NK (2020) Plasticity is key to success of *Drosophila suzukii* (Diptera: Drosophilidae) invasion. J Insect Sci 20:1–8. https://doi.org/10.1093/jisesa/ieaa034
Mazzetto F, Gonella E, Crotti E et al (2016) Olfactory attraction of *Drosophila suzukii* by symbiotic acetic acid bacteria. J Pest Sci 89:783–792. https://doi.org/10.1007/s10340-016-0754-7
Mitsui H, Beppu K, Kimura MT (2010) Seasonal life cycles and resource uses of flower- and fruit-feeding drosophilid flies (Diptera: Drosophilidae) in Central Japan. Entomol Sci 13:60–67. https://doi.org/10.1111/j.1479-8298.2010.00372.x
Mori BA, Whitener AB, Leinweber Y et al (2017) Enhanced yeast feeding following mating facilitates control of the invasive fruit pest *Drosophila suzukii*. J Appl Ecol 54:170–177. https://doi.org/10.1111/1365-2664.12688
Panel AD, Zeeman L, van der Sluis B et al (2018) Overwintered *Drosophila suzukii* are the main source for infestations of the first fruit crops of the season. Insects 9:145. https://doi.org/10.3390/insects9040145
Piñero J, Barrett BA, Bolton LG, Follett PA (2019) β-cyclocitral synergizes the response of adult *Drosophila suzukii* (Diptera: Drosophilidae) to fruit juices and isoamyl acetate in a sex-dependent manner. Sci Rep 9:10574. https://doi.org/10.1038/s41598-019-47081-z
Rendon D, Walton V, Tait G et al (2019) Interactions among morphotype, nutrition, and temperature impact fitness of an invasive fly. Ecol Evol 9:2615. https://doi.org/10.1002/ece3.4928
Renkema JM, Buitenhuis R, Hallett RH (2014) Optimizing trap design and trapping protocols for *Drosophila suzukii* (Diptera: Drosophilidae). J Econ Entomol 107:2107–2118. https://doi.org/10.1603/EC14254
Renkema JM, Iglesias LE, Bonneau P et al (2018) Trapping system comparisons for and factors affecting populations of *Drosophila suzukii* and *Zaprionus indianus* in winter-grown strawberry. Pest Manag Sci 74:2076–2088. https://doi.org/10.1002/ps.4904
Revadi S, Vitagliano S, Rossi Stacconi MV et al (2015) Olfactory responses of *Drosophila suzukii* females to host plant volatiles. Physiol Entomol 40:54–64. https://doi.org/10.1111/phen.12088
Rice KB, Short BD, Jones SK, Leskey TC (2016) Behavioral responses of *Drosophila suzukii* (Diptera: Drosophilidae) to visual stimuli under laboratory, semifield, and field conditions. Environ Entomol 45:1480–1488. https://doi.org/10.1093/ee/nvw123
Rossi-Stacconi MV, Kaur R, Mazzoni V et al (2016) Multiple lines of evidence for reproductive winter diapause in the invasive pest *Drosophila suzukii*: useful clues for control strategies. J Pest Sci 89:689–700. https://doi.org/10.1007/s10340-016-0753-8
Saeed N, Tonina L, Battisti A et al (2020) Postharvest short cold temperature treatment to preserve fruit quality after *Drosophila suzukii* damage. Int J Pest Manag 66:23–30. https://doi.org/10.1080/09670874.2018.1539531
Scheidler NH, Liu C, Hamby KA et al (2015) Volatile codes: correlation of olfactory signals and reception in *Drosophila*-yeast chemical communication. Sci Rep 5:14059. https://doi.org/10.1038/srep14059
Schöneberg T, Arsenault-Benoit A, Taylor CM et al (2020) Pruning of small fruit crops can affect habitat suitability for *Drosophila suzukii*. Agric Ecosyst Environ 294:106860. https://doi.org/10.1016/j.agee.2020.106860
Scott D (1986) Sexual mimicry regulates the attractiveness of mated *Drosophila melanogaster* females. PNAS 3:8429–8433. https://doi.org/10.1073/pnas.83.21.8429
Shaw B, Cannon MFL, Buss DS et al (2019) Comparison of extraction methods for quantifying *Drosophila suzukii* (Diptera: Drosophilidae) larvae in soft- and stone-fruits. Crop Prot 124:104868. https://doi.org/10.1016/j.cropro.2019.104868
Shearer PW, West JD, Walton VM et al (2016) Seasonal cues induce phenotypic plasticity of *Drosophila suzukii* to enhance winter survival. BMC Ecol 16. https://doi.org/10.1186/s12898-016-0070-3

Simmons FH, Bradley TJ (1997) An analysis of resource allocation in response to dietary yeast in *Drosophila melanogaster*. J Insect Phys 43:779–788. https://doi.org/10.1016/S0022-1910(97)00037-1

Snellings Y, Herrera B, Wildemann B et al (2018) The role of cuticular hydrocarbons in mate recognition in *Drosophila suzukii*. Sci Rep 8:4996. https://doi.org/10.1038/s41598-018-23189-6

Starmer WT (1981) A comparison of *Drosophila* habitats according to the physiological attributes of the associated yeast communities. Evolution 35:38–52. https://doi.org/10.2307/2407940. https://www.jstor.org/stable/2407940

Starmer WT, Aberdeen V (1990) The nutritional importance of pure and mixed cultures of yeasts in the development of *Drosophila mulleri* larvae in *Opuntia* tissues and its relationship to host plant shifts. In: Barker JSF (ed) Ecological and evolutionary genetics of *Drosophila*. Plenum Press, New York, pp 145–160

Stockton DG, Wallingford AK, Loeb GM (2018) Phenotypic plasticity promotes overwintering survival in a globally invasive crop pest, *Drosophila suzukii*. Insects 9:105. https://doi.org/10.3390/INSECTS9030105

Stockton D, Wallingford A, Rendon D et al (2019) Interactions between biotic and abiotic factors affect survival in overwintering *Drosophila suzukii* (Diptera: Drosophilidae). Environ Entomol 48:454–464. https://doi.org/10.1093/ee/nvy192

Stockton DG, Wallingford AK, Brind'Amore G et al (2020) Seasonal polyphenism of spotted-wing drosophila is affected by variation in local abiotic conditions within its invaded range, likely influencing survival and regional population dynamics. Ecol Evol 10:7669. https://doi.org/10.1002/ece3.6491

Swoboda-Bhattarai KA, Burrack HJ (2020) Diurnal and seasonal activity patterns of drosophilid species (Diptera: Drosophilidae) present in blackberry agroecosystems with a focus on spotted-wing drosophila. Environ Entomol 49:277–287. https://doi.org/10.1093/ee/nvz161

Swoboda-Bhattarai KA, McPhie DR, Burrack HJ (2017) Reproductive status of *Drosophila suzukii* (Diptera: Drosophilidae) females influences attraction to fermentation-based baits and ripe fruits. J Econ Entomol 110:1648–1652. https://doi.org/10.1093/jee/tox150

Tait G, Grassi A, Pfab F et al (2018) Large-scale spatial dynamics of *Drosophila suzukii* in Trentino, Italy. J Pest Sci 91:1213–1224. https://doi.org/10.1007/s10340-018-0985-x

Thistlewood HMA, Gill P, Beers EH et al (2018) Spatial analysis of seasonal dynamics and overwintering of *Drosophila suzukii* (Diptera: Drosophilidae) in the Okanagan-Columbia Basin, 2010–2014. Environ Entomol 47:221–232. https://doi.org/10.1093/ee/nvx178

Tochen S, Walton VM, Lee JC (2016) Impact of floral feeding on adult *Drosophila suzukii* survival and nutrient status. J Pest Sci 89:793–802. https://doi.org/10.1007/s10340-016-0762-7

Toxopeus J, Jakobs R, Ferguson LV et al (2016) Reproductive arrest and stress resistance in winter-acclimated *Drosophila suzukii*. J Insect Physiol 89:37–51. https://doi.org/10.1016/j.jinsphys.2016.03.006

Van Kerckvoorde V, Clymans R, Bangels E et al (2020) Tunnel entries and a killing agent uncover the importance of fly retention in *Drosophila suzukii* traps. Pest Manag Sci 76:3459. https://doi.org/10.1002/ps.5956

Van Timmeren S, Diepenbrock LM et al (2017) A filter method for improved monitoring of *Drosophila suzukii* (Diptera: Drosophilidae) larvae in fruit. J Integr Pest Manag 8:23. https://doi.org/10.1093/jipm/pmx019

Wallingford AK, Loeb GM (2016) Developmental acclimation of *Drosophila suzukii* (Diptera: Drosophilidae) and its effect on diapause and winter stress tolerance. Environ Entomol 45:1081–1089. https://doi.org/10.1093/ee/nvw088

Wallingford AK, Lee JC, Loeb GM (2016) The influence of temperature and photoperiod on the reproductive diapause and cold tolerance of spotted-wing drosophila, *Drosophila suzukii*. Entomol Exp Appl 159:327–337. https://doi.org/10.1111/eea.12443

Wallingford AK, Rice KB, Leskey TC et al (2018) Overwintering behavior of *Drosophila suzukii* Matsumura, and potential springtime diets for egg maturation. Environ Entomol 47:1266–1273. https://doi.org/10.1093/ee/nvy115

Wang XG, Steward TJ, Biondi A et al (2016) Population dynamics and ecology of *Drosophila suzukii* in Central California. J Pest Sci 89:701–712. https://doi.org/10.1007/s10340-016-0747-6

Weng R, Chin JSR, Yew JY et al (2013) *miR-124* controls male reproductive success in *Drosophila*. eLife 2:e00640. https://doi.org/10.7554/eLife.00640

Winkler A, Jung J, Kleinhenz B et al (2020) A review on temperature and humidity effects on *Drosophila suzukii* population dynamics. Agr Forest Entomol 22:179. https://doi.org/10.1111/afe.12381

Wittengber R, Cock MJW (2001) Early detection. In: Wittengber R, Cock MJW (eds) Invasive alien species: a toolkit of best prevention and management practices. CABI Publishing, Wallingford, pp 101–112

Witzgall P, Kirsch P, Cork A (2010) Sex pheromones and their impact on pest management. J Chem Ecol 36:80–100. https://doi.org/10.1007/s10886-009-9737-y

Wolf S, Zeisler C, Sint D et al (2018) A simple and cost-effective molecular method to track predation on *Drosophila suzukii* in the field. J Pest Sci 91:927–935. https://doi.org/10.1007/s10340-017-0948-7

Wollman J, Schlesener DCH, Vieira JGA et al (2019) Evaluation of food baits to capture *Drosophila suzukii* in the southern of Brazil. An Acad Bras Cienc 91:e20180375. https://doi.org/10.1590/0001-3765201920180375

Yee WL (2014) Comparison of the brown sugar, hot water, and salt methods for detecting Western cherry fruit fly (Diptera: Tephritidae) larvae in sweet cherry. Fla Entomol 97:422–430. https://doi.org/10.1653/024.097.0212

Yew JY, Dreisewerd K, Luftmann H et al (2009) A new male sex pheromone and novel cuticular cues for chemical communication in *Drosophila*. Curr Biol 19:1245–1254. https://doi.org/10.1016/j.cub.2009.06.037

Young Y, Buckiewicz N, Long TAF (2018) Nutritional geometry and fitness consequences in *Drosophila suzukii*, the spotted-wing Drosophila. Ecol Evol 8:2842. https://doi.org/10.1002/ece3.3849

Zawistowski S, Richmond RC (1986) Inhibition of courtship and mating of *Drosophila melanogaster* by the male-produced lipid, cis-vaccenyl acetate. J Insect Physiol 32:189–192. https://doi.org/10.1016/0022-1910(86)90057-0

Zerulla FN, Schmidt S, Streitberger M et al (2015) On the overwintering ability of *Drosophila suzukii* in South Tyrol. J Berry Res 5:41–48. https://doi.org/10.3233/JBR-150089

Ziegler AB, Berthelot-Grosjean M, Grosjean Y (2013) The smell of love in Drosophila. Front Physiol 4:72. https://doi.org/10.3389/fphys.2013.00072

Chapter 7
Chemical Control of *Drosophila suzukii*

Rady Shawer

Abstract Current effective *Drosophila suzukii* Matsumura control programs are mainly based on chemical methods although violations of maximum residue limits for specific pesticides, developing of insecticide resistance and negative impacts to beneficial arthropods. The current published data confirm the excellent activity of insecticides from four families, i.e., spinosyns (e.g., spinosad, spinetoram), pyrethroids (e.g., lambda-cyhalothrin deltamethrin bifenthrin, beta-cyfluthrin, permethrin, fenitrothion, and zeta-cypermethrin), organophosphates (e.g., dimethoate, phosmet, malathion, methidathion, and diazinon), and diamides (cyantraniliprole). The best result achieved by any of them regarding protecting fruits from damage was up to 14 days after application. While less effective insecticides provided shorter periods of fruit protection. Adding a feeding stimulant such as sugar, sugar-yeast bait, or erythritol to the insecticides, i.e., spinosad, spinetoram, acetamiprid, and cyantraniliprole, enhanced their biological performances against *D. suzukii*. The natural products including; thyme, *Leptospermum ericoides*, *L. scoparium*, erythritol+sucrose, the chitinase of *Euphorbia characias*, *Perilla aldehyde*, and the powder sulfur had promising results and should be considered in *D. suzukii* control strategy. However, to maximize the use benefits of those natural compounds, more data on their side effects on beneficials, open-field activity, and environmental impacts are currently sought after. In this chapter, we shed the light on most recent updates of chemical control of *D. suzukii*.

Keywords Spotted winged Drosophila · SWD · Chemical Based Methods · Natural Compounds · Insecticides

R. Shawer (✉)
Department of Plant Protection, Faculty of Agriculture (Saba Basha), University of Alexandria, Alexandria, Egypt
e-mail: rady.shawer@alexu.edu.eg

F. R. M. Garcia (ed.), *Drosophila suzukii Management*,
https://doi.org/10.1007/978-3-030-62692-1_7

7.1 Introduction

The new emerging pest, spotted winged Drosophila (SWD), *Drosophila suzukii* Matsumura (Diptera: Drosophilidae) (Matsumura 1931), native to Southeast Asia, is a destructive fruit pest (Delfinado and Hardy 1973; Hauser 2011). It has been recently introduced into Europe, America, and elsewhere causing a considerable economic loss in such fruits. It represents a major challenge to global production of small and thin-skinned fruits including cherry, blueberry, raspberry, grape, and strawberry (Sasaki and Sato 1995; Lee et al. 2011a, b; Walsh et al. 2011; Bellamy et al. 2013; Cuthbertson et al. 2014). As a result, it has become a key pest of soft and stone fruits in almost distribution regions and has been listed as a quarantine pest and major risk (EPPO 2013). Females *D. suzukii* oviposit in healthy fruits (Goodhue et al. 2011), as opposed to most species of drosophilids which only infest overripe, fallen, and rotting fruits. They use their strong serrated ovipositor to damage skin and lay eggs inside healthy ripening fruits (Stacconi et al. 2013). The ovipositor injury also provides a gateway for bacterial and fungal infections or other pests (Walsh et al. 2011; Shawer et al. 2018b). The hatched larvae cause the great damage by feeding on fruit core (Haye et al. 2016).

The serve infestation of SWD for cherries were observed for the first time in Japan in 1916 (Kanzawa 1935). Recently, an economic loss in sweet cherry in Japan reached sometimes up to 100% (Sasaki and Sato 1995). Since its first discovery (2009) in Italy (Cini et al. 2012), damage up to 90% in late cherry varieties was recorded (Mori et al. 2019). In northeastern Italy, an annual loss in small fruit production was determined €3.3 m/year (De Ros et al. 2013; Haye et al. 2016). In the USA, its damage to cherries was reported (26% of fruit production) for the first time in 2009 (Beers et al. 2011). And about 37% and 20% decrease in the gross revenues of raspberry and strawberry growers, respectively, was supposed (Goodhue et al. 2011).

To protect fruit effectively during the whole ripening period, the number of insecticidal applications ranges from one to eight, depending on crop and its susceptibility, pest intensity, and other environmental factors (Asplen et al. 2015; Shawer et al. 2018a; Dam et al. 2019; Shawer 2017).

Because the larvae are developed within ripening fruits, the chemical control of SWD should be mainly targeted adults, and all insecticides have to be applied on the fruits close to harvest, which of course leads to increase residues in fruits (Cini et al. 2012; CABI 2016; Shawer 2017). As a result, growers are in a great confront regarding protecting their crops with limited alternatives of effective control tactics, violations of maximum residue limits for specific pesticides, developing of insecticide resistance and negative impacts to beneficial microorganisms (Gullickson et al. 2019). Moreover, a great deal of highly efficient broad-spectrum insecticides is being progressively prevented. Therefore, additional control tactics are highly needed to decrease environmental hazards and to commit to label instructions (Haye et al. 2016). Several products were assessed for their insecticidal activity against SWD. Here in this chapter, we shed the light on most recent updates of chemical control of SWD.

7.2 Chemical-Based Methods

7.2.1 Chemical Insecticides

The severe threat of *D. suzukii* lies in that flies laid their eggs inside the healthy ripening fruit near to harvest, and then larvae spend all next life inside fruit pulp (Shawer et al. 2018b; Shawer 2017). As a result, effective adulticide-ovicide and residual approach are required while uselessness of larval control (Mori et al. 2019). The application of broad-spectrum chemical insecticides is still the main effective strategy used for managing SWD (Haye et al. 2016; Sial et al. 2017; Gullickson et al. 2019; Shawer et al. 2018a; Andreazza et al. 2017a, b), while nonchemical approaches are being improved and adopted (Sial et al. 2017). Growers principally rely on a wide range of conventional broad-spectrum insecticides (e.g., pyrethroids, carbamates, and organophosphates) which are not always compatible with integrated pest management programs (Beers et al. 2011; Haviland and Beers 2012; Van Timmeren and Isaacs 2013; Haye et al. 2016).

Neonicotinoids have been used in a limited range in control programs of SWD because their unsatisfactory performances against adults (Bruck et al. 2011; Shawer et al. 2018b; Haye et al. 2016; Shawer 2017), broad-spectrum effects and negative impacts to beneficial arthropods (He et al. 2008). However, they seem to have acceptable results in reducing egg hatching and larval development (Bruck et al. 2011; Shawer et al. 2018b) which can be attributed to neonicotinoid properties (Bruck et al. 2011; Barry and Polavarapu 2005; Wise et al. 2006; Beers et al. 2011; Shawer et al. 2018b).

An exception of this in relation to broad spectrum impacts is spinosyns (spinosad and spinetoram) (Beers et al. 2011; Haye et al. 2016; Bruck et al. 2011; Haviland and Beers 2012) which have exhibited excellent performances in controlling SWD, but for resistance management, the number of applications made per year on a given crop need to be limited (Haye et al. 2016; Shawer et al. 2018a).

In the United States, 17 active ingredients from different chemical groups including carbamates (carbaryl, methomyl), organophosphates (diazinon, malathion, phosmet), pyrethroids (bifenthrin, esfenvalerate, fenpropathrin, zeta-cypermethrin), neonicotinoids (acetamiprid, imidacloprid, imidacloprid & cyfluthrin, thiamethoxam), spinosyns (spinetoram, spinosad), pyrethrin, and pyriproxyfen are listed for use on blueberry, cranberry, strawberry, grape, and stone fruit (Fruit-Advisor 2015).

Due to the lack of effective pesticide options available for controlling SWD, it is necessary to maximize the use of what available insecticides (Haye et al. 2016) along with searching for other effective alternatives (Shawer et al. 2018a). Since the arrival of SWD, several laboratory and field screening trials have been widely carried out to identify most effective insecticides for the control of *D. suzukii* in the major of its distributed growing regions (Haye et al. 2016). Laboratory bioassays principally based on exposing flies for treated fruits or leaves and assessing fly mortality, female fecundity, oviposition, and eggs hatchability. In field experiments, crops are sprayed and sampled for managing the pest (Bruck et al. 2011; Cuthbertson et al. 2014; Shawer 2017). In Europe and the USA, several studies confirmed the efficacy of certain insecticides from various chemical groups including organophosphates, carbamates, pyrethroids, spinosyns, and diamides (Cuthbertson et al. 2014; Beers et al. 2011; Shawer et al. 2018b; Van Timmeren and Isaacs 2013; Mori et al. 2019). Bruck et al. (2011) evaluated efficacy of a wide range of insecticides against SWD. They found that insecticides including pyrethroids (bifenthrin, beta-cyfluthrin, permethrin, zeta-cypermethrin), organophosphates (malathion, diazinon), and spinosyns (spinosad, spinetoram) provided excellent control of adult *D. suzukii* following direct application (see Table 7.1). Trials conducted by Shawer et al. (2018b) confirmed that three classes of insecticides, i.e., spinosyns (e.g., spinosad and spinetoram), organophosphates (e.g., dimethoate and phosmet), and pyrethroids (e.g., lambda-cyhalothrin and deltamethrin) proved to be effective against various SWD life stages (Shawer et al. 2018a). Organophosphates, spinosyns, and cyantraniliprole showed the best performance on SWD eggs and immatures (Bruck et al. 2011; Cuthbertson et al. 2014; Beers et al. 2011; Van Timmeren and Isaacs 2013). The acute impact of tested pyrethroids (deltamethrin and lambda-cyhalothrin) as adulticides and their short residual effects were observed (Shawer et al. 2018a). The study of Diepenbrock et al. (2016) confirmed the high efficacy of organophosphates (phosmet and malathion) and pyrethroids (zeta-cypermethrin and fenpropathrin). Spinetoram, dimethoate (Profaizer et al. 2015; Shawer et al. 2018b), and lambda-cyhalothrin (Grassi et al. 2011; Cini et al. 2012) provided high efficacy against SWD in Italian cherry orchards. When *D. suzukii* females were exposed to spinetoram or lambda-cyhalothrin, number of eggs laid in cherries were reduced (Beers et al. 2011; Shawer 2017).

In other trials, spinosad and chlorantraniliprole showed a high performance (Cuthbertson et al. 2014; Shawer et al. 2018b). Same results have been recently confirmed by the trials carried out at sweet cherry orchards located in NIAB EMR (East Malling, Kent, UK) where spinosad, lambda-cyhalothrin, and cyantraniliprole protected fruits from damage up to 14 days after application (Shaw et al. 2019). The fortnightly spray applications of those products were effective for protecting fruits until harvest. Acetamiprid, lime, pyrethrin, and deltamethrin were satisfied up to day 7 after application. Andreazza et al. (2017a) evaluated the performance of different insecticides and their insecticidal baits to manage adults and larvae of both *D. suzukii* and *Zaprionus indianus*. The insecticides dimethoate, malathion, spinosad, spinetoram, lambda-cyhalothrin, and cyantraniliprole were highly effective against larvae and adults of both insects (Andreazza et al. 2017a, b). These insecticides were more effective against *D. suzukii* adults when they were added with a feeding attractant as toxic baits (Andreazza et al. 2017a, b). On the other hand, neonicotinoids (acetamiprid and thiamethoxam) and pyrolle (chlorfenapyr) showed moderate activity on adults of *D. suzukii* (40–60%); however, they caused reduction in larval infestation (Andreazza et al. 2017a, b). Azadirachtin and sulfur were less effective on adults and larvae of both pests (Andreazza et al. 2017a, b).

Sial et al. (2017) carried out laboratory assays and field trials on organic crops to evaluate efficacy of several bioinsecticides against SWD. Their results also confirmed the high activity of Spinosad. The application of azadirachtin + pyrethrins,

Table 7.1 Efficacy of tested insecticides against *Drosophila suzukii* available in most published data

IRAC MoA group	Insecticide	Efficacy	Study reference
Spinosyns (5)	Spinosad	Excellent	Beers et al. (2011); Haye et al. (2016); Bruck et al. (2011); Haviland and Beers (2012); Shawer et al. (2018a, b); Shawer (2017); Shaw et al. (2019); Sial et al. (2017); Cuthbertson et al. (2014); Hoffmann Schlesener et al. (2017); Andreazza et al. (2017a, b)
	Spinetoram	Excellent	Beers et al. (2011); Haye et al. (2016); Bruck et al. (2011); Haviland and Beers (2012); Shawer et al. (2018a, b); Shawer (2017); Andreazza et al. (2017a, b); Profaizer et al. (2015)
Organophosphates (1B)	Malathion	Excellent	Bruck et al. (2011); Hoffmann Schlesener et al. (2017); Andreazza et al. (2017a, b); Diepenbrock et al. (2016)
	Diazinon	Excellent	Bruck et al. (2011)
	Dimethoate	Excellent	Shawer et al. (2018a, b); Hoffmann Schlesener et al. (2017); Andreazza et al. (2017a, b); Profaizer et al. (2015)
	Phosmet	Excellent	(Shawer et al. (2018a); Hoffmann Schlesener et al. (2017); Diepenbrock et al. (2016)
	Fenitrothion	Excellent	Hoffmann Schlesener et al. (2017)
	Methidathion	Excellent	Hoffmann Schlesener et al. (2017)
Pyrethroids (3A)	Bifenthrin	Excellent	Bruck et al. (2011)
	Beta-cyfluthrin	Excellent	Bruck et al. (2011)
	Permethrin	Excellent	Bruck et al. (2011)
	Zeta-cypermethrin	Excellent	Hoffmann Schlesener et al. (2017); Bruck et al. (2011); Diepenbrock et al. (2016)
	Lambda-cyhalothrin	Excellent	Shawer et al. (2018b); Grassi et al. (2011); Cini et al. (2012); Shaw et al. (2019); Andreazza et al. (2017a, b)
	Deltamethrin	Excellent	Shawer et al. (2018b); Hoffmann Schlesener et al. (2017)
	Fenpropathrin	Excellent	Diepenbrock et al. (2016)
Diamides (28)	Cyantraniliprole	Excellent	Bruck et al. (2011); Cuthbertson et al. (2014); Beers et al. (2011); Van Timmeren and Isaacs (2013); Shaw et al. (2019); Shawer et al. (2018b); Andreazza et al. (2017a, b)

(continued)

Table 7.1 (continued)

IRAC MoA group	Insecticide	Efficacy	Study reference
Neonicotinoids (4A)	Thiamethoxam	Moderate	Shawer et al. (2018b); Andreazza et al. (2017a, b)
	Thiacloprid	Moderate	Shawer et al. (2018b)
	Acetamiprid	Moderate	Shawer et al. (2018b); Andreazza et al. (2017a, b); Shaw et al. (2019)
	Imidacloprid	Moderate	Shawer et al. (2018b)
Pyrolle (13)	Chlorfenapyr	Moderate	Andreazza et al. (2017a, b)
Botanical[a]	Azadirachtin + pyrethrins	Moderate	Sial et al. (2017)
	Sabadilla alkaloids	Moderate	Sial et al. (2017)
Biopesticide[a]	*Chromobacterium subtsugae*	Moderate	Sial et al. (2017)
UN[b]	Azadirachtin	Low	Andreazza et al. (2017a, b)
UN[b]	Sulfur	Low	Andreazza et al. (2017a, b)
UNF[c]	*Beauveria bassiana*	Low	Shawer et al. (2018b)
Mineral oil[a]	Paraffinic oil	Low	Shawer et al. (2018b)

[a]Not considered in IRAC MoA classification
[b]Compounds of unknown or uncertain MoA
[c]Fungal agents of unknown or uncertain MoA

Chromobacterium subtsugae and sabadilla alkaloids showed moderate activity, giving 3 days residual activity (Sial et al. 2017). Such those products can be advantageous in rotation programs to lessen dependence on spinosad and alleviate insect resistance (Sial et al. 2017).

In a laboratory study aimed at evaluation of the biological performance and ovicidal activity of different insecticides against *D. suzukii* adults, deltamethrin, dimethoate, spinosad, fenitrothion, phosmet, malathion, methidathion, and zeta-cypermethrin showed an excellent activity against adults, causing 100% mortality at 3 days after the treatment (Hoffmann Schlesener et al. 2017). Moreover, fenitrothion, malathion, and methidathion achieved an excellent ovicidal effect, disrupting all laid eggs. Phosmet and diflubenzuron caused high reduction in the emerging larvae 2 days after the treatment. Andika et al. (2019) studied the impact of simulated rainfall on SWD flies' mortality, immature survival, and residue wash-off from different plant tissues for several insecticides. The simulated rainfall adversely affected adult mortality and immature survival caused by phosmet, zeta-cypermethrin, and spinetoram (Andika et al. 2019). While, acetamiprid was the lowest insecticide affected by simulated rainfall. Residues of phosmet and spinetoram were the most sensitive to wash-off (Andika et al. 2019).

In Table 7.1, we stated the most pesticides evaluated against *D. suzukii* and their reference studies. The pesticides were classified according their efficacy based on their studies to excellent, moderate or low.

7.2.2 *Natural Compounds*

Natural compounds from different sources were assessed, and some of them seem to be promising bioinsecticides in control programs of SWD (Andreazza et al. 2017a, b). In recent studies, essential oils (EOs) extracted from the genus Piper (*P. aduncum* L.; *P. gaudichaudianum* Kunth.; *P. marginatum* L.) provided excellent adult mortality (100%) of SWD following ingestion and topical application methods. As well as, they had a repellent effect on female oviposition and negative effects on egg survival of *D. suzukii* on artificial fruits (de Souza et al. 2020). Several natural products were tested for their activity as repellents, contact or ingestion toxicants, fumigants, ovicides, or oviposition deterrents against SWD (Dam et al. 2019). The EOs of thyme or its major ingredient thymol and *Leptospermum ericoides* or *L. scoparium* showed promising results as repellents and contact toxicity, respectively. When erythritol combined with sucrose, it was a robust ingestion toxicant on flies. The chitinase of *Euphorbia characias* had an excellent larvicidal effect (100% mortality). The EO of *perilla aldehyde* (geranial and neral) had an insecticidal activity as fumigant. The powder sulfur was an efficient oviposition deterrent (76% reduction in eggs deposited into the fruits). In a laboratory trial, 11 plant species from the genera, Nepeta, and Actinidia were investigated to find effective natural products on SWD (Keesey et al. 2019). When the extracts of Actinidia were added in fruit samples, they showed an oviposition prevention.

7.2.3 *Feeding Stimulants*

Feeding stimulants are substances that can be added with insecticides to increase consumption and efficacy of the pesticide (Gullickson et al. 2019). The combination of sugar or sugar–yeast bait as phagostimulants with insecticides significantly increased biological performance of several insecticides against SWD fly (Cowles et al. 2015). Adding sucrose at a concentration of 0.1% with spinosad, spinetoram, acetamiprid, and cyantraniliprole increased adult mortality and decreased larval infestation (Cowles et al. 2015; Gullickson et al. 2019). However, more investigations regarding nontarget effects on natural enemies and fungal pathogens are needed (Cowles et al. 2015; Gullickson et al. 2019). Using the non-nutritive feeding stimulant, erythritol, combined with spinosad increased the mortality of SWD adults at 24 h after exposure of treated fruits from 62 to 96% (Gullickson et al. 2019).

7.3 Conclusion

The management strategies of SWD continue to depend basically on the chemical control methods; however, other alternatives are promising. The insecticides belonging to chemical groups, i.e., spinosyns, pyrethroids, organophosphates, and diamides, are the most effective chemicals tested against SWD (see Table 7.1). The application of insecticides (spinosad, spinetoram, acetamiprid, and cyantraniliprole) mixed with a feeding stimulant such as sugar, sugar–yeast bait, or erythritol improved their efficacy against SWD. Neonicotinoids had a limited effect in control programs because of their low performances as adulticides, broad-spectrum properties, and adverse effects on beneficials. Performance of several natural products have been investigated on SWD. Those natural products including thyme, *Leptospermum ericoides*, *L. scoparium*, erythritol + sucrose, the chitinase of *Euphorbia characias*, and *Perilla aldehyde* had promising results which will be useful in SWD control strategy.

References

Andika IP, Vandervoort C, Wise JC (2019) Rainfastness of insecticides used to control spotted-wing Drosophila in tart cherry production. Insects 10(7):203

Andreazza F, Bernardi D, Baronio CA et al (2017a) Toxicities and effects of insecticidal toxic baits to control *Drosophila suzukii* and *Zaprionus indianus* (Diptera: Drosophilidae). Pest Manag Sci 73(1):146–152

Andreazza F, Bernardi D, Santos RRS et al (2017b) *Drosophila suzukii* in southern neotropical region: current status and future perspectives. Neotrop Entomol 46(6):591–605

Asplen MK, Anfora G, Biondi A et al (2015) Invasion biology of spotted wing Drosophila (Drosophila suzukii): a global perspective and future priorities. J Pest Sci 88(3):469–494

Barry JD, Polavarapu S (2005) Feeding and survivorship of blueberry maggot flies (Diptera: Tephritidae) on protein baits incorporated with insecticides. Fla Entomol 88(3):268–277

Beers EH, Van Steenwyk RA, Shearer PW et al (2011) Developing *Drosophila suzukii* management programs for sweet cherry in the western United States. Pest Manag Sci 67(11):1386–1395

Bellamy DE, Sisterson MS, Walse SS (2013) Quantifying host potentials: indexing postharvest fresh fruits for spotted wing drosophila, *Drosophila suzukii*. PLoS One 8(4):e61227

Bruck DJ, Bolda M, Tanigoshi L et al (2011) Laboratory and field comparisons of insecticides to reduce infestation of *Drosophila suzukii* in berry crops. Pest Manag Sci 67(11):1375–1385

CABI EPPO Centre for Agricultural Bioscience International, European and Mediterranean Plant Protection Organization (2016) Distribution maps of plant pests, map 766 (1st revision). https://www.cabi.org/isc/abstract/20163203811. Accessed 19 Apr 2020

Cini A, Ioratti C, Anfora G (2012) A review of the invasion of *Drosophila suzukii* in Europe and a draft research agenda for integrated pest management. Bull Insectol 65:149–160

Cowles RS, Rodriguez-Saona C, Holdcraft R et al (2015) Sucrose improves insecticide activity against *Drosophila suzukii* (Diptera: Drosophilidae). J Econ Entomol 108(2):640–653

Cuthbertson AG, Collins DA, Blackburn LF et al (2014) Preliminary screening of potential control products against *Drosophila suzukii*. Insects 5(2):488–498

Dam D, Molitor D, Beyer M (2019) Natural compounds for controlling *Drosophila suzukii*. A review. Agron Sustain Dev 39(6):53

De Ros G, Anfora G, Grassi A et al (2013) The potential economic impact of *Drosophila suzukii* on small fruits production in Trentino (Italy). IOBC-WPRS Bull 91:317–321

de Souza MT, de Souza MT, Bernardi D et al (2020) Chemical composition of essential oils of selected species of Piper and their insecticidal activity against *Drosophila suzukii* and Trichopria anastrephae. ESPR, pp 1–10

Delfinado MD, Hardy DE (1973) A catalog of the Díptera of the oriental region. Vol. I. Suborder Nematocera. A catalog of the Díptera of the Oriental Region Vol I Suborder Nematocera

Diepenbrock LM, Rosensteel DO, Hardin JA et al (2016) Season-long programs for control of *Drosophila suzukii* in southeastern US blueberries. Crop Protect 81:76–84

EPPO Datasheet for efficacy evaluation of insecticides on *Drosophila suzukii* (2013) OEPP/EPPO Bull

Fruit-Advisor (2015) Spotted Wing Drosophila (*Drosophila suzukii*). https://ag.umass.edu/fruit/news/spotted-wing-drosophila-drosophila-suzukii. Accessed 24 Sept 2015

Goodhue RE, Bolda M, Farnsworth D et al (2011) Spotted wing drosophila infestation of California strawberries and raspberries: economic analysis of potential revenue losses and control costs. Pest Manag Sci 67(11):1396–1402

Grassi A, Giongo L, Palmieri L (2011) Drosophila (Sophophora) suzukii (Matsumura), new pest of soft fruits in Trentino (North-Italy) and in Europe. IOBC/WPRS Bull 70:121–128

Gullickson MG, Rogers MA, Burkness EC et al (2019) Efficacy of organic and conventional insecticides for *Drosophila suzukii* when combined with erythritol, a non-nutritive feeding stimulant. Crop Protect 125:104878

Hauser M (2011) A historic account of the invasion of *Drosophila suzukii* (Matsumura) (Diptera: Drosophilidae) in the continental United States, with remarks on their identification. Pest Manag Sci 67(11):1352–1357

Haviland DR, Beers EH (2012) Chemical control programs for *Drosophila suzukii* that comply with international limitations on pesticide residues for exported sweet cherries. J Integr Pest Manag 3(2):F1–F6

Haye T, Girod P, Cuthbertson A et al (2016) Current SWD IPM tactics and their practical implementation in fruit crops across different regions around the world. J Pest Sci 89(3):643–651

He LM, Troiano J, Wang A et al (2008) Environmental chemistry, ecotoxicity, and fate of lambda-cyhalothrin. In: Reviews of environmental contamination and toxicology. Springer, pp 71–91

Hoffmann Schlesener DC, Wollmann J, de Bastos PJ et al (2017) Effects of insecticides on adults and eggs of *Drosophila suzukii* (Diptera, Drosophilidae). Rev Colomb Entomol 43(2):208–214

Kanzawa T (1935) Research into the fruit-fly *Drosophila suzukii* Matsumura (preliminary report). Yamanashi Prefecture Agricultural Experiment Station Report

Keesey IW, Jiang N, Weißflog J et al (2019) Plant-based natural product chemistry for integrated pest management of *Drosophila suzukii*. J Chem Ecol 45(7):626–637

Lee JC, Bruck DJ, Curry H et al (2011a) The susceptibility of small fruits and cherries to the spotted-wing drosophila, *Drosophila suzukii*. Pest Manag Sci 67(11):1358–1367

Lee JC, Bruck DJ, Dreves AJ et al (2011b) In focus: spotted wing drosophila, *Drosophila suzukii*, across perspectives. Pest Manag Sci 67(11):1349–1351

Matsumura S (1931) A revision of the palaearctic and oriental typhlocybid-genera with descriptions of new species and new genera. Insecta Matsumurana 6(1–2):55–91

Mori N, Tonina L, Sancassani M et al (2019) Integrated pest management approaches against *Drosophila suzukii*. Italus Hortus 26:67–74

Profaizer D, Grassi A, Zadra E et al (2015) Efficacy of insecticide treatment strategies against *Drosophila suzukii* in combination with mass trapping. IOBC/WPRS Bull 109:215–218

Sasaki M, Sato R (1995) Bionomics of the cherry drosophila, *Drosophila suzukii* Matsumura (Diptera: Drosophilidae) in Fukushima Prefecture [Japan], 3: Life cycle. Annual Report of the Society of Plant Protection of North Japan (Japan)

Shaw B, Hemer S, Cannon MF et al (2019) Insecticide control of *Drosophila suzukii* in commercial sweet cherry crops under cladding. Insects 10(7):196

Shawer R (2017) Impact of traditional pesticides and new controlled release formulations on *Drosophila suzukii*. PhD thesis, Università degli Studi di Padova

Shawer R, Donati I, Cellini A et al (2018a) Insecticidal activity of *Photorhabdus luminescens* against *Drosophila suzukii*. Insects 9(4):148

Shawer R, Tonina L, Tirello P et al (2018b) Laboratory and field trials to identify effective chemical control strategies for integrated management of *Drosophila suzukii* in European cherry orchards. Crop Protect 103:73–80

Sial AA, Isaacs R, Grieshop MJ et al (2017) Development and implementation of systems-based organic management strategies for spotted wing drosophila: an OREI project update. In: Entomology ESA

Stacconi R, Grassi A, Dalton D et al (2013) First field records of *Pachycrepoideus vindemiae* as a parasitoid of *Drosophila suzukii* in European and Oregon small fruit production areas. Entomologia 1:e3. https://doi.org/10.4081/entomologia.2013.e3

Van Timmeren S, Isaacs R (2013) Control of spotted wing drosophila, *Drosophila suzukii*, by specific insecticides and by conventional and organic crop protection programs. Crop Protect 54:126–133

Walsh DB, Bolda MP, Goodhue RE et al (2011) *Drosophila suzukii* (Diptera: Drosophilidae): invasive pest of ripening soft fruit expanding its geographic range and damage potential. J Integr Pest Manag 2(1):G1–G7

Wise JC, Coombs AB, Vandervoort C et al (2006) Use of residue profile analysis to identify modes of insecticide activity contributing to control of plum curculio in apples. J Econ Entomol 99(6):2055–2064

Chapter 8
Biological Control of Spotted-Wing Drosophila: An Update on Promising Agents

Xingeng Wang, Kent M. Daane, Kim A. Hoelmer, and Jana C. Lee

Abstract Following the global invasion of *Drosophila suzukii* (spotted-wing drosophila or SWD), nearly 100 studies have explored biological control of this pest. In 2019, a review summarized 75+ papers covering 57 species of SWD parasitoids, predators, competitors, and pathogens and identified the most promising ones. This review provides an update with recent studies. Since parasitoids are promising natural enemies that can be host-specific and self-disperse, this chapter focuses on SWD parasitoids in its invaded and native ranges, and prospects for classical biological control. To date, six species have been confirmed to attack SWD in the invaded regions including three widely studied generalist pupal parasitoids, *Pachycrepoideus vindemiae*, *Trichopria drosophilae*, and *T. anastrephae*. No locally occurring larval drosophila parasitoids can develop from SWD. In contrast, foreign explorations in China, Japan, and South Korea have revealed 19 species of SWD larval parasitoids. *Asobara japonica*, *Ganaspis brasiliensis*, and *Leptopilina japonica* spp. *japonica* have been evaluated. *Ganaspis brasiliensis* is a complex of cryptic species/strains with varying host specificity, some which also occur in regions outside of Asia, but one East Asian strain was found to be the most host-specific to SWD and is currently being petitioned for introduction into North America and Europe.

Keywords *Asobara* · *Drosophila* · *Ganaspis* · *Leptopilina* · *Pachycrepoideus* · *Trichopria*

X. Wang · K. A. Hoelmer
USDA ARS Beneficial Insects Introduction Research Unit, Newark, DE, USA
e-mail: xingeng.wang@usda.gov; kim.hoelmer@usda.gov

K. M. Daane
University of California Berkeley, Berkeley, CA, USA
e-mail: kdaane@ucanr.edu

J. C. Lee (✉)
USDA ARS Horticultural Crops Research Unit, Corvallis, OR, USA
e-mail: jana.lee@usda.gov

F. R. M. Garcia (ed.), *Drosophila suzukii Management*,
https://doi.org/10.1007/978-3-030-62692-1_8

Drosophila suzukii Matsumura (Diptera: Drosophilidae) (spotted-wing drosophila or SWD) is an invasive pest in North and South America and Europe that threatens the small fruit and cherry industry (Asplen et al. 2015). Although chemical insecticides are effective, this has increased management costs (Farnsworth et al. 2017), led to resistance development (Gress and Zalom 2019), loss of markets (Haviland and Beers 2012), and impacted natural enemies (Whitehouse et al. 2018). Long-term sustainable management relies on effective biological control as well as cultural and behavioral controls. To develop biological control of invasive pests, the classical approach is to use natural enemies that are native to their countries of origin. Introduction of these specialist parasitoids has historically been preferred for controlling exotic pests because these parasitoids are generally more efficient in targeting hosts due to their long-shared history of co-adaption. However, some indigenous generalist parasitoids can adapt to exotic hosts, and such novel interactions may also play an important role in regulating the exotic pests. It is therefore important to evaluate the impacts of indigenous natural enemies while considering imported specialist parasitoids for the management of invasive pests. This chapter review will primarily focus on parasitoids since they are generally more host-specific than predators and entomopathogens, but will also summarize recent work on predators, competitors, and pathogens of SWD since the last review of SWD biological control (Lee et al. 2019). Predators and competitors of SWD are already common in crops and can be further conserved to suppress SWD populations. Pathogens are often commercially available, and if delivered effectively, provide growers additional control options.

8.1 Parasitoids of *Drosophila*

Parasitoids clearly play an important role in regulating some *Drosophila* populations, with reported levels of parasitism as high as 80–100% in Europe (e.g., Fleury et al. 2009; Janssen et al. 1988). Over 50 hymenopteran species have been reported to attack frugivorous drosophilids worldwide, including members of Braconidae (*Asobara*, *Aphaereta*, *Phaenocarpa*, *Tanycarpa*, *Aspilota*, and *Opius*), Figitidae (*Leptopilina*, *Ganaspis*, *Kleidotoma*, and *Dicerataspis*), Diapriidae (*Trichopria* and *Spilomicrus*), and Pteromalidae (*Pachycrepoideus*, *Spalangia*, *Trichomalopsis*, and *Toxomorpha*) (Carton et al. 1986). So far, no drosophilid egg or adult parasitoids have been discovered. All known *Drosophila* parasitoids are solitary koinobiont larval endoparasitoids in the families of Braconidae and Figitidae (subfamily Eucoilinae), and solitary pupal parasitoids in the families of Pteromalidae and Diapriidae. Larval *Drosophila* parasitoids attack host larvae, but all emerge from the host puparium formed from the hardened exoskeleton of the fly's last larval stage. Most known larval parasitoids belong to the genera *Asobara*, *Leptopilina*, and *Ganaspis*. Among them, *Asobara tabida* (Nees), *Leptopilina boulardi* Barbotin et al., and *L. heterotoma* (Thompson) are the three most extensively studied parasitoids attacking common drosophilids such as *Drosophila melanogaster* (Meigen) breeding in rotting fruit (reviewed in Prévost 2009). The pteromalids are ectoparasitoids that lay their eggs between the host's puparium case and the pupa, while the diapriids are endoparasitoids. *Pachycrepoideus vindemiae* (Rondani) and *Trichopria drosophilae* Perkins are two of the most common and cosmopolitan pupal drosophila parasitoids (Wang et al. 2016a, b).

The parasitoid fauna associated with SWD was poorly understood prior to its worldwide invasion. Since then, a number of studies have discovered and identified effective parasitoids both in Asia and around the world (Daane et al. 2016; Giorgini et al. 2019; Girod et al. 2018b). An earlier review article discussed the practical application potential of parasitoids for a broader audience (Lee et al. 2019). Here, we present a comprehensive review on the complexes of parasitoid species attacking SWD worldwide. We summarize the evaluations for promising parasitoids, and the diversity, dominance, and host specificity of parasitoids native to Asia. Finally, we propose future research directions for promoting parasitoids for the control of SWD by classical, augmentative, or conservation biological control.

8.2 Impact of Parasitoids in the Invaded Ranges

Surveys of locally occurring parasitoids on SWD and closely related frugivorous *Drosophila* species have been conducted in the USA (Kamiyama et al. 2019; Miller et al. 2015), Mexico (Cancino et al. 2015), Brazil (Wollmann et al. 2016), Spain (Gabarra et al. 2015), France (Kremmer et al. 2017), Switzerland (Knoll et al. 2017), Italy (Mazzetto et al. 2016; Rossi Stacconi et al. 2013), Slovenia (Modic et al. 2019), and Turkey (Zengin and Karaca 2019). These surveys used sentinel traps baited with larval or pupal SWD or *D melanogaster* in fruit or artificial diet, and occasionally with collections of infested fruits. Exposed materials in the traps were often inevitably contaminated by other drosophilid species. Thus, some reported parasitoid species, especially *L. boulardi*, may need verification in laboratory tests to confirm the host-parasitoid association (e.g., Garcia-Cancino et al. 2015; Wollmann et al. 2016; Garrido et al. 2018). The three larval parasitoids (*A. tabida*, *L. boulardi*, and *L. heterotoma*) and two pupal parasitoids (*P. vindemiae* and *T. drosophilae*) were commonly collected in North America and Europe. These larval parasitoids were, however, exclusively reared from drosophilids other than SWD, whereas the pupal parasitoids were collected from both SWD and other drosophilid species (Table 8.1). In the literature, *P. vindemiae* is sometime mentioned as *P. vindemmiae* (e.g., Wollmann et al. 2016). According to Rossi Stacconi et al. (2013), this species was originally described as *P. vindemiae* by Rondani in 1875, and the latter name should be used thereafter. In several European reports, the *Trichopria* species was reported as *T.* cf. *drosophilae* (e.g., Chabert et al. 2012; Gabarra et al. 2015; Mazzetto et al. 2016). These specimens are likely conspecific to *T. drosophilae* as reported in other studies. Other pupal parasitoids that were collected from SWD-baited sentinel traps include the pteromalid *Spalangia simplex* Perkins in Mexico, and the diapriid *Trichopria anastrephae* Lima in Brazil (Table 8.1).

Table 8.1 Parasitoid complexes and their observed parasitism on *Drosophila suzukii* (SWD) in different regions of its native and introduced ranges

Parasitoid species	Countries	Host species [2]	Parasitism (%)	References
Larval parasitoids				
Braconidae				
Asobara tabida	Japan	SWD/ Others	–	Mitsui et al. (2007)
A. brevicauda	South Korea	SWD	< 1	Daane et al. (2016)
A. japonica	Japan, South Korea	SWD/ Others	0–16.7	Mitsui et al. (2007), Daane et al. (2016)
A. leveri	China, South Korea	SWD/ DP/ Others	<1	Daane et al. (2016), Giorgini et al. (2019)
A. mesocauda	China, South Korea	SWD/ DP/DSP	–	Girod et al. (2018a), Giorgini et al. (2019)
A. pleuralis	China	SWD/ DP	–	Girod et al. (2018a)
A. sp.	Japan	SWD	<7	Kasuya et al. (2013)
A. sp. TS1	Japan	SWD	1.2	Ideo et al. (2008)
A. triangulata	South Korea	SWD	<1	Daane et al. unpub. data
A. unicolorata	China	SWD/ DP	<1	Giorgini et al. (2019)
Areotetes striatiferus	China	SWD/ DP	0.6–6.9	Girod et al. (2018a)
Tanycarpa chors	China	SWD/ DP	2.1	Girod et al. (2018a)
Figitidae				
Ganaspis brasiliensis	China, Japan, South Korea	SWD/ DP	0–47.8	Kasuya et al. (2013), Daane et al. (2016), Giorgini et al. (2019)
G. cf. *brasiliensis*	China, Japan, South Korea	SWD/ DP/DSP	0.2–75.6	Girod et al. (2018a), Daane et al. unpubl. Data
G. xanthopoda [a]	Japan	SWD/ Others	–	Mitsui et al. (2007)
Leptopilina boulardi	Argentina, Brazil, Mexico	SWD/ Others	–	Cancino et al. (2015), Wollmann et al. (2016), Garrido et al. (2018)
L. heterotoma	Italy	SWD/ Others	< 1	Miller et al. (2015)
L. j. formosana	South Korea, Japan	SWD/ Others	< 1	Novkovic et al. (2011), Daane et al. (2016)
L. j. japonica	China, Japan, South Korea	SWD/ DP	0–34.5	Novkovic et al. (2011), Kasuya et al. (2013), Daane et al. (2016), Girod et al. (2018a), Giorgini et al. (2019)

(continued)

Table 8.1 (continued)

Parasitoid species	Countries	Host species [2]	Parasitism (%)	References
L. sp.	China	SWD/ DP	7.2–35.9	Girod et al. (2018a)
Pupal parasitoids				
Diapriidae				
Trichopria cf. *drosophilae*	Italy, Spain	SWD/ Others	0–10.7	Gabarra et al. (2015), Mazzetto et al. (2016), Kremmer et al. (2017)
T. drosophilae	China, France, Italy, Mexico, Slovenia, South Korea, Switzerland, USA	SWD/ Others	0–11.1	Cancino et al. (2015), Miller et al. (2015), Rossi-Stacconi et al. (2015), Daane et al. (2016), Knoll et al. (2017), Giorgini et al. (2019), Modic et al. (2019)
T. anastrephae	Brazil	SWD	–	Wollmann et al. (2016)
Pteromalidae				
Pachycrepoideus vindemiae	China, France, Italy, Mexico, the Netherlands, South Korea, Spain, Switzerland, Turkey, USA	SWD/ Others	0–31.0	Rossi Stacconi et al. (2013), Cancino et al. (2015), Miller et al. (2015), Daane et al. (2016), Mazzetto et al. (2016), Knoll et al. (2017), Kremmer et al. (2017), Haro-Barchin et al. (2018), Zengin and Karaca (2019)
Spalangia erythromera	Italy	SWD/ Others	–	Mazzetto et al. (2016)
S. simplex	Mexico	SWD	–	Cancino et al. (2015)

[a]This species was later reassigned as *Ganaspis brasiliensis* (Nomano et al. 2017)
[b]SWD = *Drosophila suzukii*, DP = *D. pulchrella*, DSP = *D. subpulchrella*, and Others = other drosophilids when the species was either not known, not provided or not one of the above species

Various populations of *A. tabida*, *L. boulardi*, and *L. heterotoma* have been tested for their ability to attack and then develop from SWD under laboratory conditions (Table 8.2). To date, none of these larval parasitoids were able to complete development, except for a low percentage of development of *L. heterotoma*, using populations from northern Italy (Rossi-Stacconi et al. 2015) and France (Iacovone et al. 2018) (Table 8.2). The same larval parasitoid, however, successively parasitized *D. melanogaster* and other closely related drosophilids in parallel tests. The larval parasitoids' immature stages failed to develop due to a strong cellular immune response by SWD, causing the fly larvae to increase hemocyte production to encapsulate the immature parasitoids inside the host (Chabert et al. 2012; Iacovone et al. 2018; Kacsoh and Schlenke 2012; Poyet et al. 2013). Nevertheless, this species would still oviposit in SWD larvae which significantly reduced survival of SWD by up to 90%. Variation in mortality was likely due to different experimental procedures with host-parasitoid ratios and exposure times, or geographic variations of

Table 8.2 Frugivorous *Drosophila* parasitoids species evaluated for their efficiency on *Drosophila suzukii* (SWD) in laboratory or field trials

Parasitoid	Parasitoid origin	Rate of parasitism[a]	Emergence[a]	References
Larval parasitoids				
Braconidae				
Aphaereta sp.	USA	Low	None	Kacsoh and Schlenke (2012)
Asobara citri	Ivory Coast	Low	None	Kacsoh and Schlenke (2012)
A. japonica	Japan, South Korea	High	High	Ideo et al. (2008), Chabert et al. (2012), Kacsoh and Schlenke (2012), Daane et al. (2016), Girod et al. (2018b, c), Wang et al. (2018a, 2019, 2020)
A. pleuralis	Indonesia	Low	None	Kacsoh and Schlenke (2012)
A. tabida	France, Switzerland	Low	None	Chabert et al. (2012), Kacsoh and Schlenke (2012), Knoll et al. (2017)
Figitidae				
Ganaspis brasiliensis	China, South Korea	High	High	Wang et al. (2018a, 2019, 2020), Giorgini et al. (2019)
Ganaspis cf. *brasiliensis*	China, Japan	High	High	Girod et al. (2018b, c)
Ganaspis sp.	USA	Low	Low	Kacsoh and Schlenke (2012)
G. xanthopoda [b]	Japan, Uganda, USA	Low	Low	Mitsui and Kimura (2010), Kacsoh and Schlenke (2012)
Leptopilina boulardi	Congo, France, Italy, Kenya, Mexico[c], Switzerland, USA	Low	None	Chabert et al. (2012), Kacsoh and Schlenke (2012), Mazzetto et al. (2016), Gonzalez-Cabrera et al. (2020)
L. clavipes	The Netherlands	Low	None	Kacsoh and Schlenke (2012)
L. guineaensis	Cameron, South Africa	Low	None	Kacsoh and Schlenke (2012)
L. heterotoma	France, Italy, Switzerland, USA	Low	Low	Chabert et al. (2012), Kacsoh and Schlenke (2012), Rossi-Stacconi et al. (2015), Mazzetto et al. (2016), Knoll et al. (2017), Iacovone et al. (2018), Girod et al. (2018b)
L. j. japonica	China, South Korea	High	High	Girod et al. (2018b, c), Wang et al. (2018a, 2019, 2020)
L. victoriae	Philippines, USA	Low	None	Kacsoh and Schlenke (2012)
Pupal parasitoids				
Diapriidae				
Trichopria cf. *drosophilae*	France, Italy, Spain	High	High	Chabert et al. (2012), Gabarra et al. (2015), Mazzetto et al. (2016)

(continued)

Table 8.2 (continued)

Parasitoid	Parasitoid origin	Rate of parasitism[a]	Emergence[a]	References
T. drosophilae	China, Italy[c], Mexico[c], South Korea, Switzerland, USA	High	High	Rossi Stacconi et al. (Rossi-Stacconi et al. 2015, Rossi Stacconi et al. 2017, Rossi-Stacconi et al. 2018, Rossi Stacconi et al. 2019), Wang et al. (2016a, b, 2018b), Kaçar et al. (2017), Knoll et al. (2017), Gonzalez-Cabrera et al. (2019, 2020), Wolf et al. (2019), Yi et al. (2020)
T. sp.	France, USA	High	High	Kacsoh and Schlenke (2012)
T. anastrephae	Brazil	High	High	Kruger et al. (2019), Vieira et al. (2020)
Pteromalidae				
Muscidifurax raptorellus	Canada	High	High	Bonneau et al. (2019)
Pachycrepoideus sp.	USA	High	High	Kacsoh and Schlenke (2012)
P. vindemiae	Canada, China, France, Italy, South Korea, Spain, Switzerland, USA	High	High	Chabert et al. (2012), Gabarra et al. (2015), Rossi-Stacconi et al. (2015), Dancau et al. (2017), Kaçar et al. (2017), Wang et al. (2016a, b, 2018b), Knoll et al. (2017), Zhu et al. (2017), Bonneau et al. (2019), Bezerra da Silva et al. 2019a, b
Spalangia erythromera	Switzerland	High	High	Knoll et al. (2017)
Vrestovia fidenas	Switzerland	High	High	Knoll et al. (2017), Wolf et al. (2019)

[a]Studies were conducted with SWD larvae or pupae presented either in artificial diet or in contained fruit
[b]This species was later reassigned as *Ganaspis brasiliensis* (Nomano et al. 2017)
[c]Open field release

resistance and virulence among populations (Kacsoh and Schlenke 2012). Within the SWD's native range, *L. heterotoma* and *L. boulardi* have never been recorded from SWD (Daane et al. 2016; Giorgini et al. 2019; Ideo et al. 2008; Mitsui et al. 2007; Novkovic et al. 2011), while *A. tabida* has been collected from SWD in Japan possibly from a misidentification (Mitsui et al. 2007). Alternatively, some *A. tabida* populations in Japan and some *L. heterotoma* populations in Europe have locally adapted to SWD, explaining the reported parasitism by these species in those areas.

Pachycrepoideus vindemiae and *T. drosophilae* are the two most studied pupal SWD parasitoids. Although both species are cosmopolitan and sympatric in many regions, *P. vindemiae* is more widely distributed than *T. drosophilae* (Knoll et al.

2017; Miller et al. 2015; Wang et al. 2018b). *Pachycrepoideus vindemiae* is more of a generalist than *T. drosophilae*, as the former species also attacks hosts in other families of cyclorrhaphous Diptera (Wang and Messing 2004), while *T. drosophilae* attacks only Drosophilidae (Carton et al. 1986). A lack of pupal immunity against parasitoids may explain why these pupal parasitoids have broader host ranges than larval parasitoid wasps (Kacsoh and Schlenke 2012). These two pupal parasitoids have been evaluated for their efficiency, host specificity, thermal tolerance, and interspecific interactions (Kaçar et al. 2017; Rossi-Stacconi et al. 2015; Rossi Stacconi et al. 2017; Wang et al. 2016a, b, 2018b; Zhu et al. 2017; Bezerra da Silva et al. 2019a, b). Both species can locate SWD pupae in fruit or soil, but *T. drosophilae* was more efficient than *P. vindemiae* at some temperatures (Garcia-Cancino et al. 2020; Kaçar et al. 2017; Wang et al. 2018b). At 23 °C, *T. drosophilae* females from California and South Korea populations survived 27.5 and 20.2 days, respectively, and produced a total of 63.8 and 52.0 offspring, whereas *P. vindemiae* females from a California population survived 21.5 days and produced 70.0 offspring (Rossi-Stacconi et al. 2015; Wang et al. 2016a). *Pachycrepoideus vindemiae* has a wider temperature range than *T. drosophilae*, which may explain the current distribution of these species in North America (Wang et al. 2018b). Interspecific competition between these two parasitoids may reduce the overall impact on the host population. *Trichopria drosophilae* seems to have an advantage over *P. vindemiae* in laboratory tests (Wang et al. 2016b). All other tested pupal parasitoids also readily developed from SWD in laboratory tests (Table 8.2). These include the pteromalids *Vrestovia fidenas* (Walker) and *Spalangia erythromera* Förster in Europe (Knoll et al. 2017; Mazzetto et al. 2016; Wolf et al. 2019), *Muscidifurax raptorellus* Girault & Sanders in Canada (Bonneau et al. 2019), and *T. anastrephae* in Brazil (Kruger et al. 2019; Vieira et al. 2020). All four parasitoids appear to have the potential to help in the control of SWD. Naturally occurring parasitism of SWD populations by pupal parasitoids is generally low (Table 8.1), but augmentative releases may allow them to be useful. In Italy, *T. drosophilae* was commercially available, and evaluated for its host location, dispersal, and host suppression capabilities in an augmentative release in netted raspberry fields (Rossi-Stacconi et al. 2018). The parasitoid was able to locate SWD in traps up to 40 m away from the release site, and SWD emergence was significantly reduced within a radius of 10 m of the release within netting environment. Recently, the effectiveness of this parasitoid has been evaluated in releases in unmanaged vegetation surrounding cherry orchards in Italy (Rossi Stacconi et al. 2019) and in commercial berry (*Rubus fruticosus* L.) crops in Mexico (Gonzalez-Cabrera et al. 2019). In Italy, weekly release of the parasitoid at a rate of 0.33 specimens/m^2 for 7 weeks resulted in a 34% reduction in fruit infestation in the unmanaged vegetation surrounding orchards. In Mexico, semiweekly release of the parasitoid at a rate of 4.5 wasps/m^2 for 50 weeks resulted a fourfold increase in parasitism and a 50% reduction of SWD in the field. Results from these studies suggest that augmentative release of *T. drosophilae* can suppress SWD populations in the unmanaged areas surrounding crops, thus lowering the severity of pest outbreaks in the crop (Rossi Stacconi et al. 2019). While no augmentative trials have been made with *M. raptorellus*, this pupal parasitoid is commonly sold for release in livestock operations, making releases in crops potentially feasible.

A population model predicts the optimal timing for releasing *T. drosophilae* against SWD would be between late spring and early summer when the host population begins to increase (Pfab et al. 2018). Early releases would help reduce fly populations that would likely move from overwintering unmanaged vegetation into early susceptible fruit crops, and at the same time, those released parasitoids would increase their population (Pfab et al. 2018). However, the timing of release will depend on geographical region. For example, in a warm temperate climate such as Mexico, SWD populations are active year-around, and sufficient pest suppression would require repeated augmentative releases (Gonzalez-Cabrera et al. 2019).

8.3 Exploration for Parasitoids in Asia

Exploration for parasitoids native to South Korea, China, and Japan have discovered at least 19 larval parasitoids associated with SWD, including 12 *Asobara* and 7 figitids (Table 8.1). In South Korea, eight species, *Asobara japonica* Belokobylskij, *A. leveri* (Nixon), *A. brevicauda* Guerrieri & van Achterberg, *A. triangulata* van Achterberg and Guerrieri, *A. mesocauda* van Achterberg and Guerrieri, *Ganaspis brasiliensis* Ihering, *Leptopilina japonica* Novković & Kimura, and *L. j. formosana* Novković & Kimura, and the pupal parasitoid *T. drosophilae* were collected from SWD and other Drosophilidae (Daane et al. 2016). *Leptopilina japonica* is further divided into the temperate subspecies (*L. j. japonica*, thereafter, referred to *L. japonica*) and the subtropical subspecies (*L. j. formosana*) (Novkovic et al. 2011). The larval parasitoid *L. boulardi* and the pupal parasitoid *P. vindemiae* were collected from other drosophilids. *Asobara brevicauda*, *A. triangulata*, and *A. mesocauda* are newly described species (Guerrieri et al. 2016). Parasitism of SWD by these larval parasitoids varied according to geography, season, and collection methods, ranging from 0 to 28.6% (Daane et al. 2016). *Ganaspis brasiliensis* and *L. japonica* were the major parasitoids found in fresh fruits infested by SWD, whereas *A. japonica* was the major parasitoid collected from fruit bait traps infested predominantly by other drosophilids (Daane et al. 2016). A total of 3266 and 20,358 *Drosophila* puparia were collected in 2013 and 2014, respectively, from a variety of locations, and *A. japonica*, *G. brasiliensis*, and *L. japonica* accounted for 85.7% of all larval parasitoids emerged (Daane et al. 2016). In 2016, a total of 11,575 SWD puparia were collected from several wild *Rubus* fruits, and *G. brasiliensis* and *L. japonica* accounted for 87.1% of total parasitoids emerged (Daane et al. 2016).

In China, Girod et al. (2018a) conducted surveys in the provinces of Yunnan, Jilin, Beijing, Hubei, and Sichuan by collecting wild and commercial fruits, and Giorgini et al. (2019) conducted surveys in Yunnan Province using banana-baited traps and wild fruit collections. Collected fruits were often co-infested by *Drosophila. pulchrella* Tan, Hsu & Sheng in Yunnan or by *D. subpulchrella* Takamori in other regions of China. These two species are also characterized by a serrated ovipositor, like SWD, that allows them to attack fresh fruits. Because the pupae of these three *Drosophila* spp. are indistinguishable, it was impossible to determine from which host the parasitoids emerged from (Giorgini et al. 2019; Girod et al. 2018a). At least nine larval parasitoids, *A. leveri, A. mesocauda, A. unicolorata, A. pleuralis* (Ashmead), *Areotetes striatiferus* Li, *G. brasiliensis* (or *G.* cf. *brasiliensis*), *L. japonica, Tanycarpa chors* Belokobylskij, and *Leptopilina* sp., and the pupal parasitoid *T. drosophilae* were collected in China. The most abundant and frequently collected larval parasitoids were *G. brasiliensis* and *L. japonica*. For example, Giorgini et al. (2019) collected a total of 11,683 SWD and *D. pulchrella* puparia from four wild host fruits (*Rubus foliosus* Weihe, *R. niveus* Thunberg, *Fragaria moupinensis* Cardot, and *Sambucus adnate* Wallich ex de Candolle) at four different locations during 2016 in Yunnan, China. The majority of emerged parasitoids were *G. brasiliensis* (63.7%) and *L. japonica* (33.2%), accounting for 97.1% of total parasitoids. These two parasitoids also accounted for 97.8% of all larval parasitoids emerged from 1792 *D. suzukii* and *D. pulchrella* puparia (Hoelmer et al. unpubl. data). The highest parasitism by *G. brasiliensis* was 47.8% and 42.0% by *L. japonica* in the 2016 collections in Yunnan, China (Giorgini et al. 2019). The banana traps yielded mainly other Drosophilidae (>99%) and seven *Asobara* species (primarily *A. mesocauda*) and six figitids (primarily *L. japonica*) as well as *T. drosophilae* and *P. vindemiae*. Only one *A. japonica* and one *G. xanthopoda* were collected, and *G. brasiliensis* was never collected from banana traps (Giorgini et al. 2019). The surveys showed that most flies emerging from fresh fruits were SWD or the closely related *D. pulchrella* and *D. subpulchrella*. This suggests that field collection of fresh fruits is a more reliable method to collect SWD parasitoids (Daane et al. 2016; Giorgini et al. 2019) (Fig. 8.1).

In Japan, at least six larval parasitoids (*A. japonica, A. tabida, L. japonica, G. xanthopoda, T. chors*, and *Asobara* sp.) have been reported to parasitize SWD and other frugivorous *Drosophila* species (Girod et al. 2018a; Ideo et al. 2008; Kasuya et al. 2013; Mitsui and Kimura 2010; Mitsui et al. 2007; Novkovic et al. 2011). *Ganaspis* cf. *brasiliensis* was the most abundant parasitoid collected from SWD in wild fruits with parasitism of 75.6% reported in Nara, Japan (Girod et al. 2018a). Matsuura et al. (2018) showed that *G.* cf. *brasiliensis* attacked SWD larvae in fresh fruits in the tree canopy, but rarely in fruits fallen on the ground, suggesting a specific adaptation of a Japanese strain to SWD infesting fresh fruits. An *Asobara* sp. that was recorded only from SWD in wild fruits (Girod et al. 2018a; Ideo et al. 2008; Nomano et al. 2015) may be more specific; it was speculated to be *A. triangulata* based on molecular analysis of specimens (Guerrieri et al. 2016). *Asobara japonica* was the major parasitoid collected in banana traps throughout Japan (Mitsui et al. 2007). Populations of *A. japonica* in the main islands of Japan and South Korea seem to be parthenogenetic, whereas those in the south-western islands of Japan apparently reproduce sexually (Daane et al. 2016; Murata et al. 2009).

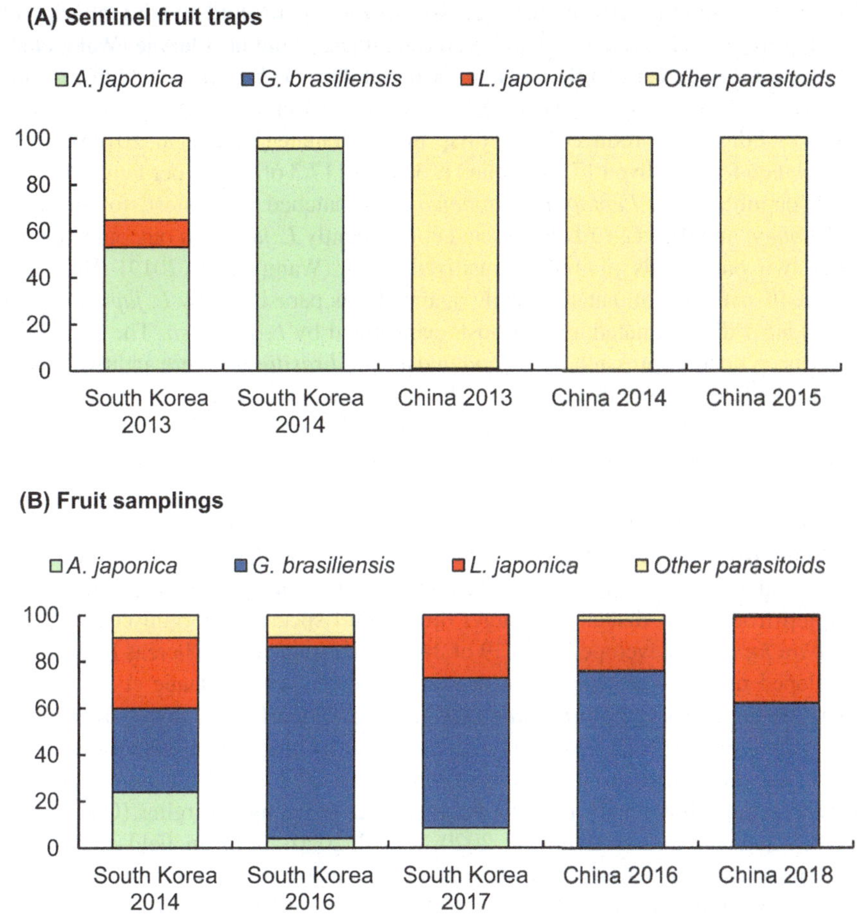

Fig. 8.1 Composition of major Asian parasitoids (*Asobara japonica*, *Ganaspis brasiliensis*, and *Leptopilina japonica*) of frugivorous Drosophilidae collected from (**a**) fruit-baited traps or (**b**) via sampling of fresh fruits in South Korea and China during 2013–2018 (for other parasitoid species, see the list on Table 8.1—parasitoid complexes). Data were compiled based on Daane et al. (2016), Giorgini et al. (2019), and unpublished data from recent collections

8.4 Prospects for Classical Biological Control

The Asian surveys suggest that *G. brasiliensis*, *L. japonica*, and *A. japonica* are the most dominant and widely distributed larval parasitoids attacking SWD (Fig. 8.1), whereas most other larval parasitoids showed a more restricted distribution and lower parasitism rates (8.1 and 8.2). These three larval parasitoids have been systematically evaluated for their efficiency, host specificity, climatic adaptability as well as potential interaction as classical biological control agents in North America and Europe (Biondi et al. 2017; Daane et al. 2016; Giorgini et al. 2019; Girod et al.

2018b, c; Wang et al. 2018a, 2019, 2020). All three parasitoids readily attack and develop from SWD (Table 8.2) and prefer to attack young host larvae (Wang et al. 2018a). At 23 °C, with SWD larvae in artificial diet, *G. brasiliensis* adult females survived 17.7 days and produced 98.3 offspring per female, and *L. japonica* survived 18.7 days and produced 107.2 offspring per female (Wang et al. 2018a), while *A. japonica* females lived 17.8 days and produced 117.3 offspring per female (Wang et al. unpubl. data). *Leptopilina japonica* eggs hatched the fastest, followed by *A. japonica* and then *G. brasiliensis* and consequently *L. japonica* outcompeted the other two parasitoids in multi-parasitized hosts (Wang et al. 2019). However, *G. brasiliensis* discriminated strongly against hosts parasitized by *L. japonica*, and *A. japonica* discriminated against hosts parasitized by *L. japonica*. The combined impacts on host suppression by *L. japonica* and *G. brasiliensis* were additive, likely due to the interspecific discrimination by *G. brasiliensis*. Indeed, both parasitoids coexist in all locations and plants sampled in China or South Korea (Daane et al. 2016; Giorgini et al. 2019), indicating they might synergistically improve the suppression of SWD.

Quarantine tests with a wide range of 24 different drosophila species showed that the South Korean and Yunnan *G. brasiliensis* populations developed from SWD and several other closely related hosts (*D. melanogaster* and *D. simulans*) but did not develop from more distant non-target drosophilid species (Giorgini et al. 2019). *Asobara japonica* developed from 19 of 24 tested host species, whereas *L. japonica* developed mainly from species in the *melanogaster* group (Daane et al. unpubl. data). By comparison, both *P. vindemiae* and *T. drosophilae* developed from all 24 tested drosophila species (Wang et al. unpubl. data). Other studies also showed that these two pupal parasitoids develop from nearly all tested hosts, preferentially attacking large hosts with correspondingly large progeny emerging (Chen et al. 2018; Wang et al. 2016a; Wolf et al. 2020; Yi et al. 2020). In Japan, field surveys and laboratory tests also found that *A. japonica* parasitized various indigenous and exotic drosophilid species (Ideo et al. 2008; Mitsui and Kimura 2010; Mitsui et al. 2007). In Switzerland, Girod et al. (2018b, c) tested six different European non-target fly species with these three larval parasitoids. Similarly, they found that *A. japonica* developed from all tested drosophilids, and *L. japonica* successfully parasitized *D. melanogaster* and *D. subobscura*. A Japanese population of *G.* cf. *brasiliensis* collected from SWD was strictly specific to SWD as reported by Kasuya et al. (2013), whereas another population from China parasitized SWD and *D. melanogaster* and sporadically parasitized *D. subobscura*. Thus, *A. japonica* is more of a generalist, whereas *L. japonica* appears to be a specialist on *melanogaster* species group. Currently, *G. brasiliensis* is considered as the first candidate for classical biological control of SWD due to its demonstrated specificity.

8.5 Diversity of the *Ganaspis brasiliensis* "Complex"

Buffington and Forshage (2016) first described *G. brasiliensis* as a new combination based on the specimens collected from SWD in South Korea (Daane et al. 2016) and historical specimens from the Neotropical region. Previously in Japan, Mitsui and Kimura (2010) reported that *Ganaspis* collected from *Drosophila lutescens* Okada readily parasitized *D. lutescens* and other drosophilids tested (>90% parasitism) but rarely accepted SWD (only 3.3% parasitism). These *Ganaspis* were initially assigned the name *G. xanthopoda* (Table 8.1). However, Kasuya et al. (2013) showed that SWD was the only drosophilid species infesting fresh wild cherries in Tokyo area, and *Ganaspis* individuals were the major parasitoids attacking SWD in wild cherry fruits. They reported that this *Ganaspis* population did not parasitize SWD in *Drosophila* medium and other *Drosophila* spp. in fresh cherries; and they identified the population as the *D. suzukii*-associated *G. xanthopoda* type. *Ganaspis* specimens previously assigned as *G. xanthopoda* are morphologically similar to specimens that were collected from South Korea and identified as *G. brasiliensis* by Buffington and Forshage (2016) and were thus reassigned to *G. brasiliensis* (Nomano et al. 2017).

Subsequent molecular analyses of different individuals based on nucleotide sequences of the mitochondrial cytochrome oxidase subunit 1 (CO1) gene, and the inter-transcribed spacers 1 and 2 (ITS1 and ITS2) suggest that individuals thus far morphologically identified as *G. brasiliensis* could be subdivided into five lineages (Nomano et al. 2017): G1, including individuals collected from SWD from Sendi and Tokyo in Japan; G2, including individuals from a subtropical Japanese island parasitizing *Drosophila ficusphila* Kikkawa & Peng; G3, including individuals from temperate regions of Japan and high mountains of Southeast Asia (Indonesia, Malaysia) parasitizing different species of *Drosophila*; G4, including individuals from Indonesia parasitizing *Drosophila eugracilis* Bock & Wheeler; G5, including individuals previously reported as *G. xanthopoda* or *Ganaspis* sp. from Thailand and the Philippines (Schilthuizen et al. 1998), Hawaii and Uganda (Kacsoh and Schlenke 2012), Indonesia (Kimura and Suwito 2012, 2015), Malaysia (Nomano et al. 2017), Benin, Puerto Rico and the Caribbean Sea (Carton et al. 1986), Brazil (Buffington and Forshage 2016), and Mexico (Gonzalez-Cabrera et al. 2020).

Phylogenetic analysis of COI sequences revealed that the *G. brasiliensis* specimens collected in Yunnan, China (Giorgini et al. 2019), consisted of 77% G1 and 23% G3. Similarly, the *G. brasiliensis* specimens collected in South Korea in 2017 (and similar sites reported in Daane et al. 2016) consisted of 65% G1 and 35% G3. These results suggest that these two lineages (G1 and G3) appear to be widely distributed in East Asia. They coexist in many locations and attack SWD and the closely related *D. pulchrella* and *D. subpulchrella* inhabiting fresh fruits, and have thus been considered sufficiently specific to SWD based on field collections and quarantine evaluations (Daane et al. 2016; Giorgini et al. 2019; Girod et al. 2018b, c; Kasuya et al. 2013). The host range of other lineages is unclear, and they have not been collected from SWD in fresh fruits nor tested in the laboratory with SWD, except that some G5 individuals from Hawaii and Uganda have a capacity to parasitize SWD in laboratory tests but with no or low development (Kacsoh and Schlenke 2012). Thus, *G. brasiliensis* appears to be a complex of several cryptic strains with varying host specificity and distributions. Given the Asian origin of SWD and the common ancestor of different lineages likely occurs in Asia, the species have likely been introduced to the Neotropics and Africa (Buffington and Forshage 2016; Nomano et al. 2017).

Recent studies further suggest that G1 (called *G.* cf. *brasiliensis* in Girod et al. 2018a) and G3 may be two different species. Reeve and Seehausen (2019) compared the acid-soluble insect protein spectra among three different G1 populations collected from Tokyo, Japan, and Dali and Ximing, China and a G1 population collected from Hasuike, Japan, and found that the G3 is significantly different from all G1 specimens. Other ongoing studies indicate the absence of positive crossing between G1 and G3, and different host-searching behaviors. G1 prefers hosts infesting fruits, whereas G3 prefers hosts in rotting substrates (M. Kenis, personal comm.). Further research combining multiple gene analyses and crossing-mating experiments across geographical populations or lineages is clearly needed to fully understand the ecological and genetic diversity of the *G. brasiliensis* complex.

8.6 Predicted Geographical Ranges of *Ganaspis brasiliensis*

The CLIMEX model (Kriticos et al. 2015) has been used to predict the potential geographical range of *G. brasiliensis* based on the current known distribution of G1 and G3 lineages in Asia (Daane et al. unpubl. data). Geographical coordinates of 37 collection sites where parasitoids were found in China, South Korea, and Japan were obtained (Kasuya et al. 2013; Daane et al. 2016; Nomano et al. 2017; Matsuura et al. 2018; Giorgini et al. 2019). The model parameters were repetitively adjusted and the function "Compare location," which describes the potential geographical distribution of species, as controlled by weather variables was subsequently run until the estimated potential *G. brasiliensis* range coincided best with the known distributions of the species in East Asia. The model predicted that *G. brasiliensis* would likely establish in the western, southeastern, and east coastal states in North America and most southern European countries where SWD is a major concern of small fruit crops (Fig. 8.2). Indeed, a recent survey in British Columbia, Canada, found that G1 has established in the Vancouver area, possibly through accidental introduction (P. Abram, personal comm.). It remains to be discovered whether *G. brasiliensis* will be able to colonize all invaded regions by SWD or whether it will be limited by climatic constraints. A comparative study on thermal performance between two populations originally from Yunnan Province of China and Gyeongsangnam-do Province of South Korea revealed the occurrence of a facultative diapause in *G. brasiliensis* below 17.2 °C (Hougardy et al. 2019). This cold temperature response varied between the populations: South Korean populations entered diapause at 17.2 °C, whereas only a proportion of its Chinese counterpart

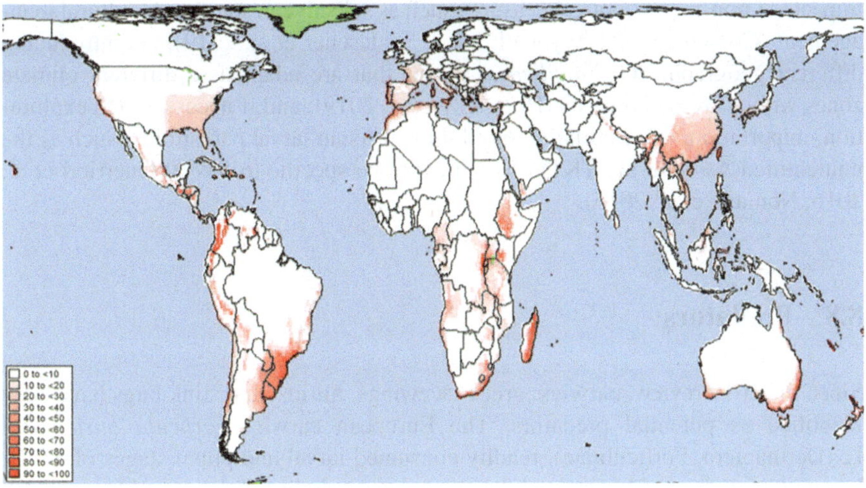

Fig. 8.2 Predicted distribution of *Ganapsis brasiliensis* worldwide based on CLIMEX climatic suitability indices. EI: <10 is not suitable; 10–50 moderate level of suitability; and >50 highly suitable for long-term survival

entered diapause at the same temperature. This suggests that some populations could be a better match for colder climates, or that mixing of populations from different origins could increase plasticity in response to cold seasons.

8.7 Future Directions with Parasitoids

Biological control using parasitoids could be a key component of areawide management programs for SWD by reducing fly populations at the landscape level. To date, three resident pupal parasitoids (*P. vindemiae*, *T. drosophilae*, and *T. anastrephae*) and one Asian larval parasitoid species or species complex (*G. brasiliensis* or *G.* cf. *brasiliensis*) have been identified as potentially promising biological control agents for SWD. A petition for release of the Asian *G. brasiliensis* in North America and Europe has been submitted, and a regulatory decision is currently pending. The resident pupal parasitoids already adapted to local ecological conditions and which can readily attack SWD could be manipulated either through conservation or augmentation to contribute to SWD suppression. However, the most effective and permanent biological control will likely be achieved by the introduction and augmentation of *G. brasiliensis*.

Future studies may include (1) the genetic improvement of natural enemies by selecting biological traits among different populations for selection or breeding that are important for effective biological control (Kruitwagen et al. 2018); (2) developing optimal rearing and release strategies for promising parasitoids to maximize establishment potential in different regions; (3) developing strategies to reduce the impacts of non-target control measures such as selective pesticides or cultural management (Cossentine and Ayyanath 2017; Schlesener et al. 2019), (4) introducing different geographic *G. brasiliensis* strains that are adapted to different climate zones within invaded regions (Hougardy et al. 2019); and if necessary (5) exploration, importation, and evaluation of additional Asian larval parasitoids (such as the unidentified *Asobara* sp. TK1) that appear to be specific to SWD (Guerrieri et al. 2016; Nomano et al. 2015).

8.8 Predators

Since the 2019 review, earwigs, green lacewings, mirids, and stink bugs have been identified as potential predators. The European earwig, *Forficula auricularia* L. (Dermaptera, Forficulidae), readily consumed larval and pupal stages of SWD, but could not effectively catch adults in laboratory studies (Englert and Herz 2019). Similarly, *F. auricularia* reduced the emergence of SWD adults by 45% when confined to infested cherries in a growth chamber, and the reduction was likely due to predation on developing SWD and not removal of parental SWD (Bourne et al. 2019). Green lacewing larvae, *Chrysoperla carnea* (Stephens) (Neuroptera: Chrysopidae), have reduced emergence of SWD from infested raspberries by 32% (Bonneau et al. 2019), and from infested cherries by 33% (Englert and Herz 2019). The mirid *Dicyphushesperus* (Knight) (Heteroptera: Miridae) preyed on exposed SWD eggs, and the stink bug *Podisus maculiventris* (Say) (Hemiptera: Pentatomidae) preyed on exposed larvae that were placed on leaves (Bonneau et al. 2019). Recent studies with minute pirate bugs further support previous work, as *Orius insdiosus* (Say) (Hemiptera: Anthocoridae) reduced emergence from infested fruit by 49% (Bonneau et al. 2019), and *O. majusculus* (Reuter) (Hemiptera: Anthocoridae) reduced emergence by 31% (Englert and Herz 2019).

Previous assessments of field predation have revealed 61–100% removal of sentinel pupae on or below the soil surface, and a 19–49% reduction in emerging SWD from infested fruit in fields in Oregon and Maine, USA (Ballman et al. 2017; Woltz and Lee 2017). A recent survey in organic raspberry fields in Wisconsin documented 1–4% predation on sentinel pupae (Kamiyama et al. 2019). Therefore, actual predation levels can vary from field to field depending on the conditions. Predation has also been assessed by molecular analysis as first demonstrated by Wolf et al. (2018) which surveyed predators in organic farms in Germany. In Georgia blueberries, 0.4% of the 1600 collected predators tested positive for predation by molecular analysis (Schmidt et al. 2019). These included hunting spiders, a web-building spider, and one mantid. Both studies using molecular surveys reveal that generalist spiders prey on SWD.

To date, most work on parasitoids and predators has focused on field surveys, measuring natural predation and parasitism rates, or studied the efficacy of agents in enclosed arenas or field releases. More work remains to be done on conserving these natural enemies, and whether specific habitat manipulations will benefit natural enemies and increase SWD control. A landscape-level analysis of blueberry fields revealed that organic systems and fields with vegetation between rows harbored more natural enemies (Schmidt et al. 2019). In their analysis, landscapes with greater composition of non-crop habitats also had higher SWD populations. This may be expected since SWD reproduces on many different wild hosts (Kenis et al. 2016; Lee et al. 2015), and SWD can move from wild fruit to nearby crops (Leach et al. 2018).

8.9 Entomopathogens

In 2019–2020, more studies were published on entomopathogenic nematodes than other pathogens. The recently discovered *Oscheius onirici* Torrini et al. (Nematoda: Rhabditidae) was sprayed on infested blueberries, reducing pupation by 78% in laboratory trials (Foye and Steffan 2020). Contrary to previous reports (Garriga et al. 2017; Hübner et al. 2017), SWD pupae have appeared to be more susceptible to nematode infections than larvae. For example, *Heterorhabditis bacteriophora* (Poinar) (Nematoda: Heterorhabditidae) and *Steinernema feltiae* (Filipjev) (Nematoda: Steinernematidae) caused 72% mortality among SWD pupae, and 20% mortality among larvae in Petri dish assays (Ibouh et al. 2019). Newly tested *Heterorhabditis amazonensis* (Andalo) and *H. indica* (Poinar) (Nematoda: Heterorhabditidae), as well as *S. carpocapsae* (Weiser) and *S. feltiae*, caused 35, 26, 13, and 43% mortality among SWD pupae, respectively (Brida et al. 2019).

Additional work has supported the effectiveness of *S. carpocapsae,* including recent assays with adult SWD. Adults exposed to *S. carpocapsae* had an infection rate of 65% compared to 4% by *S. feltiae* and *H. bacteriophora* (Garriga et al. 2020b). Moreover, when soil with buried SWD pupae was treated with *S. carpocapsae*, 89% of emerging adults were infected. Teneral adults may be especially vulnerable to infection, and 59% could not move up their plastic cylinder arena. In lab arenas, 21% of infected adults were able to fly, and the authors suggested that this may help with nematode dispersion. As with any pathogen, infected hosts can have defensive responses, and studies of SWD larvae infected with *S. carpocapsae* and its symbiont bacteria *Xenorhabdus nematophila* Thomas & Poinar revealed that the pathogen avoided cellular defenses and depressed humoral responses (Garriga et al. 2020a).

The fungal pathogens *Beauveria bassiana* (Bals.) Vuill and *Metarhizium anisopliae* (Metch.) Sorok. have been the most widely studied (reviewed in Lee et al. 2019). Recent work has shown them to cause 38% mortality of larvae, and 32–64% of adults when sprayed on SWD in Petri dishes (Ibouh et a. 2019). Interestingly, when grape berries were dipped in fungal suspensions, oviposition by SWD was reduced by 80% compared to the controls. Thus, while fungal treatments may not always directly contact adults when sprayed in the field, and require several days to induce adult mortality, the sprays provide additional protection to the fruit. Assays conducted by Ibouh et al. (2019) exposed flies to grapes for 5 days under standard laboratory conditions, and the duration of fruit protection has still to be determined.

To find bacterial pathogens, Hiebert et al. (2020) collected SWD from infested fruits in the field, isolated, and screened the associated bacteria. Seven isolates were detrimental including the Gram-positive bacteria *Brevibacterium frigoritolerans* Delaporte & Sasson, *Bacillus simplex* (exMeyer and Gottheil), *Bacillus altitudinis* Schivaji et al., *Leuconostoc pseudomesenteroides* Farrow et al., *Paenibacillus dongdonensis* Son et al. and *Paenibacillus odorifer*, and the Gam-negative bacterium *Tatumella terrea* (Kageyama et al.). The mode of action was explored; *Paenibacillus dongdonensis* and *L. pseudomesenteroides* appeared to reduce food uptake in SWD larvae.

8.10 Competitors

Previous laboratory and greenhouse work has shown *Drosophila melanogaster* to be a promising competitor of SWD. The presence of *D. melanogaster* is not expected to pose a threat to harvested fruit since it attacks overripe or damaged fruit and could foreseeably compete with SWD during the late season when dropped fruit remains on the ground. The African fig fly, *Zaprionus indianus* (Gupta), was recently shown to compete with SWD in grapes in laboratory studies, and induce higher SWD mortality (Shrader et al. 2020). *Zaprionus indianus* generally does not lay eggs in intact fruit but can use the oviposition sites of SWD to lay eggs (Bernardi et al. 2017). Whether co-infestations occur often in the field or could be advantageous for IPM remains to be studied.

8.11 Compatibility of Biological Control

Recent work has investigated the compatibility of biological controls with other control approaches, especially with pesticides commonly used in SWD management. Organophosphates, pyrethroids, and neonicotinoids cause high mortality in the parasitoids *T. anastrephae* and *P. vindemiae* in lab bioassays (Schlesener et al. 2019). Spinosad is a commonly used organic insecticide which is unfortunately detrimental to *P. vindemiae* adults, and female wasps are unable to avoid treated SWD pupae (Cossentine and Ayyanath 2017). The same study also determined that the larval stage of *P. vindemiae* was susceptible to spinosad when SWD pupae were treated 1 week post-parasitization, but they survived better at the pupal stage when treated 2 weeks post-parasitization. A variety of organic insecticides were tested on two generalist predators of SWD; the green lacewing *Chrysoperla rufilabris* (Burmeister) was susceptible to spinosad, and the minute pirate bug *Orius insidiosus* was susceptible to fresh and aged residues of spinosad and sabadilla alkaloids

(Sarkar et al. 2019). Moreover, sublethal effects of insecticide exposure resulted in reduced egg hatch of *O. insidiosus*.

With the variety of pathogens being tested for SWD control, more work is needed to assess compatibility of pathogens with predators and parasitoids, especially if releases are anticipated. Recently, *T. drosophilae* was unaffected when parasitized SWD pupae were exposed to treatments of *B. bassiana*, *M. anisopliae*, *H. bacteriophora*, or *S. feltiae* and parasitoid emergence was subsequently monitored in the laboratory (Ibouh et al. 2019). Likewise, adults of *T. drosophilae* and rove beetle, *D. coriaria* (Kraatz), were unaffected by *H. bacteriophora*, *S. feltiae*, and *S. carpocapsae* in Petri dish assays (Garriga et al. 2019). However, the predator *O. laevigatus* (Fieber) experienced reduced survival from exposure to *S. carpocapsae* in Petri dish assays but not when nematodes were applied to a plant. This suggests that this predator would escape harmful effects in a field situation. Mulching and floor management have been examined as cultural practices to control SWD (Rendon et al. 2020; Rendon and Walton 2019), and specifically target SWD as they often wander to pupate in the soil (Woltz and Lee 2017). Such ground practices to make the soil less hospitable to SWD may however be incompatible with soil drench treatments with nematodes where a moist soil environment is necessary for infective juveniles to survive and find hosts.

8.12 Summary

Many researchers have been dedicated to advancing biological control of SWD as demonstrated by the nearly 100 publications at the time of writing this review. A longer-term approach relies on importing the parasitoid, *Ganapsis brasiliensis*, to invaded regions. With this parasitoid, there is a need to: breed more effective traits, develop efficient rearing and release strategies, and use geographic strains adapted to various climates. A variety of endemic predators in the field prey on SWD. Augmentative releases of predators have not yet been recommended since their cost-effectiveness and efficacy need determination. As for pathogens, new nematode species have been tested in the laboratory, and nematodes can affect teneral SWD adults as they emerge from treated soil. Moreover, several nematode and fungal pathogens appear to be compatible with common SWD parasitoids and predators. This is promising since many commonly used insecticides for SWD are harmful to these parasitoids and predators. Since most pathogen research has been conducted in the laboratory, field trials are required to develop recommendations. As more information becomes available with biological control agents, additional work is needed to integrate them into SWD management programs.

Acknowledgments Funding was provided by the National Institute of Food and Agriculture, USDA Specialty Crops Research Initiative under Agreement No. 2015-51181-24252, USDA Organic Research and Extension Initiative under agreement No. 2018-51300-28434, the California

Cherry Board, and the University of California Agricultural and Natural Research Grant, and base funds USDA CRIS 8010-22000-028-00D and 2072-22000-040-00D.

References

Asplen MK, Anfora G, Biondi A et al (2015) Invasion biology of spotted wing drosophila (*Drosophila suzukii*): a global perspective and future priorities. J Pest Sci 88:469–494. https://doi.org/10.1007/s10340-015-0681-z

Ballman ES, Collins JA, Drummond FA (2017) Pupation behavior and predation on *Drosophila suzukii* (Diptera: Drosophilidae) pupae in Maine wild blueberry fields. J Econ Entomol 110:2308–2317. https://doi.org/10.1093/jee/tox233

Bernardi D, Andreazza F, Botton M et al (2017) Susceptibility and interactions of *Drosophila suzukii* and *Zaprionus indianus* (Diptera: Drosophilidae) in damaging strawberry. Neotrop Entomol 46:1–7. https://doi.org/10.1007/s13744-016-0423-9

Bezerra da Silva CS, Price BE, Soohoo-Hui A et al (2019a) Factors affecting the biology of *Pachycrepoideus vindemmiae* (Hymenoptera: Pteromalidae), a parasitoid of spotted-wing drosophila (*Drosophila suzukii*). PLoS One 14:19326203

Bezerra da Silva CH, Price BE, Walton VM (2019b) Water-deprived parasitic wasps (*Pachycrepoideus vindemmiae*) kill more pupae of a pest (*Drosophila suzukii*) as a water-intake strategy. Sci Rep 9:3592

Biondi A, Wang XG, Miller JC et al (2017) Innate olfactory responses of *Asobara japonica* toward fruits infested by the invasive spotted wing drosophila. J Insect Behav 30:495–506. https://doi.org/10.1007/s10905-017-9636-y

Bonneau P, Renkema J, Fournier V et al (2019) Ability of *Muscidifurax raptorellus* and other parasitoids and predators to control *Drosophila suzukii* populations in raspberries in the laboratory. Insects 10:68. https://doi.org/10.3390/insects10030068

Bourne A, Fountain MT, Wijnen H, Shaw B (2019) Potential of the European earwig (Forficula auricularia) as a biocontrol agent of the soft and stone fruit pest Drosophila suzukii. Pest Manag Sci 75:3340–3345

Brida A, Wilcken SRS, Garrigós Leite L et al (2019) Virulence of entomopathogenic nematode to pupae and adults of *Drosophila suzukii* in laboratory. Revista Científica Rural 21:126–138. https://doi.org/10.30945/rcr-v21i2.2725

Buffington ML, Forshage M (2016) Redescription of *Ganaspis brasiliensis* (Ihering, 1905), new combination (Hymenoptera: Figitidae), a natural enemy of the invasive *Drosophila suzukii* (Matsumura, 1931) (Diptera: Drosophilidae). Proc Entomol Soc Wash 118:1–13. https://doi.org/10.4289/0013-8797.118.1.1

Cancino MDG, Hernandez AG, Cabrera JG et al (2015) Parasitoids of *Drosophila suzukii* (Matsumura) (Diptera: Drosophilidae) in Colima, Mexico. Southwest Entomol 40:855–858. https://doi.org/10.3958/059.040.0418

Carton Y, Boulétreau B, van Alphen JJM et al(1986) The *Drosophila* parasitic wasps. In: Ashburner M, Carson HL, Thompson JN (eds) The genetics and biology of *Drosophila*, vol 3e. Academic, London, pp. 347–394

Chabert S, Allemand R, Poyet M et al (2012) Ability of European parasitoids (Hymenoptera) to control a new invasive Asiatic pest, *Drosophila suzukii*. Biol Control 63:40–47. https://doi.org/10.1016/j.biocontrol.2012.05.005

Chen JN, Zhou SC, Wang Y et al (2018) Biocontrol characteristics of the fruit fly pupal parasitoid *Trichopria drosophilae* (Hymenoptera: Diapriidae) emerging from different hosts. Sci Rep 8:e13323. https://doi.org/10.1038/s41598-018-31718-6

Cossentine JE, Ayyanath MM (2017) Limited protection of the parasitoid *Pachycrepoideus vindemiae* from *Drosophila suzukii* host-directed spinosad suppression. Entomol Exp Appl 164:78–86. https://doi.org/10.1111/eea.12592

Daane KM, Wang XG, Biondi A et al (2016) First exploration of parasitoids of *Drosophila suzukii* in South Korea as potential classical biological agents. J Pest Sci 89:823–835. https://doi.org/10.1007/s10340-016-0740-0

Dancau T, Stemberger TLM, Clarke P, Gillespie DR (2017) Can competition be superior to parasitism for biological control? The case of spotted wing Drosophila (*Drosophila suzukii*), Drosophila melanogaster and Pachycrepoideus vindemmiae. Biocontrol Sci Tech 27:3–16

Englert C, Herz A (2019) Acceptability of *Drosophila suzukii* as prey for common predators occurring in cherries and berries. J Appl Entomol 143:387–396. https://doi.org/10.1111/jen.12613

Farnsworth D, Hamby KA, Bolda M et al (2017) Economic analysis of revenue losses and control costs associated with the spotted wing drosophila, *Drosophila suzukii* (Matsumura), in the California raspberry industry. Pest Manag Sci 73:1083–1090. https://doi.org/10.1002/ps.4497

Fleury F, Gibert P, Ris N et al (2009) Ecology and life history evolution of frugivorous *Drosophila* parasitoids. In: Prévost G (ed) Advances in Parasitology: Parasitoids of *Drosophila*, Advances in parasitology, vol 70. Academic, London, pp 3–44. https://doi.org/10.1016/s0065-308x(09)70001-6

Foye S, Steffan SA (2020) A rare, recently discovered nematode, *Oscheius onirici* (Rhabditida: Rhabditidae), kills *Drosophila suzukii* (Diptera: Drosophilidae) within fruit. J Econ Entomol 113:1047–1051. https://doi.org/10.1093/jee/toz365

Gabarra R, Riudavets J, Rodríguez GA et al (2015) Prospects for the biological control of *Drosophila suzukii*. Biocontrol 60:331–339

Garcia-Cancino MD, Gonzalez-Cabrera J, Sanchez-Gonzalez JA et al (2020) Biological and population parameters, as well as oviposition preference, of two pupal parasitoids of *Drosophila suzukii* (Diptera: Drosophilidae) in Mexico. J Entomol Sci 55:87–97. https://doi.org/10.18474/0749-8004-55.1.87

Garrido SA, Cichón LI, Lago JD et al (2018) First record of *Leptopilina boulardi* (Hymenoptera: Figitidae) associated to *Drosophila suzukii* (Diptera: Drosophilidae) in Alto Valle of Río Negro and Neuquén, Patagonia, Argentina. Revista de la Sociedad Entomológica Argentina 77:22–27

Garriga A, Morton A, Garcia-del-Pino F (2017) Is *Drosophila suzukii* as susceptible to entomopathogenic nematodes as *Drosophila melanogaster*? J Pest Sci 91:789–798. https://doi.org/10.1007/s10340-017-0920-6

Garriga A, Morton A, García-López D, García-del-Pino F (2019) Compatibility of entomopathogenic nematodes with natural enemies for horticultural pest control. Biol Control 138:104050

Garriga A, Mastore M, Morton A et al (2020a) Immune response of *Drosophila suzukii* larvae to infection with the nematobacterial complex *Steinernema carpocapsae-Xenorhabdus nematophila*. Insects 11:210. https://doi.org/10.3390/insects11040210

Garriga A, Morton A, Ribes A, Garcia-del-Pino F (2020b) Soil emergence of Drosophila suzukii adults: a susceptible period for entomopathogenic nematodes infection. J Pest Sci 93(2):639–646

Giorgini M, Wang XG, Wang Y et al (2019) Exploration for native parasitoids of *Drosophila suzukii* in China reveals a diversity of parasitoid species and narrow host range of the dominant parasitoid. J Pest Sci 92:509–522. https://doi.org/10.1007/s10340-018-01068-3

Girod P, Borowiec N, Buffington M et al (2018a) The parasitoid complex of *D. suzukii* and other fruit feeding *Drosophila* species in Asia. Sci Rep 8. https://doi.org/10.1038/s41598-018-29555-8

Girod P, Lierhmann O, Urvois T et al (2018b) Host specificity of Asian parasitoids for potential classical biological control of *Drosophila suzukii*. J Pest Sci 91:1241–1250. https://doi.org/10.1007/s10340-018-1003-z

Girod P, Rossignaud L, Haye T et al (2018c) Development of Asian parasitoids in larvae of *Drosophila suzukii* feeding on blueberry and artificial diet. J Appl Entomol 142:483–494. https://doi.org/10.1111/jen.12496

Gonzalez-Cabrera J, Moreno-Carrillo G, Sanchez-Gonzalez JA et al (2019) Single and combined release of *Trichopria drosophilae* (Hymenoptera: Diapriidae) to control *Drosophila suzukii* (Diptera: Drosophilidae). Neotrop Entomol 48:949–956. https://doi.org/10.1007/s13744-019-00707-3

Gonzalez-Cabrera J, Cordoba-Urtiz EG, Moreno-Carrillo G et al (2020) First report of the parasitoid *Ganaspis brasiliensis* Ihering (Hymenoptera: Figitidae) in Mexico. Entomol News 129:67–70. https://doi.org/10.3157/021.129.0110

Gress BE, Zalom FG (2019) Identification and risk assessment of spinosad resistance in a California population of *Drosophila suzukii*. Pest Manag Sci 75:1270–1276. https://doi.org/10.1002/ps.5240

Guerrieri E, Giorgini M, Cascone P et al (2016) Species diversity in the parasitoid genus *Asobara* (Hymenoptera: Braconidae) from the native area of the fruit fly pest *Drosophila suzukii* (Diptera: Drosophilidae). PLoS One 11. https://doi.org/10.1371/journal.pone.0147382

Haviland DR, Beers EH (2012) Chemical control programs for *Drosophila suzukii* that comply with international limitations on pesticide residues for exported sweet cherries. J Integr Pest Manag 3:1. https://doi.org/10.1603/ipm11034

Haro-Barchin E, Scheper J, Ganuza C, De Groot GA, Colombari F, van Kats R, Kleijn D (2018) Landscape-scale forest cover increases the abundance of Drosophila suzukii and parasitoid wasps. Basic Appl Ecol 31:33–43

Hiebert N, Carrau T, Bartling M et al (2020) Identification of entomopathogenic bacteria associated with the invasive pest *Drosophila suzukii* in infested areas of Germany. J Invertebr Pathol 173:107389. https://doi.org/10.1016/j.jip.2020.107389

Hougardy E, BN Hogg, X Wang, Daane KM (2019) Comparsion of thermal performances of two Asian larval parasitoids of Droophila suzukii. Biol Control 136:10400

Hübner A, Englert C, Herz A (2017) Effect of entomopathogenic nematodes on different developmental stages of *Drosophila suzukii* in and outside fruits. Biocontrol 62:669–680. https://doi.org/10.1007/s10526-017-9832-x

Iacovone A, Ris N, Poirie M et al (2018) Time-course analysis of *Drosophila suzukii* interaction with endoparasitoid wasps evidences a delayed encapsulation response compared to *D. melanogaster*. PLoS One 13:e0201573. https://doi.org/10.1371/journal.pone.0201573

Ibouh K, Oreste M, Bubici G et al (2019) Biological control of *Drosophila suzukii*: Efficacy of parasitoids, entomopathogenic fungi, nematodes and deterrents of oviposition in laboratory assays. Crop Prot 125:104897. https://doi.org/10.1016/j.cropro.2019.104897

Ideo S, Watada M, Mitsui H et al (2008) Host range of *Asobara japonica* (Hymenoptera : Braconidae), a larval parasitoid of drosophilid flies. Entomol Sci 11:1–6. https://doi.org/10.1111/j.1479-8298.2007.00244.x

Janssen A, Driessen G, Dehaan M et al (1988) The impact of parasitoids on natural populations of temperate woodland *Drosophila*. Neth J Zool 38:61–73

Kaçar G, Wang XG, Biondi A et al (2017) Linear functional response by two pupal *Drosophila* parasitoids foraging within single or multiple patch environments. PLoS One 12:e0183525. https://doi.org/10.1371/journal.pone.0183525

Kacsoh BZ, Schlenke TA (2012) High hemocyte load is associated with increased resistance against parasitoids in *Drosophila suzukii*, a relative of *D. melanogaster*. PLoS One 7:e34721. https://doi.org/10.1371/journal.pone.0034721

Kamiyama MT, Schreiner Z, Guedot C (2019) Diversity and abundance of natural enemies of *Drosophila suzukii* in Wisconsin, USA fruit farms. Biocontrol 64:665–676. https://doi.org/10.1007/s10526-019-09966-w

Kasuya N, Mitsui H, Ideo S et al (2013) Ecological, morphological and molecular studies on *Ganaspis* individuals (Hymenoptera: Figitidae) attacking *Drosophila suzukii* (Diptera: Drosophilidae). Appl Entomol Zool 48:87–92. https://doi.org/10.1007/s13355-012-0156-0

Kenis M, Tonina L, Eschen R et al (2016) Non-crop plants used as hosts by *Drosophila suzukii* in Europe. J Pest Sci 89:735–748. https://doi.org/10.1007/s10340-016-0755-6

Kimura MT, Suwito A (2012) Diversity and abundance of frugivorous drosophilids and their parasitoids in Bogor, Indonesia. J Nat Hist 46:1947–1957. https://doi.org/10.1080/00222933.2012.707239

Kimura MT, Suwito A (2015) Altitudinal patterns of abundances and parasitism in frugivorous drosophilids in west Java, Indonesia. J Nat Hist 49:1627–1639. https://doi.org/10.1080/00222933.2015.1005709

Knoll V, Ellenbroek T, Romeis J et al (2017) Seasonal and regional presence of hymenopteran parasitoids of *Drosophila* in Switzerland and their ability to parasitize the invasive *Drosophila suzukii*. Sci Rep 7:e40697. https://doi.org/10.1038/srep40697

Kremmer L, Thaon M, Borowiec N et al (2017) Field monitoring of *Drosophila suzukii* and associated communities in south eastern France as a pre-requisite for classical biological control. Insects 8:124. https://doi.org/10.3390/insects8040124

Kriticos DJ, Maywald GF, Yonow T et al (2015) CLIMEX Version 4: Exploring the effects of climate on plants, animals and diseases. CSIRO, Canberra, 156 pp

Kruger AP, Scheunemann T, Vieira JGA et al (2019) Effects of extrinsic, intraspecific competition and host deprivation on the biology of *Trichopria anastrephae* (Hymenoptera: Diapriidae) reared on *Drosophila suzukii* (Diptera: Drosophilidae). Neotrop Entomol 48:957. https://doi.org/10.1007/s13744-019-00705-5

Kruitwagen A, Beukeboom LW, Wertheim B (2018) Optimization of native biocontrol agents, with parasitoids of the invasive pest *Drosophila suzukii* as an example. Evol Appl 11:1473–1497. https://doi.org/10.1111/eva.12648

Leach H, Hagler JR, Machtley SA et al (2018) Spotted wing drosophila (*Drosophila suzukii*) utilization and dispersal from the wild host Asian bush honeysuckle (Lonicera spp.). Agri For Entomol 21:149–158. https://doi.org/10.1111/afe.12315

Lee JC, Dreves AM, Cave AM et al (2015) Infestation of wild and ornamental noncrop fruits by *Drosophila suzukii* (Diptera: Drosophilidae). Ann Entomol Soc Am 108:117–129. https://doi.org/10.1093/aesa/sau014

Lee JC, Wang XG, Daane KM et al (2019) Biological control of spotted-wing drosophila (Diptera: Drosophilidae): current and pending tactics. J Integr Pest Manag 10:13. https://doi.org/10.1093/jipm/pmz012

Matsuura A, Mitsui H, Kimura MT (2018) A preliminary study on distributions and oviposition sites of *Drosophila suzukii* (Diptera: Drosophilidae) and its parasitoids on wild cherry tree in Tokyo, central Japan. Appl Entomol Zool 53:47–53. https://doi.org/10.1007/s13355-017-0527-7

Mazzetto F, Marchetti E, Amiresmaeili N et al (2016) *Drosophila* parasitoids in northern Italy and their potential to attack the exotic pest *Drosophila suzukii*. J Pest Sci 89:837–850. https://doi.org/10.1007/s10340-016-0746-7

Miller B, Anfora G, Buffington M et al (2015) Seasonal occurrence of resident parasitoids associated with *Drosophila suzukii* in two small fruit production regions of Italy and the USA. Bull Insect 68:255–263

Mitsui H, Kimura MT (2010) Distribution, abundance and host association of two parasitoid species attacking frugivorous drosophilid larvae in central Japan. Eur J Entomol 107:535–540

Mitsui H, Van Achterberg K, Nordlander G et al (2007) Geographical distributions and host associations of larval parasitoids of frugivorous Drosophilidae in Japan. J Nat Hist 41:1731–1738. https://doi.org/10.1080/00222930701504797

Modic S, Žigon P, Razinger J (2019) *Trichopria drosophilae* (Diapriidae) and *Leptopilina heterotoma* (Figitidae), native parasitoids of *Drosophila suzukii*, confirmed in Slovenia. Acta Agri Slovenica 113:181–185

Murata Y, Ideo S, Watada M et al (2009) Genetic and physiological variation among sexual and parthenogenetic populations of *Asobara japonica* (Hymenoptera: Braconidae), a larval parasitoid of drosophilid flies. Eur J Entomol 106:171–178. https://doi.org/10.14411/eje.2009.020

Nomano FY, Mitsui H, Kimura MT (2015) Capacity of Japanese *Asobara* species (Hymenoptera; Braconidae) to parasitize a fruit pest *Drosophila suzukii* (Diptera; Drosophilidae). J Appl Entomol 139:105–113. https://doi.org/10.1111/jen.12141

Nomano FY, Kasuya N, Matsuura A et al (2017) Genetic differentiation of *Ganaspis brasiliensis* (Hymenoptera: Figitidae) from East and Southeast Asia. Appl Entomol Zool 52:429–437. https://doi.org/10.1007/s13355-017-0493-0
Novkovic B, Mitsui H, Suwito A et al (2011) Taxonomy and phylogeny of *Leptopilina* species (Hymenoptera: Cynipoidea: Figitidae) attacking frugivorous drosophilid flies in Japan, with description of three new species. Entomol Sci 14:333–346. https://doi.org/10.1111/j.1479-8298.2011.00459.x
Pfab F, Rossi-Stacconi MV, Anfora G et al (2018) Optimized timing of parasitoid release: a mathematical model for biological control of *Drosophila suzukii*. Theor Ecol 11:489–501. https://doi.org/10.1007/s12080-018-0382-3
Poyet M, Havard S, Prevost G et al (2013) Resistance of *Drosophila suzukii* to the larval parasitoids *Leptopilina heterotoma* and *Asobara japonica* is related to haemocyte load. Physiol Entomol 38:45–53. https://doi.org/10.1111/phen.12002
Prévost G (2009) Advances in parasitology, Parasitoids of Drosophila, vol 70. Academic, London
Rendon D, Walton VM (2019) Drip and overhead sprinkler irrigation in blueberry as cultural control for *Drosophila suzukii* (Diptera: Drosophilidae) in Northwestern United States. J Econ Entomol 112:745–752. https://doi.org/10.1093/jee/toy395
Rendon D, Hamby KA, Arsenault-Benoit AL et al (2020) Mulching as a cultural control strategy for Drosophila suzukii in blueberry. Pest Manag Sci 76:55–66. https://doi.org/10.1002/ps.5512
Reeve MA, Seehausen ML (2019) Discrimination between Asian populations of the parasitoid wasp Ganaspis cf. brasiliensis using a simple MALDI-TOF MS-based method for use with insects. Biology Methods and Protocols 4:bpz002. https://doi.org/10.1093/biomethods/bpz002
Rossi Stacconi MV, Grassi A, Dalton DT et al (2013) First field records of *Pachycrepoideus vindemiae* as a parasitoid of *Drosophila suzukii* in European and Oregon small fruit production areas. Entomologia 1:1–15
Rossi Stacconi MV, Panel A, Baser N et al (2017) Comparative life history traits of indigenous Italian parasitoids of *Drosophila suzukii* and their effectiveness at different temperatures. Biol Control 112:20–27. https://doi.org/10.1016/j.biocontrol.2017.06.003
Rossi Stacconi MV, Grassi A, Ioriatti C et al (2019) Augmentative releases of *Trichopria drosophilae* for the suppression of early season *Drosophila suzukii* populations. Biocontrol 64:9–19. https://doi.org/10.1007/s10526-018-09914-0
Rossi-Stacconi MV, Buffington M, Daane KM et al (2015) Host stage preference, efficacy and fecundity of parasitoids attacking *Drosophila suzukii* in newly invaded areas. Biol Control 84:28–35. https://doi.org/10.1016/j.biocontrol.2015.02.003
Rossi-Stacconi MV, Amiresmaeili N, Biondi A et al (2018) Host location and dispersal ability of the cosmopolitan parasitoid *Trichopria drosophilae* released to control the invasive spotted wing drosophila. Biol Control 117:188–196. https://doi.org/10.1016/j.biocontrol.2017.11.013
Sarkar N, Rhodes EM, Spies J et al (2019) Evaluation of non-target effects of OMRI-listed insecticides for management of *Drosophila suzukii* Matsumura in berry crops. J Appl Entomol 144:12–25. https://doi.org/10.1111/jen.12713
Schilthuizen M, Nordlander G, Stouthamer R et al (1998) Morphological and molecular phylogenetics in the genus *Leptopilina* (Hymenoptera: Cynipoidea: Eucoilidae). Syst Entomol 23:253–264. https://doi.org/10.1046/j.1365-3113.1998.00049.x
Schlesener DCH, Wollmann J, Pazini JB et al (2019) Insecticide toxicity to *Drosophila suzukii* (Diptera: Drosophilidae) parasitoids: *Trichopria anastrephae* (Hymenoptera: Diapriidae) and *Pachycrepoideus vindemmiae* (Hymenoptera: Pteromalidae). J Econ Entomol 112:1197–1206. https://doi.org/10.1093/jee/toz033
Schmidt JM, Whitehouse TS, Green K et al (2019) Local and landscape-scale heterogeneity shape spotted wing drosophila (*Drosophila suzukii*) activity and natural enemy abundance: implications for trophic interactions. Agri Eco Environ 272:86–94. https://doi.org/10.1016/j.agee.2018.11.014

Shrader M, Burrack HJ, Pfeiffer D (2020) Effects of interspecific larval competition on developmental parameters in nutrient sources between *Drosophila suzukii* (Diptera: Drosophilidae) and *Zaprionus indianus*. J Econ Entomol 113:230–238. https://doi.org/10.1093/jee/toz297

Vieira JGA, Kruger AP, Scheuneumann T et al (2020) Some aspects of the biology of *Trichopria anastrephae* (Hymenoptera: Diapriidae), a resident parasitoid attacking *Drosophila suzukii* (Diptera: Drosophilidae) in Brazil. J Econ Entomol 113:81–87. https://doi.org/10.1093/jee/toz270

Wang XG, Messing RH (2004) The ectoparasitic pupal parasitoid, *Pachycrepoideus vindemmiae* (Hymenoptera: Pteromalidae), attacks other primary tephritid fruit fly parasitoids: host expansion and potential non-target impact. Biol Control 31:227–236. https://doi.org/10.1016/j.biocontrol.2004.04.019

Wang XG, Kaçar G, Biondi A et al (2016a) Foraging efficiency and outcomes of interactions of two pupal parasitoids attacking the invasive spotted wing drosophila. Biol Control 96:64–71. https://doi.org/10.1016/j.biocontrol.2016.02.004

Wang XG, Kaçar G, Biondi A et al (2016b) Life-history and host preference of *Trichopria drosophilae*, a pupal parasitoid of spotted wing drosophila. Biocontrol 61:387–397. https://doi.org/10.1007/s10526-016-9720-9

Wang XG, Nance AH, Jones JML et al (2018a) Aspects of the biology and reproductive strategy of two Asian larval parasitoids evaluated for classical biological control of *Drosophila suzukii*. Biol Control 121:58–65. https://doi.org/10.1016/j.biocontrol.2018.02.010

Wang XG, Serrato MA, Son Y et al (2018b) Thermal performance of two indigenous pupal parasitoids attacking the invasive *Drosophila suzukii* (Diptera: Drosophilidae). Environ Entomol 47:764–772. https://doi.org/10.1093/ee/nvy053

Wang XG, Hogg BN, Hougardy E et al (2019) Potential competitive outcomes among three solitary larval endoparasitoids as candidate agents for classical biological control of *Drosophila suzukii*. Biol Control 130:18–26

Wang XG, Biondi A, Daane KM (2020) Functional responses of three candidate Asian larval parasitoids evaluated for classical biological control of *Drosophila suzukii*. J Econ Entomol 113:73–80

Whitehouse TS, Sial AA, Schmidt JM (2018) Natural enemy abundance in southeastern blueberry agroecosystems: distance to edge and impact of management practices. Environ Entomol 47:32–38. https://doi.org/10.1093/ee/nvx188

Wolf S, Zeisler C, Sint D et al (2018) A simple and cost-effective molecular method to track predation on *Drosophila suzukii* in the field. J Pest Sci 91:927–935. https://doi.org/10.1007/s10340-017-0948-7

Wolf S, Baur H, Collatz J (2019) Life history of *Vrestovia fidenas*, a potential control agent of *Drosophila suzukii*. Biocontrol 64:263–275. https://doi.org/10.1007/s10526-019-09933-5

Wolf S, Boycheva-Woltering S, Romeis J et al (2020) *Trichopria drosophilae* parasitizes *Drosophila suzukii* in seven common non-crop fruits. J Pest Sci 93:627–638. https://doi.org/10.1007/s10340-019-01180-y

Wollmann J, Schlesener DCH, Ferreira MS et al (2016) Parasitoids of Drosophilidae with potential for parasitism on *Drosophila suzukii* in Brazil. Drosophila Inform Serv 99:38–42

Woltz JM, Lee JC (2017) Pupation behavior and larval and pupal biocontrol of *Drosophila suzukii* in the field. Biol Control 110:62–69. https://doi.org/10.1016/j.biocontrol.2017.04.007

Yi CD, Cai PM, Lin J et al (2020) Life history and host preference of *Trichopria drosophilae* from Southern China, one of the effective pupal parasitoids on the *Drosophila* species. Insects 11:103. https://doi.org/10.3390/insects11020103

Zengin E, Karaca I (2019) Dynamics of trapped adult populations of *Drosophila suzukii* Matsumura (Diptera: Drosophilidae) and its parasitoids in Uak Province, Turkey. Egypt J Biol Pest Control 29. https://doi.org/10.1186/s41938-019-0147-3

Zhu CJ, Li J, Wang H et al (2017) Demographic potential of the pupal parasitoid *Trichopria drosophilae* (Hymenoptera: Diapriidae) reared on *Drosophila suzukii* (Diptera: Drosophilidae). J Asia Pac Entomol 20:747–751. https://doi.org/10.1016/j.aspen.2017.04.008

Chapter 9
Sterile Insect Technique and Incompatible Insect Technique for the Integrated *Drosophila suzukii* Management

Fabiana Sassù, Katerina Nikolouli, Christian Stauffer, Kostas Bourtzis, and Carlos Cáceres

Abstract The spotted wing *Drosophila* (SWD) vinegar fly, *Drosophila suzukii*, has become a significant global pest of a wide variety of commercial soft fruits. The sterile insect technique (SIT) is a species-specific method of population control that has been successfully used for the suppression or local eradication of several economically important insect pests. Repetitive releases of mass-produced sterile insects in the target area lead progressively to the decline of the pest population since the mating of sterile males with wild females results in no offspring. As part of an area-wide integrated pest management programme, the SIT can also be combined with the incompatible insect technique (IIT). The combined SIT/IIT approach allows for the use of flies that are infected with *Wolbachia* strains that can induce cytoplasmic incompatibility, to be irradiated with lower doses compared to the ones required for SIT as a stand-alone method. Both the SIT and the SIT/IIT concepts are overruled by strengths and weaknesses when it comes to their application for *D. suzukii* management in confined locations. In this chapter, we are discussing the requirements and the challenges of SIT and IIT, and we review the progress achieved on these fields for *D. suzukii* so far.

F. Sassù · C. Stauffer
Department of Forest and Soil Sciences, Boku, University of Natural Resources and Life Sciences, Vienna, Austria
e-mail: christian.stauffer@boku.ac.at

K. Nikolouli (✉)
Department of Forest and Soil Sciences, Boku, University of Natural Resources and Life Sciences, Vienna, Austria

Insect Pest Control Laboratory, Joint FAO/IAEA Programme of Nuclear Techniques in Food and Agriculture, A-1400 Vienna, Austria
e-mail: k.nikolouli@iaea.org

K. Bourtzis · C. Cáceres
Insect Pest Control Laboratory, Joint FAO/IAEA Programme of Nuclear Techniques in Food and Agriculture, A-1400 Vienna, Austria
e-mail: k.bourtzis@iaea.org; C.E.Caceres-Barrios@iaea.org

F. R. M. Garcia (ed.), *Drosophila suzukii Management*,
https://doi.org/10.1007/978-3-030-62692-1_9

Keywords Pest management · Mass-rearing · Quality control · *Wolbachia*

9.1 The Sterile Insect Technique for *Drosophila suzukii* Management

The sterile insect technique (SIT) is a species-specific and environment-friendly method for the population control of agricultural, livestock and human disease vector insect pests. It relies on the large-scale rearing of individuals from the targeted pest species that are sterilized and released in a delimitated geographical area (Knipling 1955). The reproductive sterility is achieved by exposing the insects to ionizing radiation that causes germ cells atrophy, dominant lethal mutations in the sperm and complete ovary atrophy in females (Robinson 2005). Then, sterile insects are released in the field at overflooding numbers. Once in the field, the sterile males will inseminate the wild females, thus leading to zygotes that will die during early embryogenesis and reduction of the wild population (Knipling 1979).

The SIT should be combined with other control methods to combat insect pests (Enkerlin 2005; Vreysen et al. 2006). The theoretical basis of SIT fits perfectly to the context of the area-wide integrated pest management programmes (AW-IPM) which is defined as the management of the total pest population within a delimited area (Hendrichs et al. 2007). Depending on the AW-IPM strategy, the amount of and the frequency with which the sterile insects are released can prevent their establishment, and also engage the containment, suppression or even the eradication of the wild pest population from the selected area (Hendrichs et al. 2005).

The first application of the SIT as part of an AW-IPM programme was tested against the New World Screwworm, *Cochliomyia hominivorax* (Coquerel) (Diptera: Calliphoridae), which was eradicated in the 1966 in North America and in subsequent years in other Central America countries (Baumhover et al. 1955, 1959; Baumhover 2002; Coppedge et al. 1978; Vargas 1984). After these successful programmes, SIT was also adopted in other AW-IPM programmes to manage agricultural insect pests, such as several Lepidoptera (Marec and Vreysen 2019), tephritid flies (Fisher 1996; Hendrichs et al. 1983; Yosiaki et al. 2003) and disease vector species such as tsetse flies (Vreysen et al. 2000; De Beer et al. 2017) and mosquitoes to reduce the spread of vector-borne diseases (Bourtzis et al. 2016; Lees et al. 2015). This strategy is currently applied in 19 countries on more than 27 different species (Enkerlin et al. 2017; IDIDAS n.d.).

The settlement of an AW-IPM programme with an SIT component on a newly targeted insect pest is generally determined by economic and political considerations (Enkerlin et al. 2015), and its implementation strongly depends on technical factors (Green et al. 2007). Some of the most important baseline data required for a SIT programme are as follows (Hendrichs et al. 2007):

- Extensive studies on the target area for the presence, distribution and abundance of the native population;
- Selection of a delimitated geographical area to avoid possible re-entry or introduction of wild populations;
- Understanding of the mating system of the target species;
- Cost-benefit evaluation and programme goal definition;
- Coordination planning with the integration of other control approaches;
- Public relation campaigns for the general public;
- Long-term surveillance and monitoring of the programme.

In addition to these points, large research effort is dedicated to the establishment of functional SIT protocols: viable mass-rearing protocols, irradiation sterility dose, quality control assessments, storage, shipment and release procedures (Pereira et al. 2013). Those protocols are major components of the "SIT package", and they must be specifically developed and adapted for the target pest.

The spotted wing *Drosophila* (SWD), *Drosophila suzukii* (Matsumura 1931) (Diptera, Drosophilidae), is an invasive pest native to Asia (Kanzawa 1936). Since its first reports outside of Asia (Calabria et al. 2012; Hauser 2011), it has widely colonized the Americas and Europe where it has rapidly become a major pest of berries, and also grapes and stone fruits causing great economic losses (De Ros et al. 2015; Mazzi et al. 2017; Dos Santos et al. 2017). In the last decade, the urgent need for a solution to its invasiveness has elevated SWD into a case study for pest management research programmes (Atallah et al. 2014). The ongoing progress on alternative biological methods to manage this pest has brought promising results, but yet the infested areas are mostly subjected to chemical-dependent management strategies (Sial et al. 2019; Van Timmeren and Isaacs 2013). The demand for alternative environmentally sound and sustainable pest management has prompted interest in using the SIT as part of an AW-IPM approach to control the infestation of SWD populations in confined areas such as greenhouses (Krüger et al. 2018; Lanouette et al. 2017; Sassù et al. 2019b).

The SIT shows specific benefits when it comes to its application on SWD:

- The SIT is a species-specific control method; thus, it has no impact on non-target species.
- The major concern regarding SWD control is that its damage on crops is extremely close to the harvest time (Haviland and Beers 2012). The SIT can be safely applied as a pre-harvest activity and eliminate any risk on human and environmental health.
- The SIT can be easily integrated with other biological control strategies (Klassen 2005).
- In greenhouses, the SIT can be integrated without any impact on the ongoing biocontrol programmes against other pests, contrary to the insecticide application that will ruin all the biocontrol scheme in place.

The SIT utilizes ionizing radiation to prevent insect reproduction either partly or completely (Proverbs 1969). Irradiation sources such as Cobalt-60 (^{60}Co) or

Caesium-137 (^{137}Cs) are the most common methods to induce reproduction sterility in programmes that release sterile insects (Helinski et al. 2009; Mastrangelo and Walder 2011). The tolerance to irradiation is conditioned by several factors, e.g. radiation dose, insect life stage, age, sex, ambient conditions during exposure (hypoxia, anoxia or normoxia), type of radiation source and dose rate (Bakri et al. 2005; Yamada et al. 2019). Consequently, a dose–response curve must be assessed on each insect species in order to establish the dose that achieves the full sterility (Mastrangelo et al. 2010). However, the 100% sterility dose is not often applied in SIT programmes because male mating competitiveness can be decreased as the dose of radiation increases (Bloem et al. 1999; De Beer et al. 2017; Toledo et al. 2004). Therefore, the optimal irradiation dose should induce high sterility with minimal effects on the male quality (Collins et al. 2008).

The effects of radiation on reproduction sterility have been evaluated on SWD (Fig. 9.1). Gamma radiation doses ranging from 30 to 240 Gy were tested on SWD pupae by three different studies (Krüger et al. 2018; Lanouette et al. 2017; Sassù et al. 2019b). Although there were variations in intervals and irradiation doses tested by the different studies, the dose–response curve based on egg hatch showed comparable results. Dose of about 200 Gy produced near-complete sterility in males (>99%) and therefore it has been proposed as the optimal dose for SIT releases (Krüger et al. 2018; Sassù et al. 2019b). SWD females irradiated with a dose of 75 Gy showed complete sterility (Krüger et al. 2018; Lanouette et al. 2017; Sassù et al. 2019b) and ovarian atrophy (Krüger et al. 2018). Thus, females revealed more radiosensitivity than males, as observed in other insects (Bakri et al. 2005; Mastrangelo and Walder 2011), ensuring their complete sterility following the irradiation with any of the male potential dose. Usually, the most convenient stage for induction of reproductive sterilization through the use of ionizing radiation in insects is late pupae/pharate adults (Robinson 2005). The pupal eye colour changes

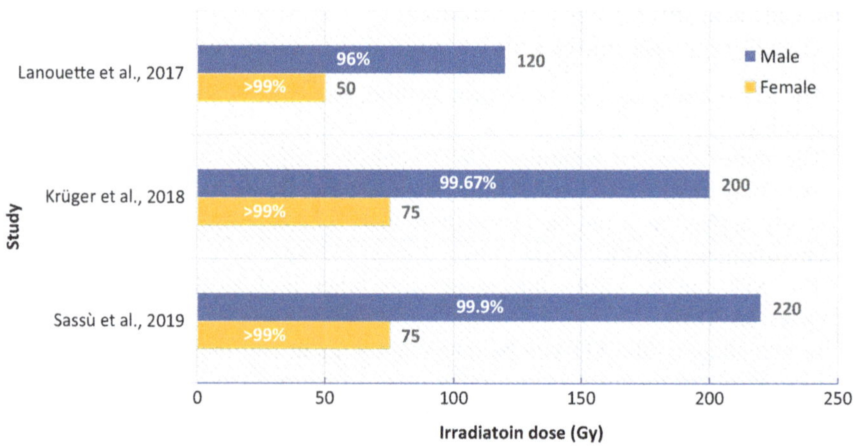

Fig. 9.1 Irradiation doses for *Drosophila suzukii* male and female sterility suggested by different studies. Percentages of egg hatch sterility are indicated in each bar

during the development, and this can be used in SIT programmes for synchronizing and timing the sterilization of pupae (Resilva and Pereira 2014; Ruhm and Calkins 1981). Due to the fact that the duration of the SWD pupal development (about 7–8 days at 23 ± 1 °C) is shorter compared to other tephritid fruit flies (Resilva and Pereira 2014), irradiation of SWD pupae should be as close as possible (≤24 h before) to their emergence time (Krüger et al. 2018, 2019; Lanouette et al. 2017; Nikolouli et al. 2020; Sassù et al. 2019b).

Radiological sensitivity of SWD to gamma irradiation has also been tested under low-oxygen atmospheric conditions (hypoxia) (Sassù et al. 2019b). Sterility of SWD treated under hypoxia required higher doses to obtain the same level of sterility compared with the pupae irradiated under ambient conditions thus confirming prior findings (Balock et al. 2015; Condon et al. 2017; Langley et al. 1974; Robinson 2006). Irradiation of pupae under hypoxia can be advantageous for a SIT application since inducing hypoxia right before irradiation will allow the irradiated pupae to be kept for additional time inside the transportation containers (e.g. plastic bags) until they arrive at the holding and emergence facility. An additional benefit of hypoxia is that cells irradiated in low oxygen conditions tend to produce less free radicals which induce significant somatic cell damages (Nestel et al. 2007; López-Martínez and Hahn 2012). Due to this benefit, irradiation under hypoxia is a routine practice in fruit flies SIT programmes (Nestel et al. 2007).

Despite the fundamental knowledge acquired on the SIT from its successful application on several tephritid species, caution should be exercised before developing a genetic control method with an SIT component for SWD. The high fecundity of SWD and its ability for recurrent establishments in non-confined crop areas might compromise the SIT feasibility. In addition, the absence of an adequate sexing system that imposes bisexual releases, the short generation time that requires intensive releases to avoid rapid population recovery and the limitations of the utilization of SIT for large field populations collectively signify the need for combining the SIT with additional control methods in greenhouses and other confined locations.

9.2 Combination of the Sterile and the Incompatible Insect Techniques (SIT/IIT) for *Drosophila suzukii* Management

The exploitation of insects' symbiotic microorganisms has resurged research interest towards the development of environmentally and economically benign approaches for insect pest management (Iturbe-Ormaetxe et al. 2011; Zabalou et al. 2004, 2009). *Wolbachia*, a widespread endosymbiont in arthropods and filarial nematodes, is a maternally inherited alphaproteobacterium (Hilgenboecker et al. 2008). The so-called sex parasite is mainly localized in the reproductive tissues of arthropods, and it can act as both a parasite and a mutualist (Fytrou et al. 2006; Kambris et al. 2010; Martinez et al. 2014; Cattel et al. 2016). *Wolbachia* is particularly known for its ability to induce reproductive distortions in its hosts (Hiroki et al. 2002;

Jiggins et al. 2001; Turelli and Hoffmann 1991; Vavre et al. 2002; Weeks and Breeuwer 2001; Zeh et al. 2005). Cytoplasmic incompatibility (CI) is one of these reproductive abnormalities used by *Wolbachia* to enhance the production of infected females and spread rapidly through populations (Stevens and Wade 1990; Turelli and Hoffmann 1991). CI is a sperm–egg incompatibility expressed when infected males mate with females with different *Wolbachia* infection status (non-infected or infected with different and incompatible strains) hence resulting in embryonic mortality. This phenotype enables the transfer of *Wolbachia* to the next generation and confers a selective advantage to infected female hosts (Toomey et al. 2013; Turelli 1994).

The CI induction by *Wolbachia* was early recognized as a means of manipulating the host reproductive system (Turelli and Hoffmann 1991) and included in the framework of the incompatible insect technique (IIT) (Zabalou et al. 2004). The IIT is based on immense releases of incompatible males in targeted areas that will mate with wild females leading to incompatible crosses and embryo mortality (Zabalou et al. 2004). Several semi-field and field trials employed the CI mechanism in an effort to control mosquito species via inundative releases of *Wolbachia*-infected male mosquitoes (Atyame et al. 2015; Crawford et al. 2020; Laven 1967; Mains et al. 2019; O'Connor et al. 2012). O'Connor et al. (2012) performed an open field study using *Wolbachia*-infected *Aedes polynesiensis* Marks (Diptera: Culicidae) male mosquitoes that were released in isolated islands for 30 weeks. The results showed that the released males were competitive mates under field conditions and thus the application of an IIT approach at a larger scale against additional disease vectors gained momentum. Despite the promising outcomes, the fact that the IIT claims for strict male release cannot be overlooked. Any accidental female release would result in the replacement of the targeted population by a population carrying the *Wolbachia* infection and the IIT success would be compromised (Dobson et al. 2002; Xi et al. 2005). Application of the IIT as a stand-alone control method is facing the utmost requirement of a solid and sufficient sex separation system (ideally in the form of a genetic sexing strain) to ensure male-only releases (Crawford et al. 2020). In the absence of a robust sex separation system, the IIT is self-restrained from any field application for SWD.

Coupling the SIT with the IIT has been considered as an alternative approach that can overcome the "strict-male release" barrier. The strategy profits from the increased sensitivity to radiation demonstrated by female insects compared to males in terms of sterility and therefore, the identification of a minimum radiation dose that sterilizes completely the females is facilitated (Bourtzis and Robinson 2006; Zhang et al. 2015a, b, 2016). In such a system, a *Wolbachia* strain that induces CI is irradiated with a low irradiation dose to ensure female sterility. As a result, bisexual releases can be applied since the *Wolbachia*-infected females are sterile, and the risk of population replacement is avoided. In addition, the released males are partially or completely sterile due to the low irradiation dose and able to confer sterility in the wild population through the CI mechanism (Lees et al. 2015). The mosquito vector *Aedes albopictus* (Skuse) (Diptera: Culicidae) has served as the core of the experimental quest towards the deployment of a combined SIT/IIT approach (Bourtzis et al. 2014, 2016; Lees et al. 2015; Zhang et al. 2015a, b, 2016; Zheng et al. 2019). In particular, a natural *Ae. albopictus* population originated from China and carrying a native double *Wolbachia* infection (*w*AlbA and *w*AlbB) was used for the establishment of a triple-infected line (*w*AlbA, *w*AlbB and *w*Pip). Males of the new strain were able to induce high levels of cytoplasmic incompatibility when mated with wild-type females (Zhang et al. 2015b). Zheng et al. (2019) provided a proof-of-concept study under field conditions, and the aftermath showed a near-elimination of a wild-type *Ae. albopictus* field population and prevention of its replacement by the released triple-infected strain. Substantial research towards the development of a combined SIT/IIT protocol is also carried out for *Aedes aegypti* (L.) (Diptera: Culicidae) mosquitoes (Carvalho et al. 2020; Kittayapong et al. 2018, 2019). In the frame of a pilot population suppression trial in Thailand, Kittayapong et al. (2018) reported that both male and female *Wolbachia*-infected *Ae. aegypti* were completely sterile at 70 Gy, while the male survival and longevity were not affected. The pilot trial occurred in a semi-rural setting and lasted for 6 months. Reduction of wild *Ae. aegypti* females and low egg hatch rates in the treated area evidenced the effectiveness of the combined SIT/IIT approach in suppressing natural populations of *Ae. aegypti* (Kittayapong et al. 2019).

Apart from mosquito species, the combination of the SIT with the IIT is an appealing approach that has also been suggested as an alternate route for the biological control of SWD (Nikolouli et al. 2020). However, the abundance of *Wolbachia* strains in insects does not necessarily mean that natural *Wolbachia* infections are present in all species or, even if present, that they induce CI in their hosts. Trans-infections using embryo microinjections have assisted in constructing novel and stable *Wolbachia*-infected hosts capable of inducing CI (Hughes and Rasgon 2014; Zabalou et al. 2004). An example is the trans-infection of the Mediterranean fruit fly, *Ceratitis capitata* (Wied.) (Diptera: Tephritidae), a naturally uninfected species, with two *Wolbachia* strains from *Rhagoletis cerasi* (L.) (Diptera: Tephritidae) that led to the expression of high CI levels (Zabalou et al. 2004). In their study, Cattel et al. (2018) applied the trans-infection approach in a SWD population that was naturally infected by *w*Suz, a *Wolbachia* strain that does not exhibit in its natural host. Two *Wolbachia* strains (*w*Ha and *w*Tei) acquired by other *Drosophila* species were used to trans-infect SWD and develop stable lines. Following the establishment of the lines and the calculation of the CI rates, a combined SIT/IIT protocol was developed (Nikolouli et al. 2020). A low range of irradiation doses (45–90 Gy) was tested, and results indicated complete sterility both for *w*Ha and *w*Tei females at all doses. The quality parameters tested at 45 Gy did not suggest any negative effect which encouraged the continuation of research to develop a combined SIT/IIT strategy (Nikolouli et al. 2020).

The low irradiation dose required to achieve complete female sterility in case that a combined SIT/IIT protocol is used is an asset for this approach as it is expected that any effects on the male mating competitiveness will be minimum or even zero (Carvalho et al. 2020; Zhang et al. 2015a, b, 2016; Nikolouli et al. 2020). On the other hand, the host nuclear background as well as the different *Wolbachia* strains are crucial factors for the CI expression. The *Wolbachia* phenotype and density may vary across different genomic backgrounds of the same host species which in turn might also affect the fitness and sexual behaviour of the host (Dean 2006; Mouton et al. 2007). This type of interactions might lead to negative effects on the host sexual competitiveness and fitness traits, and therefore the importance of rearing insects that have the same genomic background with the one in the target field population has long been recognized (Bourtzis et al. 2014; Carvalho et al. 2020; Fraser et al. 2017).

9.3 Requirements for the Development of the SIT Package for *Drosophila suzukii*

Before the application of any SIT-based programme against SWD, a series of standard procedures needs to be developed and validated which will allow production of high-quality insects (FAO/IAEA/USDA 2019). These procedures should reflect the SWD biology and include routine and periodic quality control tests to prevent any adverse effects on the field programme. According to the flowchart presented in the "Product quality control for sterile mass-reared and released tephritid flies" of 2019 (FAO/IAEA/USDA 2019), the following requirements will be discussed here: mass-rearing, sterilization, quality controls and insect supply.

9.3.1 Mass-Rearing

The SIT relies on the release of large numbers of sterile insects that are reared on an industrial scale and systematically produced in specifically designed facilities. Mass-rearing facilities stand out because of their large-scale capacity that can reach several billion of insects per week (Hendrichs et al. 2002). To achieve that, mass-rearing facilities have controlled environments, processes are automatized as much as possible and the technical choices are based on optimizing the production and balancing cost/quality (Cortes Ortiz et al. 2016; Parker 2005). The establishment of mass-rearing protocols for a new target species depends at first on the possibility of adapting the species to artificial conditions. Once the species has been adapted, a second important step is to develop and implement economically viable mass-rearing methods to scale up the production of insects that must be capable of competing in the field with fertile males for mating with fertile females. To achieve large-scale production of SWD, one of the first challenges to overcome is the adaptation of the colony to the artificially controlled environment. Previous studies on fruit flies have shown that insect populations can adapt to the mass-rearing environment within some generations since their initial introduction in the laboratory (Gilchrist et al. 2012; Gilligan and Frankham 2003; Parreño et al. 2014; Zygouridis et al. 2013). During the laboratory adaptation, the genotype–environment interactions could lead to fitness loss and significant changes of the phenotype. These changes are attributed to bottleneck effects and selective pressures posed by artificial rearing and can alter several life history traits including fertility, fecundity and lifespan, and also stress resistance and developmental time (Gilchrist et al. 2012; Hoffmann et al. 2001; Raphael et al. 2014). It is evident that any of these changes can be a potential barrier to the efficient implementation of an SIT-based programme. However, the implementation of routine quality control tests and protocols allows monitoring the performance and quality of the released insects (FAO/IAEA/USDA 2019).

Symbiotic communities can also be affected by the colonization process although their structure is predominantly shaped by the new artificial environment provided for feeding and oviposition (Augustinos et al. 2019). Diet can be a driving factor in shaping the bacterial communities of *Drosophila* species in the laboratory environment since it can define which symbionts can thrive in the new artificial environment (Staubach et al. 2013). Changes in the symbiotic profile of the digestive tract have been proved to play a crucial role in the biology, ecology and developmental processes of insect hosts including, but not limited to, sex ratio, nutrition and mating behaviour (Augustinos et al. 2015; Koskinioti et al. 2019; Koukou et al. 2006; Miller et al. 2010; Sharon et al. 2010). These alterations should be assessed with respect to rearing efficiency and insect quality parameters to continuously verify the production of high-quality sterile insects for SIT applications at a reduced cost.

An efficient egg collection system and an artificial larval rearing diet that will yield a sufficient number of pupae and adults constitute fundamental components for the industrial-scale SWD production (Ahmad et al. 2016; Franz 2005). In addition, monitoring the quality of mass-reared insects should be a continuous process to ensure that the success of any SIT-based technique is not jeopardized.

9.3.1.1 Cage Design and Oviposition System

Appropriate adult holding cages are an essential part of mass-rearing technologies. The biological traits may be influenced by the size and structure of the cage, i.e. number of adults hold, likelihood of mating, feeding and survivorship with overall consequences on quality and quantity of colony's production (Liedo et al. 2007; Aluja et al. 2009; Orozco-Dávila et al. 2014). Additionally, rearing cages should have a functional design to facilitate the daily activities and restrain operative costs (Schwarz et al. 1985; Vargas 1984).

In the setup of a mass-rearing cage, careful attention is usually paid for the development of the section served as oviposition site for females (Prokopy and Boller 1970; Szentesi et al. 1979). Female oviposition behaviour must be adapted to the artificial oviposition devices allowing easy egg collection (Boller 1972).

Small-scale rearing of SWD laboratory strains lay eggs directly in the adult diet in which larvae feed, develop and pupate. This system allows minimal maintenance efforts because the whole cycle is completed in the same setting, but it does not provide an estimation of the colony's egg production. Estimation of egg production requires first and foremost an artificial oviposition system that will be adopted by the females thus allowing for high egg productivity. An artificial egg oviposition system was recently developed and evaluated for SWD (Sassù et al. 2019a). The so-called wax panel suggested by Sassù et al. (2019a) is based on the similar principle followed for the olive fly, *Bactrocera oleae* (Rossi) (Diptera: Tephritidae) and *C. capitata* (Ahmad et al. 2016; Vargas 1984). This method permits the easy egg collection outside of the cage avoiding diet contamination and enabling the estimate of total egg production. Therefore, this method could be recommended to initiate mass-rearing of SWD for small pilot projects. However, factors like light, resting places, texture of larval diet, height and colour of the ovipositional site have been shown to influence female oviposition preference of mass-reared insects which consequently seeks for more studies on SWD (Ekesi and Mohamed 2011; McLay et al. 2017; Tsitsipis 1977; Zhang et al. 2010).

9.3.1.2 Larval Diet

Like the majority of Drosophilidae, SWD colonies are easily grown on standard laboratory fly food mainly consisting of yeast and sugar where both adults and larvae can complete their development (Jaramillo et al. 2015). On the contrary, in mass-rearing facilities of tephritid fruit flies and other reared insect species, adult and larval diets usually differ. Adult diet is constantly available or regularly provided to adults inside the rearing cages while eggs are removed from the cages and separately placed in a larval diet specifically prepared for the appropriate development of immature stages. Diets should be accepted by the insects, secure adequately their nutritional requirements to produce fit insects, guarantee consistency in the purchase and allow for acceptable costs. Deficit in one of these factors may lead to a decline in colony production and/or quality (Jácome et al. 1999; Liedo et al. 2007, 2012).

Fresh fruits are often used as an initial substrate to adapt a fruit fly species to artificial rearing and increase colony size (Cooley et al. 1986). Slices of banana were used to produce a large number of SWD adults for the rearing of their parasitoids (Gonzalez-Cabrera et al. 2018b). In most countries, bananas are available all year-round, and they are an easy-to-adapt substrate for the rearing of SWD. However, in large-scale rearing, the use of fresh fruits as an oviposition substrate is not sustainable because it increases the maintenance labour and the risk of possible microbial contaminations (Parker 2005; Saldanha and Silva 1999). Moreover, bananas are often naturally infested by *Drosophila melanogaster* (Meigen) (Diptera: Drosophilidae) that can be easily established in a SWD colony and cause its "crash" (Dancau et al. 2016). Adult SWD and *D. melanogaster* females were shown to compete for accessing a common oviposition resource, and this competition had a strong impact on SWD progeny numbers. *D. melanogaster* presence significantly reduced the SWD offspring emergence in a laboratory setting, a phenomenon that could be attributed to larval competition for resources or egg laying disruption of SWD (Dancau et al. 2016). Therefore, the use of fresh fruits in mass-rearing facilities is discouraged since it involves the apparent risk of *D. melanogaster* infestation that will completely outcompete SWD.

The role that the relative concentration of proteins and carbohydrates (P:C) plays on different insect life stages has essential implications on fitness-related traits (Behmer et al. 2003; Roeder and Behmer 2014). The contribution of P:C on SWD development has been recently investigated (Bellutti et al. 2018; Hardin et al. 2015; Schlesener et al. 2018; Rendón et al. 2019; Young et al. 2018). Independently of the medium, all authors indicate that SWD larval performance is greater in high protein environments. Understanding the nutritional requirements of the SWD immature stages is a key factor for its larval diet formulation. The viability of a cheap and suitable artificial diet is a major concern for mass-rearing (Cáceres et al. 2014; Rajabpour et al. 2018). At present, only few studies have compared the composition of artificial diets for the mass-rearing of SWD by considering both production efficiency and cost (Aceituno-Medina et al. 2020; Gonzalez-Cabrera et al. 2018a). Results from these studies showed that diets composed mainly of brewer's yeast, sugar, and either coconut fibres or wheat germ offered higher yield and nutritional balance for the development of SWD. Brewer's yeast and sugar proved to be the principal components for the correct development of SWD as demonstrated by previous studies (Bellutti et al. 2018; Lewis and Hamby 2019), while coconut fibres and wheat germ were successfully used as bulking agents to reduce the economic cost and provide a moist retention texture of the diet that facilitates larval movement and bioavailability of nutrients (Raulston and Shaver 1970; Reinecke 1985; Vera et al. 2014). The texture of the larval diet is important for the larval movement and also for enabling the separation of pupae from the medium. In some species, e.g. Mediterranean fruit fly and melon fly, third instar larvae are self-removed from their diet by jumping out of the trays to pupate in moist sterile sand (Balock et al. 2015; Baumhover 1966; Suenega et al. 1992). This trait speeds up the process since mass-reared pupae can be easily collected and irradiated without further handling. In other mass-reared species such as the West Indian fruit fly, *Anastrepha obliqua* (Macquart) (Diptera: Tephritidae), and the black soldier fly, *Hermetia illucens* (L.) (Diptera: Stratyomidae), larvae usually pupate in the diet requiring special procedures to remove or wash them out from it (Bosch et al. 2019; Orozco-Dávila et al. 2014).

A field study on SWD reported that larvae might either pupate inside the fruit or crawl out to pupate on the ground (Woltz and Lee 2017). In the artificial diet, SWD pupae are separated from the diet residues by placing them in a fine mesh plastic colander and rinsing them under running tap water (Sassù F, personal communication). Other systems that are currently under investigation use a centrifuge to separate pupae and rearing diet since pupae tend to be less dense than the larval diet components (Taret G, personal communication). Alternatively, liquid or gel larval diets that could simplify pupae collection could be also tested (Anato et al. 2017; Chang et al. 2006; Ekesi and Mohamed 2011; Mainali et al. 2019).

9.3.1.3 Sex Separation System

The development of efficient sexing methods that allow for strict male releases has fuelled several scientific studies in fruit flies and mosquitoes SIT programmes (Franz 2005; Rendón et al. 2004; Robinson 2002; Rössler 1979). Sex sorting for pest control programmes can be achieved through various techniques, like phenotypic sorting based on sexual dimorphism at the pupal or adult stage, sex ratio distortion, and sexing systems based on classical genetic or molecular methods (Franz 2005; Robinson 2002; Gong et al. 2005; Kandul et al. 2020).

The development of the Mediterranean fruit fly genetic sexing strains that addressed the requirement of male-only releases marked an era that revolutionized the SIT application programmes (Franz 2005; Rendón et al. 2009). Although bisexual releases were initially in place to combat insect pests, it soon became evident that male-only releases would accelerate the level of induced sterility and reduce the operational costs (Rendón et al. 2000). The Mediterranean fruit fly genetic sexing strain was developed using classical genetic approaches and resulted in a strain that has two selectable markers (pupal colour and a conditionally lethal mutation), and both of these markers are linked to the male sex, i.e. the wild-type alleles are linked to the Y chromosome (Franz 2005). However, the development of novel genetic sexing strains for pest insects through classical genetics requires the identification of suitable genetic markers that could be used for the construction of a genetic sexing mechanism. The whole process is time-consuming as the induction of mutations through chemicals or irradiation is random, requires extensive knowledge on the genetics and cytogenetics of the species and large-scale screening for the isolation of suitable marker(s) (Fisher and Cáceres 2000). To the authors' knowledge, no markers for the construction of genetic sexing strains by classical genetics is currently available for the SWD.

The CRISPR/Cas9 system is a genome engineering method used for the targeted mutagenesis of specific genes (Taning et al. 2017). Utilizing this system, gene-specific changes can be introduced in the genome leading to mutant phenotypes of interest (Kalajdzic and Schetelig 2017). Both classical genetic sexing strains and genetic sexing strains developed by modern molecular biology-based approaches require extensive testing before any long-term application in open field SIT programmes, while in addition, transgenic strains require regulatory approval (Donovan 2009; Wozniak 2007).

9.3.2 *Quality Control*

The selection of optimal protocols is primarily based on the consequences generated on adult quality. Natural traits such as body size, survival, pupal weight, adult emergence, longevity, flight ability, fecundity, fertility and mating ability can be altered by the SIT and IIT processes (Calkins and Parker 2005; Ekesi and Mohamed 2011; Kyritsis et al. 2019; Sarakatsanou et al. 2011). Manuals for product control procedures were developed to evaluate the standard quality of mass-reared fruit flies (FAO/IAEA/USDA 2019). Following the same procedures used for fruit fly species, several biological attributes of SWD were tested in relation to irradiation exposure. Emergence, longevity and sex ratio of SWD flies were not affected by exposure to radiation in any of the studies available to date (Gutierrez-Palomares et al. 2019; Krüger et al. 2018; Lanouette et al. 2017; Nikolouli et al. 2020; Sassù et al. 2019b) nor was any increase in the rate of deformities recorded (Lanouette et al. 2017). Likewise, an irradiation dose of 200 Gy did not negatively affect the flight ability of SWD adults (Krüger et al. 2018; Gutierrez-Palomares et al. 2019; Sassù et al. 2019b). However, it is important to mention that a specific device for measuring the flight ability of SWD has not yet been developed, and that the current method used for tephritid flies might not be appropriate (Collins and Taylor 2010). The mating ability is of great importance for preserving the success of the sterile males (Rull et al. 2005). Without effective matings, the SIT-based programmes cannot be implemented as a pest control management method (Calkins and Ashley 1989; Gallardo-Ortiz et al. 2018a; Hendrichs et al. 2002).

Irradiation, mass-rearing and insect supply procedures can alter natural behaviours and the mating propensity, compatibility and competitiveness of the adult males (Calkins and Parker 2005). The ability of the reared and sterilized adult SWD to mate has been the objective of a recent study (Krüger et al. 2019). Krüger and colleagues observed that sterile males had shorter copula duration, but similar likelihood to mate, re-mate and time to initiate the mating (latency period) than fertile males. Performance parameters under mass-rearing such as pupae recovery, weight of pupae, emergence, sex ratio and others have also been applied as parameters to assess the profile of SWD reared in different artificial diets (Aceituno-Medina et al. 2020; Gonzalez-Cabrera et al. 2018a; Schlesener et al. 2018).

At present, there has been a lack of mating tests using natural or semi-natural conditions and/or using wild SWD males obtained from the field population. The latter will be required to ensure that the competitiveness of the reared sterile males is as close as possible to that of the wild males present in the targeted area (Gallardo-Ortiz et al. 2018b; Mudavanhu et al. 2016; Rempoulakis et al. 2016; Virginio et al. 2017). Furthermore, an estimate of the sperm transferred and related mechanisms of sperm competitiveness of SWD sterile males have not yet been explored (Esfandi et al. 2019; Pérez-Staples et al. 2013; Seo et al. 1990; Sirot et al. 2011).

9.3.3 *Insect Supply*

Packaging, shipping, handling and release protocols are main components of the SIT package that need to be correctly addressed to guarantee insect quality (Cáceres et al. 2014; Culbert et al. 2017) and operational safety (FAO/IAEA/USDA 2019; Rull et al. 2012). However, these components are still under development for SWD. Since the beginning of its invasion, a significant proportion of research has been concentrated on the ability of SWD to adapt in cold conditions by overwintering in a dormant reproductive period (Grassi et al. 2018; Enriquez and Colinet 2017; Enriquez et al. 2018b; Rossi-Stacconi et al. 2016; Thistlewood et al. 2018). Defining the temperatures that SWD can tolerate and understanding the biological and physiological reactions can nevertheless be an important starting point for most procedures that may require a chilling phase, i.e. handling, package and shipment (Mutika et al. 2019). Storage of pupae at low temperatures delays the emergence thus making it possible to ship the sterile insects in infested areas far from the production facility and even in long distance shipments across nations as in the case of Mediterranean fruit fly, *Ceratitis capitata* (Wied.); Mexican fruit fly, *Anastrepha ludens* (Loew); Codling moth, *Cydia pomonella* (L.) and tsetse, *Glossina palpalis* spp. campaigns (Blomefield et al. 2011; Diallo et al. 2019; Enkerlin et al. 2017; Seck et al. 2015). Chilling fruit fly pupae is also a common practice to immobilize adults preventing them from moving and damaging during transport, loading and releasing (FAO/IAEA/USDA 2019).

The extensive literature published on SWD thermal requirements, in particular on the low-temperature thresholds, may be of great use to further studies on SWD pre- and post-irradiation protocols (Enriquez and Colinet 2017; Enriquez et al. 2018a, b; Hamby et al. 2016; Sánchez-Ramos et al. 2019). Likewise, understanding the SWD diapause can improve the management of mass-rearing by a more efficient colony storage reducing the labour and costs (Rossi-Stacconi et al. 2016; Zhai et al. 2016; Leopold 2007).

As mentioned before, low-oxygen atmospheric conditions, i.e. hypoxia and anoxia, are also known to prevent insect's development (Hoback and Stanley 2001). In SIT programmes, packaging, handling and shipping are often realized under hypoxia, low temperatures or combination of both (Benelli et al. 2019; Diallo et al. 2019; FAO/IAEA/USDA 2019). Reactions to low-oxygen atmospheres have been well-studied on insect pests (Basson and Terblanche 2010; Calkins and Parker 2005; Hallman and Hellmich 2010) and *D. melanogaster* (Burggren et al. 2017; Colinet and Renault 2012; Frazier et al. 2001; Haddad et al. 1997; Zhao and Haddad 2011). Although there have been some studies that have used hypoxia for irradiation protocols on SWD, the effects of low-oxygen atmospheric conditions on this pest are poorly known (Sassù et al. 2019b; Vacas et al. 2019). Determining the appropriate operational procedures of SWD packaging, shipping, handling and releasing will also depend on the level of production, shipping distance, release methodology and availability of resources.

9.4 Conclusions

The rapid spread of SWD across the world and the considerable yield and economic losses it has caused call for the development of a control strategy to mitigate its effects. Studies on the insect biology revealed exceptional biological traits including the short generation time, the high fertility rate of the species and its ability to recurrently infest non-confined areas. An AW-IPM programme would be the most competent and economically viable solution to combat SWD. We discussed here the development of SIT-based techniques as part of an AW-IPM approach. The SIT has an acknowledged record of success as part of eradication and suppression programmes of several insect pests, and therefore its development for SWD is expected to contribute significantly in suppressing the populations in confined areas. The adequate experience gained on SIT applications from other pests and the recent developments on the irradiation and mass-rearing of SWD have paved the way towards this goal (Krüger et al. 2018, 2019; Lanouette et al. 2017; Nikolouli et al. 2020; Sassù et al. 2019a, b). The release of irradiated *Wolbachia*-infected males that has been recently applied on *Ae. albopictus* and *Ae. aegypti* and the encouraging field results presented by these studies (Zheng et al. 2019), as well as the combined SIT/IIT protocol developed from the ground up for SWD (Nikolouli et al. 2020) have put the cornerstone for the exploitation of *Wolbachia* as part of a SIT application. In such a case, significantly reduced levels of radiation are required, and the radiation-induced sterility complements the one induced by *Wolbachia* infection. Nonetheless, before any small- or large-scale application of SIT or SIT/IIT, there are various challenges to be addressed that mainly concern the quality control protocols, including the male mating competitiveness in the field. To this aim, future research should be pursued on further improving the current protocols and also acquiring essential knowledge under semi-field conditions.

References

Aceituno-Medina M, Ordoñez A, Carrasco M et al (2020) Mass rearing, quality parameters, and bioconversion in *Drosophila suzukii* (Diptera: Drosophilidae) for sterile insect technique purposes. J Econ Entomol 113:1097–1104. https://doi.org/10.1093/jee/toaa022

Ahmad S, ul Haq I, Rempoulakis P et al (2016) Artificial rearing of the olive fruit fly *Bactrocera oleae* (Rossi) (Diptera: Tephritidae) for use in the sterile insect technique: improvements of the egg collection system. Int J Ind Entomol 33:15–23. https://doi.org/10.7852/ijie.2016.33.1.15

Aluja M, Sivinski J, Ovruski S et al (2009) Colonization and domestication of seven species of native new world hymenopterous larval-prepupal and pupal fruit fly (Diptera: Tephritidae) parasitoids. Biocontrol Sci Technol 19:49–79. https://doi.org/10.1080/09583150802377373

Anato FM, Bokonon-Ganta AH, Gnanvossou D et al (2017) Assessment of a liquid larval diet for rearing *Dacus* species and *Bactrocera dorsalis* (Diptera: Tephritidae). J Appl Entomol 141:860–865. https://doi.org/10.1111/jen.12419

Atallah J, Teixeira L, Salazar R et al (2014) The making of a pest: the evolution of a fruit-penetrating ovipositor in *Drosophila suzukii* and related species. Proc R Soc B Biol Sci 281:20132840. https://doi.org/10.1098/rspb.2013.2840

Atyame CM, Cattel J, Lebon C et al (2015) *Wolbachia*-based population control strategy targeting *Culex quinquefasciatus* mosquitoes proves efficient under semi-field conditions. PLoS One 10(3):e0119288. https://doi.org/10.1371/journal.pone.0119288

Augustinos AA, Kyritsis GA, Papadopoulos NT et al (2015) Exploitation of the medfly gut microbiota for the enhancement of sterile insect technique: use of *Enterobacter sp.* in larval diet-based probiotic applications. PLoS One 10:1–17. https://doi.org/10.1371/journal.pone.0136459

Augustinos AA, Tsiamis G, Cáceres C et al (2019) Taxonomy, diet, and developmental stage contribute to the structuring of gut-associated bacterial communities in tephritid pest species. Front Microbiol 10:2004. https://doi.org/10.3389/fmicb.2019.02004

Bakri A, Mehta K, Lance DR (2005) Sterilizing insects with ionizing radiation. In: Dyck VA, Hendrichs J, Robinson AS (eds) Sterile insect technique: principles and practice in area-wide integrated pest management. Springer, Dordrecht, pp 233–269

Balock JW, Burditt AK, Christenson LD (2015) Effects of gamma radiation on various stages of three fruit fly species. J Econ Entomol 56:42–46. https://doi.org/10.1093/jee/56.1.42

Basson CH, Terblanche JS (2010) Metabolic responses of *Glossina pallidipes* (Diptera: Glossinidae) puparia exposed to oxygen and temperature variation: implications for population dynamics and subterranean life. J Insect Physiol 56:1789–1797. https://doi.org/10.1016/j.jinsphys.2010.07.010

Baumhover AH (1966) Eradication of the screwworm fly. JAMA 196:240. https://doi.org/10.1001/jama.1966.03100160090027

Baumhover AH (2002) A personal account of developing the sterile insect technique to eradicate the screwworm from Curacao, Florida and the southeastern United States. Florida Entomol 85:666–673. https://doi.org/10.1653/0015-4040(2002)085[0666:apaodt]2.0.co;2

Baumhover AH, Graham AJ, Bitter BA et al (1955) Screw-worm control through release of sterilized flies. J Econ Entomol 48:462–466. https://doi.org/10.1093/jee/48.4.462

Baumhover AH, Husman CN, Skipper CC, New WD (1959) Field observations on the effects of releasing sterile screw-worms in Florida. J Econ Entomol 52:1202–1206. https://doi.org/10.1093/jee/52.6.1202

Behmer ST, Cox E, Raubenheimer D, Simpson SJ (2003) Food distance and its effect on nutrient balancing in a mobile insect herbivore. Anim Behav 66:665–675. https://doi.org/10.1006/anbe.2003.2243

Bellutti N, Gallmetzer A, Innerebner G et al (2018) Dietary yeast affects preference and performance in *Drosophila suzukii*. J Pest Sci (2004) 91:651–660. https://doi.org/10.1007/s10340-017-0932-2

Benelli M, Ponton F, Lallu U et al (2019) Cool storage of Queensland fruit fly pupae for improved management of mass production schedules. Pest Manag Sci 75:3184–3192. https://doi.org/10.1002/ps.5436

Bloem S, Bloem KA, Carpenter JE, Calkins CO (1999) Inherited sterility in codling moth (Lepidoptera: Tortricidae): effect of substerilizing doses of radiation on field competitiveness. Environ Entomol 28:669–674. https://doi.org/10.1093/ee/28.4.669

Blomefield T, Blomefield T, Carpenter JE, Vreysen MJB (2011) Quality of mass-reared codling moth (Lepidoptera: Tortricidae) after long-distance transportation: 1. Logistics of shipping procedures and quality parameters as measured in the laboratory. J Econ Entomol 104:814–822. https://doi.org/10.1603/EC10238

Boller E (1972) Behavioral aspects of mass-rearing of insects. Entomophaga 17:9–25. https://doi.org/10.1007/BF02371070

Bosch G, Oonincx DGAB, Jordan HR et al (2019) Standardisation of quantitative resource conversion studies with black soldier fly larvae. J Insects Food Feed 6:1–16. https://doi.org/10.3920/jiff2019.0004

Bourtzis K, Robinson AS (2006) Insect pest control using *Wolbachia* and/or radiation. In: Bourtzis K, Miller TA (eds) Insect symbiosis, vol 2. CRC Press, Boca Raton, pp 225–246

Bourtzis K, Dobson SL, Xi Z et al (2014) Harnessing mosquito-*Wolbachia* symbiosis for vector and disease control. Acta Trop 132:S150–S163. https://doi.org/10.1016/j.actatropica.2013.11.004

Bourtzis K, Lees RS, Hendrichs J, Vreysen MJB (2016) More than one rabbit out of the hat: radiation, transgenic and symbiont-based approaches for sustainable management of mosquito and tsetse fly populations. Acta Trop 157:115–130. https://doi.org/10.1016/j.actatropica.2016.01.009

Burggren W, Souder BM, Ho DH (2017) Metabolic rate and hypoxia tolerance are affected by group interactions and sex in the fruit fly (*Drosophila melanogaster*): new data and a literature survey. Biol Open 6:471–480. https://doi.org/10.1242/bio.023994

Cáceres C, Hendrichs J, Vreysen MJB (2014) Development and improvement of rearing techniques for fruit flies (Diptera: Tephritidae) of economic importance. Int J Trop Insect Sci 34:S1–S12. https://doi.org/10.1017/S1742758414000034

Calabria G, Máca J, Bächli G et al (2012) First records of the potential pest species *Drosophila suzukii* (Diptera: Drosophilidae) in Europe. J Appl Entomol 136:139–147. https://doi.org/10.1111/j.1439-0418.2010.01583.x

Calkins CO, Ashley TR (1989) The impact of poor quality of mass-reared Mediterranean fruit flies on the sterile insect technique used for eradication. J Appl Entomol 108:401–408. https://doi.org/10.1111/j.1439-0418.1989.tb00474.x

Calkins CO, Parker AG (2005) Sterile insect quality. In: Dyck VA, Hendrichs J, Robinson AS (eds) Sterile insect technique: principles and practice in area-wide integrated pest management. Springer, Dordrecht, pp 269–296

Carvalho DO, Torres-Monzon JA, Koskinioti P et al (2020) *Aedes aegypti* lines for combined sterile insect technique and incompatible insect technique applications: the importance of host genomic background. Entomol Exp Appl 168:560–572. https://doi.org/10.1111/eea.12892

Cattel J, Martinez J, Jiggins F et al (2016) *Wolbachia*-mediated protection against viruses in the invasive pest *Drosophila suzukii*. Insect Mol Biol 25(5):595–603. https://doi.org/10.1111/imb.12245

Cattel J, Nikolouli K, Andrieux T et al (2018) Back and forth *Wolbachia* transfers reveal efficient strains to control spotted wing *Drosophila* populations. J Appl Ecol 55:2408–2418. https://doi.org/10.1111/1365-2664.13101

Chang CL, Caceres C, Jang EB (2006) A novel liquid larval diet and its rearing system for Melon fly, *Bactrocera cucurbitae* (Diptera: Tephritidae). Ann Entomol Soc Am 97:524–528. https://doi.org/10.1603/0013-8746(2004)097[0524:anllda]2.0.co;2

Colinet H, Renault D (2012) Metabolic effects of CO2 anaesthesia in *Drosophila melanogaster*. Biol Lett 8:1050–1054. https://doi.org/10.1098/rsbl.2012.0601

Collins SR, Taylor PW (2010) Flight ability procedures for mass-reared Queensland fruit flies, *Bactrocera tryoni*: an assessment of some variations. Entomol Exp Appl 136:308–311. https://doi.org/10.1111/j.1570-7458.2010.01030.x

Collins SR, Weldon CW, Banos C, Taylor PW (2008) Effects of irradiation dose rate on quality and sterility of Queensland fruit flies, *Bactrocera tryoni* (Froggatt). J Appl Entomol 132:398–405. https://doi.org/10.1111/j.1439-0418.2008.01284.x

Condon CH, White S, Meagher RL et al (2017) Effects of low-oxygen environments on the radiation tolerance of the Cabbage looper moth (Lepidoptera: Noctuidae). J Econ Entomol 110:80–86. https://doi.org/10.1093/jee/tow273

Cooley SS, Prokopy RJ, McDonald PT, Wong TTY (1986) Learning in oviposition site selection by *Ceratitis capitata* flies. Entomol Exp Appl 40:47–51. https://doi.org/10.1111/j.1570-7458.1986.tb02154.x

Coppedge JR, Goodenough JL, Broce AB et al (1978) Evaluation of the screwworm adult suppression system (SWASS) on the island of Curacao. J Econ Entomol 71:579–584. https://doi.org/10.1093/jee/71.4.579

Cortes Ortiz JA, Ruiz AT, Morales-Ramos JA et al (2016) Insect mass production technologies. In: Insects as sustainable food ingredients. Elsevier, San Diego, pp 153–201

Crawford JE, Clarke DW, Criswell V et al (2020) Efficient production of male *Wolbachia*-infected *Aedes aegypti* mosquitoes enables large-scale suppression of wild populations. Nat Biotechnol 38:482–492. https://doi.org/10.1038/s41587-020-0471-x

Culbert NJ, Lees RS, Vreysen MJB et al (2017) Optimised conditions for handling and transport of male *Anopheles arabiensis*: effects of low temperature, compaction, and ventilation on male quality. Entomol Exp Appl 164:276–283. https://doi.org/10.1111/eea.12610

Dancau T, Stemberger TLM, Clarke P, Gillespie DR (2016) Can competition be superior to parasitism for biological control? The case of spotted wing *Drosophila* (*Drosophila suzukii*), *Drosophila melanogaster* and *Pachycrepoideus vindemmiae*. Biocontrol Sci Technol 27:3–16. https://doi.org/10.1080/09583157.2016.1241982

De Beer CJ, Moyaba P, Boikanyo SNB et al (2017) Evaluation of radiation sensitivity and mating performance of *Glossina brevipalpis* males. PLoS Negl Trop Dis 11:e0005473. https://doi.org/10.1371/journal.pntd.0005473

De Ros G, Conci S, Pantezzi T, Savini G (2015) The economic impact of invasive pest *Drosophila suzukii* on berry production in the Province of Trento, Italy. J Berry Res 5:89–96. https://doi.org/10.3233/JBR-150092

Dean MD (2006) A *Wolbachia*-associated fitness benefit depends on genetic background in *Drosophila simulans*. Proc R Soc B Biol Sci 273:1415–1420. https://doi.org/10.1098/rspb.2005.3453

Diallo S, Seck MT, Rayaissé JB et al (2019) Chilling, irradiation and transport of male *Glossina palpalis gambiensis* pupae: effect on the emergence, flight ability and survival. PLoS One 14:e0216802. https://doi.org/10.1371/journal.pone.0216802

Dobson SL, Fox CW, Jiggins FM (2002) The effect of *Wolbachia*-induced cytoplasmic incompatibility on host population size in natural and manipulated systems. Proc R Soc B Biol Sci 269:437–445. https://doi.org/10.1098/rspb.2001.1876

Donovan MJ (2009) Genetically modified insects: why do we need them and how will they be regulated? Mo Envtl L & Pol'y Rev 17:62

Dos Santos LA, Mendes MF, Krüger AP et al (2017) Global potential distribution of *Drosophila suzukii* (Diptera Drosophilidae). PLoS One 12(3):e0174318. https://doi.org/10.1371/journal.pone.0174318

Ekesi S, Mohamed S (2011) Mass rearing and quality control parameters for tephritid fruit flies of economic importance in Africa. In: Akyar I (ed) Wide spectra of quality control. InTech, p 387–410

Enkerlin W (2005) Impact of fruit fly control programmes using the sterile insect technique. In: Dyck VA, Hendrichs J, Robinson AS (eds) Sterile insect technique. Principles and practice in area-wide integrated pest management. Springer, Dordrecht, pp 651–676

Enkerlin W, Gutiérrez-Ruelas JM, Cortes AV et al (2015) Area freedom in Mexico from Mediterranean fruit fly (Diptera: Tephritidae): a review of over 30 years of a successful containment program using an integrated area-wide SIT approach. Florida Entomol 98:665–681. https://doi.org/10.1653/024.098.0242

Enkerlin WR, Gutiérrez Ruelas JM, Pantaleon R et al (2017) The Moscamed regional programme: review of a success story of area-wide sterile insect technique application. Entomol Exp Appl 164:188–203. https://doi.org/10.1111/eea.12611

Enriquez T, Colinet H (2017) Basal tolerance to heat and cold exposure of the spotted wing *Drosophila*, *Drosophila suzukii*. PeerJ 5:e3112. https://doi.org/10.7717/peerj.3112

Enriquez T, Renault D, Charrier M, Colinet H (2018a) Cold acclimation favors metabolic stability in *Drosophila suzukii*. Front Physiol 9:1–17. https://doi.org/10.3389/fphys.2018.01506

Enriquez T, Ruel D, Charrier M, Colinet H (2018b) Effects of fluctuating thermal regimes on cold survival and life history traits of the spotted wing *Drosophila* (*Drosophila suzukii*). Insect Sci 27:317–335. https://doi.org/10.1111/1744-7917.12649

Esfandi K, He XZ, Wang Q (2019) Sperm allocation strategies in a sperm heteromorphic insect. Curr Zool 66(3):285–292. https://doi.org/10.1093/cz/zoz048

FAO/IAEA/USDA (2019) Product quality control for sterile mass-reared and released tephritid fruit flies, version 7.0. http://www-naweb.iaea.org/nafa/ipc/public/QCV7.pdf

Fisher K (1996) Queensland fruit fly (*Bactrocera tryoni*): eradication from Western Australia. In: McPheron BA, Steck GJ (eds) Fruit fly pests. A world assessment of their biology and management. St. Lucie Press, Delray Beach, pp 535–541

Fisher K, Caceres C (2000) A filter rearing system for mass reared genetic sexing strains of Mediterranean fruit fly (Diptera: Tephritidae). In: Tan K-H (ed) Area-wide control of fruit flies and other insect pests. Penerbit Universiti Sains Malaysia, Penang, pp 543–550

Franz G (2005) Genetic sexing strains in mediterranean fruit fly, an example for other species amenable to large-scale rearing for the sterile insect technique. In: Dyck VA, Hendrichs J, Robinson AS (eds) Sterile insect technique. Principles and practice in area-wide integrated pest management. Springer, Dordrecht, pp 427–451

Fraser JE, De Bruyne JT, Iturbe-Ormaetxe I et al (2017) Novel *Wolbachia*-transinfected *Aedes aegypti* mosquitoes possess diverse fitness and vector competence phenotypes. PLoS Pathog 13(12):e1006751. https://doi.org/10.1371/journal.ppat.1006751
Frazier MR, Woods HA, Harrison JF (2001) Interactive effects of rearing temperature and oxygen on the development of *Drosophila melanogaster*. Physiol Biochem Zool 74(5):641–650. https://doi.org/10.1086/322172
Fytrou A, Schofield PG, Kraaijeveld AR, Hubbard SF (2006) *Wolbachia* infection suppresses both host defence and parasitoid counter-defence. Proc R Soc B Biol Sci 273:791–796. https://doi.org/10.1098/rspb.2005.3383
Gallardo-Ortiz U, Pérez-Staples D, Liedo P, Toledo J (2018a) Sexual competitiveness, field survival, and dispersal of *Anastrepha obliqua* (Diptera: Tephritidae) fruit flies irradiated at different doses. J Econ Entomol 111:761–769. https://doi.org/10.1093/jee/tox326
Gallardo-Ortiz U, Pérez-Staples D, Liedo P, Toledo J (2018b) Sexual competitiveness of fertile and sterile, wild and mass-reared males of *Anastrepha obliqua*. Entomol Exp Appl 166:771–777. https://doi.org/10.1111/eea.12709
Gilchrist AS, Cameron EC, Sved JA (2012) Genetic consequences of domestication and mass rearing of pest fruit fly *Bactrocera tryoni* (Diptera: Tephritidae). J Econ Entomol 105(3):1051–1056. https://doi.org/10.1603/ec11421
Gilligan DM, Frankham R (2003) Dynamics of genetic adaptation to captivity. Conserv Genet 4:189–197. https://doi.org/10.1023/A:1023391905158
Gong P, Epton MJ, Fu G, Scaife S, Hiscox A, Condon KC (2005) A dominant lethal genetic system for autocidal control of the Mediterranean fruit fly. Nat Biotechnol 23:453–456. https://doi.org/10.1038/nbt1071
Gonzalez-Cabrera J, Contreras-Bermudez Y, Sanchez-Gonzalez JA et al (2018a) Optimization of a wheat germ diet for mass rearing *Drosophila suzukii*. Entomol Exp Appl 166:925–931. https://doi.org/10.1111/eea.12745
Gonzalez-Cabrera J, Garcia-Cancino MD, Moreno-Ca Rrillo G et al (2018b) Fresh banana as an alternative host for mass rearing *Drosophila suzukii*. Bull Insectol 71:65–70
Grassi A, Gottardello A, Dalton DT et al (2018) Seasonal reproductive biology of *Drosophila suzukii* (Diptera: Drosophilidae) in temperate climates. Environ Entomol 47(1):166–174. https://doi.org/10.1093/ee/nvx195
Green KK, Potgieter FT, Vreysen MJB (2007) A strategy for an area-wide control campaign with an SIT component to establish a tsetse (*Glossina austeni* and *Glossina brevipalpis*) free South Africa. In: Vreysen MJB, Robinson AS, Hendrichs J (eds) Area-wide control of insect pests: from research to field implementation. Springer, Dordrecht, pp 309–323
Gutierrez-Palomares VM, Cibrian-Tovar J, Alatorre-Rosas R, Quezada-Salinas A (2019) Effect of irradiation on quality and fertility parameters of *Drosophila suzukii* in Mexico. Southwest Entomol 44:617. https://doi.org/10.3958/059.044.0307
Haddad GG, Sun Y-A, Wyman RJ, Xu T (1997) Genetic basis of tolerance to O_2 deprivation in *Drosophila melanogaster*. Genetics 94:10809–10812. https://doi.org/10.1073/pnas.94.20.10809
Hallman GJ, Hellmich RL (2010) Modified atmosphere storage may reduce efficacy of irradiation phytosanitary treatments. In IX international controlled atmosphere research conference. p 159–162
Hamby KA, Bellamy ED, Chiu JC et al (2016) Biotic and abiotic factors impacting development, behavior, phenology, and reproductive biology of *Drosophila suzukii*. J Pest Sci (2004) 89:605–619. https://doi.org/10.1007/s10340-016-0756-5
Hardin JA, Kraus DA, Burrack HJ (2015) Diet quality mitigates intraspecific larval competition in *Drosophila suzukii*. Entomol Exp Appl 156:59–65. https://doi.org/10.1111/eea.12311
Hauser M (2011) A historic account of the invasion of *Drosophila suzukii* (Matsumura) (Diptera: Drosophilidae) in the continental United States, with remarks on their identification. Pest Manag Sci 67:1352–1357. https://doi.org/10.1002/ps.2265

Haviland DR, Beers EH (2012) Chemical control programs for *Drosophila suzukii* that comply with international limitations on pesticide residues for exported sweet cherries. J Integr Pest Manag 3:1–6. https://doi.org/10.1603/ipm11034

Helinski MEH, Parker AG, Knols BGJ (2009) Radiation biology of mosquitoes. Malar J 13:1–13. https://doi.org/10.1186/1475-2875-8-S2-S6

Hendrichs J, Ortiz G, Liedo P, Schwarz A (1983) Six years of successful medfly program in Mexico and Guatemala. In: Cavalloro R (ed), Proceedings, symposium: fruit flies of economic importance. CEC/IOBC International Symposium, 16–19 November 1982, Athens, Greece. A. A. Balkema, Rotterdam, p 353–365

Hendrichs J, Robinson AS, Cayol JP, Enkerlin W (2002) Medfly area-wide sterile insect technique programmes for prevention, suppression or eradication: the importance of mating behavior studies. Florida Entomol 85:1–13. https://doi.org/10.1653/0015-4040(2002)085[0001:masit p]2.0.co;2

Hendrichs J, Vreysen MJB, Enkerlin WR, Cayol JP (2005) Strategic options in the use of the sterile insect technique. In: Dyck VA, Hendrichs J, Robinson AS (eds) Sterile in-sect technique. Principles and practice in area-wide integrated pest management. Springer, Dordrecht, pp 563–600

Hendrichs J, Kenmore P, Robinson AS, Vreysen MJB (2007) Area-wide integrated pest management (AW-IPM): principles, practice and prospects. In: Vreysen MJB, Robinson A.S., Hendrichs J. (eds) Area-wide control of insect pests. Springer, Dordrecht, p 3–33

Hilgenboecker K, Hammerstein P, Schlattmann P et al (2008) How many species are infected with *Wolbachia*?—a statistical analysis of current data. FEMS Microbiol Lett 281:215–220. https://doi.org/10.1111/j.1574-6968.2008.01110.x

Hiroki M, Kato Y, Kamito T, Miura K (2002) Feminization of genetic males by a symbiotic bacterium in a butterfly, *Eurema hecabe* (Lepidoptera: Pieridae). Naturwissenschaften 89:167–170. https://doi.org/10.1007/s00114-002-0303-5

Hoback WW, Stanley DW (2001) Insects in hypoxia. J Insect Physiol 47:533–542. https://doi.org/10.1016/s0022-1910(00)00153-0

Hoffmann AA, Hallas R, Sinclair C, Partridge L (2001) Rapid loss of stress resistance in *Drosophila melanogaster* under adaptation to laboratory culture. Evolution 55:436–438. https://doi.org/10.1554/0014-3820(2001)055[0436:RLOSRI]2.0.CO;2

Hughes GL, Rasgon JL (2014) Transinfection: a method to investigate *Wolbachia*-host interactions and control arthropod-borne disease. Insect Mol Biol 23:141–151. https://doi.org/10.1111/imb.12066

IDIDAS International Database on Insect Disinfestation and Sterilization (n.d.). http://www--ididas.iaea.org/IDIDAS/start.htm%0Afile:///y:/8934.pdf

Iturbe-Ormaetxe I, Walker T, O'Neill SL (2011) *Wolbachia* and the biological control of mosquito-borne disease. EMBO Rep 12:508–518. https://doi.org/10.1038/embor.2011.84

Jácome I, Aluja M, Liedo P (1999) Impact of adult diet on demographic and population parameters of the tropical fruit fly *Anastrepha serpentina* (Diptera: Tephritidae). Bull Entomol Res 89:165–175. https://doi.org/10.1017/s0007485399000255

Jaramillo SL, Mehlferber E, Moore PJ (2015) Life-history trade-offs under different larval diets in *Drosophila suzukii* (Diptera: Drosophilidae). Physiol Entomol 40:2–9. https://doi.org/10.1111/phen.12082

Jiggins FM, Hurst GDD, Schulenburg JHGVD, Majerus ME (2001) Two male-killing *Wolbachia* strains coexist within a population of the butterfly *Acraea encedon*. Heredity (Edinb) 86(Pt 2):161–166. https://doi.org/10.1046/j.1365-2540.2001.00804.x

Kalajdzic P, Schetelig MF (2017) CRISPR/Cas-mediated gene editing using purified protein in *Drosophila suzukii*. Entomol Exp Appl 164:350–362. https://doi.org/10.1111/eea.12599

Kambris Z, Blagborough AM, Pinto SB et al (2010) *Wolbachia* stimulates immune gene expression and inhibits plasmodium development in *Anopheles gambiae*. PLoS Pathog 6:e1001143. https://doi.org/10.1371/journal.ppat.1001143

Kandul NP, Liu J, Hsu AD et al (2020) A drug-inducible sex-separation technique for insects. Nat Commun 11:2106. https://doi.org/10.1038/s41467-020-16020-2

Kanzawa T (1936) Studies on Drosophila suzukii Mats. J Plant Prot (Tokyo) 23:66–70, 127–132, 183–191. In: Rev Appl Entomol 24:315

Kittayapong P, Kaeothaisong NO, Ninphanomchai S, Limohpasmanee W (2018) Combined sterile insect technique and incompatible insect technique: sex separation and quality of sterile *Aedes aegypti* male mosquitoes released in a pilot population suppression trial in Thailand. Parasites and Vectors 11:73–83. https://doi.org/10.1186/s13071-018-3214-9

Kittayapong P, Ninphanomchai S, Limohpasmanee W et al (2019) Combined sterile insect technique and incompatible insect technique: the first proof-of-concept to suppress *Aedes aegypti* vector populations in semi-rural settings in Thailand. PLoS Negl Trop Dis 13(10):e0007771. https://doi.org/10.1371/journal.pntd.0007771

Klassen W (2005) Area-wide integrated pest management and the sterile insect technique. In: Dyck VA, Hendrichs J, Robinson AS (eds) Sterile insect technique. Principles and practice in area-wide integrated pest management. Springer, Dordrecht, pp 39–68

Knipling EF (1955) Possibilities of insect control or eradication through the use of sexually sterile males. J Econ Entomol 48:459–462. https://doi.org/10.1093/jee/48.4.459

Knipling EF (1979) The basic principles of insect population suppression and management. Agriculture handbook. USDA No. 512. pp ix + 659 pp

Koskinioti P, Ras E, Augustinos AA et al (2019) The effects of geographic origin and antibiotic treatment on the gut symbiotic communities of *Bactrocera oleae* populations. Entomol Exp Appl 167:197–208. https://doi.org/10.1111/eea.12764

Koukou K, Pavlikaki H, Kilias G, Werren JH, Bourtzis K, Alahiotis SN (2006) Influence of antibiotic treatment and *Wolbachia* curing on sexual isolation among *Drosophila melanogaster* cage populations. Evolution 60(1):87–96. https://doi.org/10.1111/j.0014-3820.2006.tb01084.x

Krüger AP, Schlesener DCH, Martins LN et al (2018) Effects of irradiation dose on sterility induction and quality parameters of *Drosophila suzukii* (Diptera: Drosophilidae). J Econ Entomol 111:741–746. https://doi.org/10.1093/jee/tox349

Krüger AP, Schlesener DCH, Martins LN et al (2019) Radiation effects on *Drosophila suzukii* (Diptera: Drosophilidae) reproductive behaviour. J Appl Entomol 143:88–94. https://doi.org/10.1111/jen.12563

Kyritsis GA, Augustinos AA, Livadaras I et al (2019) Medfly-*Wolbachia* symbiosis: genotype x genotype interactions determine host's life history traits under mass rearing conditions. BMC Biotechnol 19:96. https://doi.org/10.1186/s12896-019-0586-7

Langley PA, Curtis CF, Brady J (1974) The viability, fertility and behaviour of tsetse flies (*Glossina morsitans*) sterilized by irradiation under various conditions. Entomol Exp Appl 17:97–111. https://doi.org/10.1111/j.1570-7458.1974.tb00323.x

Lanouette G, Brodeur J, Fournier F et al (2017) The sterile insect technique for the management of the spotted wing *Drosophila*, *Drosophila suzukii*: establishing the optimum irradiation dose. PLoS One 12:e0180821. https://doi.org/10.1371/journal.pone.0180821

Laven H (1967) A possible model for speciation by cytoplasmic isolation in the *Culex pipiens* complex. Bull World Health Organ 37(2):263. https://doi.org/10.1101/SQB.1959.024.01.017

Lees RS, Gilles JR, Hendrichs J et al (2015) Back to the future: the sterile insect technique against mosquito disease vectors. Curr Opin Insect Sci 10:156–162. https://doi.org/10.1016/j.cois.2015.05.011

Leopold RA (2007) Colony maintenance and mass-rearing: using cold storage technology for extending the shelf-life of insects. In: Vreysen MJB, Robinson AS, Hendrichs J (eds) Area-wide control of insect pests: from research to field implementation. Springer, Dordrecht, pp 149–162

Lewis MT, Hamby KA (2019) Differential impacts of yeasts on feeding behavior and development in larval *Drosophila suzukii* (Diptera:Drosophilidae). Sci Rep 9:13370. https://doi.org/10.1038/s41598-019-48863-1

Liedo P, Salgado S, Oropeza A, Toledo J (2007) Improving mating performance of mass-reared sterile Mediterranean fruit flies (Diptera: Tephritidae) through changes in adult holding conditions: demography and mating competitiveness. Florida Entomol 90:33–40. https://doi.org/10.1653/0015-4040(2007)90[33:impoms]2.0.co;2

Liedo P, Carey JR, Ingram DK, Zou S (2012) The interplay among dietary fat, sugar, protein and açai (*Euterpe oleracea* Mart.) pulp in modulating lifespan and reproduction in a tephritid fruit fly. Exp Gerontol 47(7):536–539. https://doi.org/10.1016/j.exger.2012.05.001

López-Martínez G, Hahn DA (2012) Short-term anoxic conditioning hormesis boosts antioxidant defenses, lowers oxidative damage following irradiation and enhances male sexual performance in the Caribbean fruit fly, *Anastrepha suspensa*. J Exp Biol 215:2150–2161. https://doi.org/10.1242/jeb.065631

Mainali BP, Moadeli T, Ponton F et al (2019) Comparison of gel larval diet with traditional lucerne chaff and carrot solid diets for rearing of Queensland fruit fly (Diptera: Tephritidae). J Econ Entomol 112:2278–2286. https://doi.org/10.1093/jee/toz140

Mains JW, Kelly PH, Dobson KL et al (2019) Localized control of *Aedes aegypti* (Diptera: Culicidae) in Miami, FL, via inundative releases of *Wolbachia*-infected male mosquitoes. J Med Entomol 56:1296–1303. https://doi.org/10.1093/jme/tjz051

Marec F, Vreysen MJB (2019) Advances and challenges of using the sterile insect technique for the management of pest Lepidoptera. Insects 10:371. https://doi.org/10.3390/insects10110371

Martinez J, Longdon B, Bauer S et al (2014) Symbionts commonly provide broad spectrum resistance to viruses in insects: a comparative analysis of *Wolbachia* strains. PLoS Pathog 10(9):e1004369. https://doi.org/10.1371/journal.ppat.1004369

Mastrangelo T, Walder J (2011) Use of radiation and isotopes in insects. In: Singh N (ed) Radioisotopes—applications in bio-medical science. IntechOpen. p 67–92

Mastrangelo T, Parker AG, Jessup A et al (2010) A new generation of X ray irradiators for insect sterilization. J Econ Entomol 103:85–94. https://doi.org/10.1603/ec09139

Mazzi D, Bravin E, Meraner M et al (2017) Economic impact of the introduction and establishment of *Drosophila suzukii* on sweet cherry production in Switzerland. Insects 8:18. https://doi.org/10.3390/insects8010018

Matsumura S (1931) 6000 illustrated insects of Japan-empire. The Toko-Shoin, Tokyo, Japan

McLay LK, Green MP, Jones TM (2017) Chronic exposure to dim artificial light at night decreases fecundity and adult survival in *Drosophila melanogaster*. J Insect Physiol 100:15–20. https://doi.org/10.1016/j.jinsphys.2017.04.009

Miller WJ, Ehrman L, Schneider D (2010) Infectious speciation revisited: impact of symbiont-depletion on female fitness and mating behavior of *Drosophila paulistorum*. PLoS Pathog 6(12):e1001214. https://doi.org/10.1371/journal.ppat.1001214

Mouton L, Henri H, Charif D et al (2007) Interaction between host genotype and environmental conditions affects bacterial density in *Wolbachia* symbiosis. Biol Lett 3:210–213. https://doi.org/10.1098/rsbl.2006.0590

Mudavanhu P, Addison P, Carpenter JE, Conlong DE (2016) Mating compatibility and co mpetitiveness between wild and laboratory strains of *Eldana saccharina* (Lepidoptera: Pyralidae) after radiation treatment. Florida Entomol 99:54–65. https://doi.org/10.1653/024.099.sp108

Mutika GN, Parker AG, Vreysen MJB, Chaudhury M (2019) Tolerance to a combination of low temperature and sterilizing irradiation in male *Glossina palpalis gambiensis* (Diptera: Glossinidae): simulated transport and release conditions. J Insect Sci 19:1–6. https://doi.org/10.1093/jisesa/iez082

Nestel D, Nemny-Lavy E, Mohammad Islam S et al (2007) Effect of pre-irradiation conditioning of medfly pupae (Diptera: Tephritidae): hypoxia and quality of sterile males. Florida Entomol 90:80–87. https://doi.org/10.1653/0015-4040(2007)90[80:eopcom]2.0.co;2

Nikolouli K, Sassù F, Mouton L et al (2020) Combining sterile and incompatible insect techniques for the population suppression of *Drosophila suzukii*. J Pest Sci 93:647–661. https://doi.org/10.1007/s10340-020-01199-6

O'Connor L, Plichart C, Sang AC et al (2012) Open release of male mosquitoes infected with a *Wolbachia* biopesticide: field performance and infection containment. PLoS Negl Trop Dis 6(11):e1797. https://doi.org/10.1371/journal.pntd.0001797

Orozco-Dávila D, Artiaga-López T, Del Refugio Hernández M et al (2014) *Anastrepha obliqua* (Diptera: Tephritidae) mass-rearing: effect of relaxed colony management. Int J Trop Insect Sci 34:S19–S27. https://doi.org/10.1017/S1742758414000058

Parker AG (2005) Mass-rearing for sterile insect release. In: Dyck VA, Hendrichs J, Robinson AS (eds) Sterile insect technique. Principles and practice in area-wide integrated pest management. Springer, Dordrecht, pp 209–232

Parreño MA, Scannapieco AC, Remis MI et al (2014) Dynamics of genetic variability in *Anastrepha fraterculus* (Diptera: Tephritidae) during adaptation to laboratory rearing conditions. BMC Genet 15:S14. https://doi.org/10.1186/1471-2156-15-S2-S14

Pereira R, Yuval B, Liedo P et al (2013) Improving sterile male performance in support of programmes integrating the sterile insect technique against fruit flies. J Appl Entomol 137:178–190. https://doi.org/10.1111/j.1439-0418.2011.01664.x

Pérez-Staples D, Shelly TE, Yuval B (2013) Female mating failure and the failure of "mating" in sterile insect programs. Entomol Exp Appl 146:66–78. https://doi.org/10.1111/j.1570-7458.2012.01312.x

Prokopy RJ, Boller EF (1970) Artificial egging system for the European cherryfruit fly. J Econ Entomol 63:1413–1417. https://doi.org/10.1093/jee/63.5.1413

Proverbs D (1969) Induced sterilization and control of insects. Annu Rev Entomol 14:81–102

Rajabpour A, Seraj AA, Mehrnia A (2018) Developing an artificial diet for rearing *Orius albidipennis* reuter (Het., Anthocoridae). J Plant Prot Res 58:276–281. https://doi.org/10.24425/jppr.2018.124635

Raphael KA, Shearman DC, Gilchrist AS et al (2014) Australian endemic pest tephritids: genetic, molecular and microbial tools for improved sterile insect technique. BMC Genet 15:S9. https://doi.org/10.1186/1471-2156-15-S2-S9

Raulston JR, Shaver TN (1970) A low agar casein-wheat germ diet for rearing Tobacco budworms. J Econ Entomol 63:1743–1744. https://doi.org/10.1093/jee/63.6.1743

Reinecke JP (1985) Nutrition: artificial diets. In: Kerkut GA, Gilbert LI (eds) Comprehensive insect physiology, biochemistry and pharmacology, vol 4. Pergamon Press, Oxford, pp 391–419

Rempoulakis P, Nemny-Lavy E, Castro R, Nestel D (2016) Mating behaviour of *Dacus ciliatus* (Loew) [Diptera:Tephritidae]: comparisons between a laboratory and a wild population. J Appl Entomol 140:250–260. https://doi.org/10.1111/jen.12252

Rendón P, McInnis D, Lance D, Stewart J (2000). Comparison of medfly male-only and bisexual releases in large scale field trials. In Area-wide control of fruit flies and other insect pests. Joint proceedings of the international conference on area-wide control of insect pests, 28 May–2 June, 1998 and the fifth international symposium on fruit flies of economic importance, Penang, Malaysia, 1-5 June, 1998 (pp 517-525). Penerbit Universiti Sains Malaysia

Rendón P, McInnis D, Lance D, Stewart J (2004) Medfly (Diptera:Tephritidae) genetic sexing: large-scale field comparison of males-only and bisexual sterile fly releases in Guatemala. J Econ Entomol 97:1547–1553. https://doi.org/10.1603/0022-0493-97.5.1547

Rendón D, Walton V, Tait G et al (2019) Interactions among morphotype, nutrition, and temperature impact fitness of an invasive fly. Ecol Evol 9:2615–2628. https://doi.org/10.1002/ece3.4928

Rendón P, McInnis D, Lance D, Stewart J (2009) Medfly (Diptera:Tephritidae) Genetic Sexing: Large-Scale Field Comparison of Males-Only and Bisexual Sterile Fly Releases in Guatemala. J Econ Entomol. https://doi.org/10.1603/0022-0493-97.5.1547

Resilva SS, Pereira R (2014) Age- and temperature-related pupal eye colour changes in various tephritid fruit fly species with a view to optimizing irradiation timing. J Trop Insect Sci 34:59–65. https://doi.org/10.1017/S1742758414000095

Robinson AS (2002) Mutations and their use in insect control. Mutat Res Rev Mutat Res 511:113–132

Robinson AS (2005) Genetic basis of the sterile insect technique. In: Dyck VA, Hendrichs J, Robinson AS (eds) Sterile insect technique. Principles and practice in area-wide integrated pest management. Springer, Dordrecht, pp 95–114

Robinson AS (2006) Influence of anoxia during gamma irradiation on the fertility and competitiveness of the adult male codling moth, *Laspeyresia pomonella* (L.). Radiat Res 61:526. https://doi.org/10.2307/3574127

Roeder KA, Behmer ST (2014) Lifetime consequences of food protein-carbohydrate content for an insect herbivore. Funct Ecol 28:1135–1143. https://doi.org/10.1111/1365-2435.12262

Rossi-Stacconi MV, Kaur R, Mazzoni V et al (2016) Multiple lines of evidence for reproductive winter diapause in the invasive pest *Drosophila suzukii*: useful clues for control strategies. J Pest Sci 89:689–700. https://doi.org/10.1007/s10340-016-0753-8

Rössler Y (1979) The genetics of the Mediterranean fruit fly: a "white pupae" mutant. Ann Entomol Soc Am 72:583–585. https://doi.org/10.1093/aesa/72.5.583

Ruhm ME, Calkins CO (1981) Eye-color changes in *Ceratitis capitata* pupae, a technique to determine pupal development. Entomol Exp Appl 29:237–240. https://doi.org/10.1111/j.1570-7458.1981.tb03064.x

Rull J, Brunel O, Mendez ME (2005) Mass rearing history negatively affects mating success of male *Anastrepha ludens* (Diptera: Tephritidae) reared for sterile insect technique programs. J Econ Entomol 98:1510–1516. https://doi.org/10.1603/0022-0493-98.5.1510

Rull J, Birke A, Ortega R et al (2012) Quantity and safety vs. quality and performance: conflicting interests during mass rearing and transport affect the efficiency of sterile insect technique programs. Entomol Exp Appl 142:78–86. https://doi.org/10.1111/j.1570-7458.2011.01196.x

Saldanha LA, Silva NM (1999) Semi-artificial rearing of the larvae of *Anastrepha obliqua* (Diptera: Tephritidae) in Manaus, Amazonas-Brazil. Florida Entomol 82:82. https://doi.org/10.2307/3495838

Sánchez-Ramos I, Fernández CE, González-Núñez M (2019) Comparative analysis of thermal performance models describing the effect of temperature on the preimaginal development of *Drosophila suzukii*. J Pest Sci 92:523–541. https://doi.org/10.1007/s10340-018-1030-9

Sarakatsanou A, Diamantidis AD, Papanastasiou SA, Bourtzis K, Papadopoulos NT (2011) Effects of *Wolbachia* on fitness of the Mediterranean fruit fly (Diptera: Tephritidae). J Appl Entomol 135:554–563. https://doi.org/10.1111/j.1439-0418.2011.01610.x

Sassù F, Nikolouli K, Caravantes S et al (2019a) Mass-rearing of *Drosophila suzukii* for sterile insect technique application: evaluation of two oviposition systems. Insects 10(12):448. https://doi.org/10.3390/insects10120448

Sassù F, Nikolouli K, Pereira R et al (2019b) Irradiation dose response under hypoxia for the application of the sterile insect technique in *Drosophila suzukii*. PLoS One 14:e0226582. https://doi.org/10.1371/journal.pone.0226582

Schlesener DC, Wollmann J, Krüger AP et al (2018) Biology and fertility life table of *Drosophila suzukii* on artificial diets. Entomol Exp Appl 166:932–936. https://doi.org/10.1111/eea.12736

Schwarz AJ, Zambada A, Orozco DHS, Zavala JL (1985) Mass production of the Mediterranean fruit fly at Metapa, Mexico. Florida Entomol 68:467–477

Seck MT, Pagabeleguem S, Bassene MD et al (2015) Quality of sterile male tsetse after long distance transport as chilled, irradiated pupae. PLoS Negl Trop Dis 9:e0004229. https://doi.org/10.1371/journal.pntd.0004229

Seo ST, Vargas RI, Gilmore JE et al (1990) Sperm transfer in normal and gamma-irradiated, laboratory-reared Mediterranean fruit flies (Diptera: Tephritidae). J Econ Entomol 83:1949–1953. https://doi.org/10.1093/jee/83.5.1949

Sharon G, Segal D, Ringo JM et al (2010) Commensal bacteria play a role in mating preference of *Drosophila melanogaster*. Proc Natl Acad Sci U S A 107:20051–20056. https://doi.org/10.1073/pnas.1009906107

Sial AA, Roubos CR, Gautam BK et al (2019) Evaluation of organic insecticides for management of spotted wing *Drosophila* (*Drosophila suzukii*) in berry crops. J Appl Entomol 143:593–608. https://doi.org/10.1111/jen.12629

Sirot LK, Wolfner MF, Wigby S (2011) Protein-specific manipulation of ejaculate composition in response to female mating status in *Drosophila melanogaster*. Proc Natl Acad Sci U S A 108:9922–9926. https://doi.org/10.1073/pnas.1100905108

Staubach F, Baines JF, Kü Nzel S et al (2013) Host species and environmental effects on bacterial communities associated with *Drosophila* in the laboratory and in the natural environment. PLoS One 8(8):e70749. https://doi.org/10.1371/journal.pone.0070749

Stevens L, Wade MJ (1990) Cytoplasmically inherited reproductive incompatibility in *Tribolium* flour beetles: the rate of spread and effect on population size. Genetics 124:367–372

Suenega H, Kamiwada H, Tanaka A, Chishaki N (1992) Difference in the timing of larval jumping behavior of mass-reared and newly-colonized strains of the Melon fly, *Dacus cucurbitae* Coquillett (Diptrera: Tephritidae). Appl Entomol Zool 27:177–183

Szentesi A, Greany PD, Chambers DL (1979) Oviposition behavior of laboratory-reared and wild Caribbean fruit flies (*Anastrepha suspensa*; Diptera: Tephritidae): I. selected chemical influences. Entomol Exp Appl 26:227–238. https://doi.org/10.1111/j.1570-7458.1979.tb02923.x

Taning CNT, Van Eynde B, Yu N et al (2017) CRISPR/Cas9 in insects: applications, best practices and biosafety concerns. J Insect Physiol 98:245–257. https://doi.org/10.1016/j.jinsphys.2017.01.007

Thistlewood HMA, Gill P, Beers EH et al (2018) Spatial analysis of seasonal dynamics and overwintering of *Drosophila suzukii* (Diptera: Drosophilidae) in the Okanagan-Columbia Basin, 2010–2014. Environ Entomol 47:221–232. https://doi.org/10.1093/ee/nvx178

Toledo J, Rull J, Oropeza A et al (2004) Irradiation of *Anastrepha obliqua* (Diptera: Tephritidae) revisited: optimizing sterility induction. J Econ Entomol 97:383–389. https://doi.org/10.1093/jee/97.2.383

Toomey ME, Panaram K, Fast EM et al (2013) Evolutionarily conserved *Wolbachia*-encoded factors control pattern of stem-cell niche tropism in *Drosophila* ovaries and favor infection. Proc Natl Acad Sci U S A 110:10788–10793. https://doi.org/10.1073/pnas.1301524110

Tsitsipis JA (1977) An improved method for the mass rearing of the olive fruit fly, *Dacus oleae* (Gmel.) (Diptera, Tephritidae). Zeitschrift für Angew Entomol 83:419–426. https://doi.org/10.1111/j.1439-0418.1977.tb02419.x

Turelli M (1994) Evolution of incompatibility-inducing microbes and their hosts. Evolution 48:1500–1513. https://doi.org/10.1111/j.1558-5646.1994.tb02192.x

Turelli M, Hoffmann AA (1991) Rapid spread of an inherited incompatibility factor in California *Drosophila*. Nature 353:440–442. https://doi.org/10.1038/353440a0

Vacas S, Primo J, Manclús JJ et al (2019) Survey on *Drosophila suzukii* natural short-term dispersal capacities using the mark-release-recapture technique. Insects 10(9):268. https://doi.org/10.3390/insects10090268

Van Timmeren S, Isaacs R (2013) Control of spotted wing drosophila, *Drosophila suzukii*, by specific insecticides and by conventional and organic crop protection programs. Crop Prot 54:126–133. https://doi.org/10.1016/j.cropro.2013.08.003

Vargas RI (1984) Alternative egg collection system for mass production of Mediterranean fruit fly (Diptera: Tephritidae). J Econ Entomol 77:1064–1069. https://doi.org/10.1093/jee/77.4.1064

Vavre F, Fleury F, Varaldi J et al (2002) Infection polymorphism and cytoplasmic incompatibility in hymenoptera-*Wolbachia* associations. Heredity (Edinb) 88:361–365. https://doi.org/10.1038/sj.hdy.6800063

Vera MT, Oviedo A, Abraham S et al (2014) Development of a larval diet for the south American fruit fly *Anastrepha fraterculus* (Diptera: Tephritidae). Int J Trop Insect Sci 34:S73–S81. https://doi.org/10.1017/s1742758414000150

Virginio JF, Gómez M, Pinto AM et al (2017) Male sexual competitiveness of two *Ceratitis capitata* strains, tsl Vienna 8 and OX3864A transgenics, in field cage conditions. Entomol Exp Appl 164:318–326. https://doi.org/10.1111/eea.12615

Vreysen MJB, Saleh KM, Ali MY, Abdulla AM, Zhu ZR et al (2000) *Glossina austeni* (Dip-tera: Glossinidae) eradicated on the island of Unguja, Zanzibar, using the sterile insect technique. J Econ Entomol 93(1):123–135. https://doi.org/10.1603/0022-0493-93.1.123

Vreysen MJB, Hendrichs J, Enkerlin WR (2006) The sterile insect technique as a component of sustainable area-wide integrated pest management of selected horticultural insect pests. Res Inst Pomol Floric J Fruit Ornam Plant Res 14:107–132

Weeks AR, Breeuwer JAJ (2001) *Wolbachia*-induced parthenogenesis in a genus of phytophagous mites. Proc R Soc B Biol Sci 268:2245–2251. https://doi.org/10.1098/rspb.2001.1797

Woltz JM, Lee JC (2017) Pupation behavior and larval and pupal biocontrol of *Drosophila suzukii* in the field. Biol Control 110:62–69. https://doi.org/10.1016/j.biocontrol.2017.04.007

Wozniak C (2007) Regulatory impact on insect biotechnology and pest management. Entomol Res 37:221–230. https://doi.org/10.1111/j.1748-5967.2007.00118.x

Xi Z, Khoo CCH, Dobson SL (2005) *Wolbachia* establishment and invasion in an *Aedes aegypti* laboratory population. Science 310(5746):326–328. https://doi.org/10.1126/science.1117607

Yamada H, Maiga H, Juarez J et al (2019) Identification of critical factors that significantly affect the dose-response in mosquitoes irradiated as pupae. Parasit Vectors 12:1–13. https://doi.org/10.1186/s13071-019-3698-y

Yosiaki ITO, Hiroyuki K, Masaaki Y, Tsuguo K (2003) Eradication of the melon fly, *Bactrocera cucurbitae*, from Okinawa, Japan, by means of the sterile insect technique, with special emphasis on the role of basic studies. J Asia Pac Entomol 6(2):119–129. https://doi.org/10.1016/S1226-8615(08)60177-6

Young Y, Buckiewicz N, Long TAF (2018) Nutritional geometry and fitness consequences in *Drosophila suzukii*, the spotted wing *Drosophila*. Ecol Evol 8:2842–2851. https://doi.org/10.1002/ece3.3849

Zabalou S, Riegler M, Theodorakopoulou M et al (2004) *Wolbachia*-induced cytoplasmic incompatibility as a means for insect pest population control. Proc Natl Acad Sci U S A 101:15042–15045. https://doi.org/10.1073/pnas.0403853101

Zabalou S, Apostolaki A, Livadaras I et al (2009) Incompatible insect technique: incompatible males from a *Ceratitis capitata* genetic sexing strain. Entomol Exp Appl 132:232–240. https://doi.org/10.1111/j.1570-7458.2009.00886.x

Zeh DW, Zeh JA, Bonilla MM (2005) *Wolbachia*, sex ratio bias and apparent male killing in the harlequin beetle riding pseudoscorpion. Heredity (Edinb) 95:41–49. https://doi.org/10.1038/sj.hdy.6800666

Zhai Y, Lin Q, Zhang J et al (2016) Adult reproductive diapause in *Drosophila suzukii* females. J Pest Sci 89:679–688. https://doi.org/10.1007/s10340-016-0760-9

Zhang J, Huang L, He J et al (2010) An artificial light source influences mating and oviposition of black soldier flies, *Hermetia illucens*. J Insect Sci 10:1–7. https://doi.org/10.1673/031.010.20201

Zhang D, Lees RS, Xi Z et al (2015a) Combining the sterile insect technique with *Wolbachia*-based approaches: II—A safer approach to *Aedes albopictus* population suppression programmes, designed to minimize the consequences of inadvertent female release. PLoS One 10:1–14. https://doi.org/10.1371/journal.pone.0135194

Zhang D, Zheng X, Xi Z, Bourtzis K, Gilles JR (2015b) Combining the sterile insect technique with the incompatible insect technique: I-impact of Wolbachia infection on the fitness of triple- and double-infected strains of Aedes albopictus. PLoS One 10(4):e0121126. https://doi.org/10.1371/journal.pone.0121126

Zhang D, Lees RS, Xi Z et al (2016) Combining the sterile insect technique with the incompatible insect technique: III-robust mating competitiveness of irradiated triple *Wolbachia*-infected *Aedes albopictus* males under semi-field conditions. PLoS One 11:1–15. https://doi.org/10.1371/journal.pone.0151864

Zhao HW, Haddad GG (2011) Review: hypoxic and oxidative stress resistance in *Drosophila melanogaster*. Placenta 32:S104–S108. https://doi.org/10.1016/j.placenta.2010.11.017

Zheng X, Zhang D, Li Y et al (2019) Incompatible and sterile insect techniques combined eliminate mosquitoes. Nature 572:56–61. https://doi.org/10.1038/s41586-019-1407-9

Zygouridis NE, Argov Y, Augustinos AA et al (2013) Genetic changes during laboratory domestication of an olive fly SIT strain. J Appl Entomol 138:423–432. https://doi.org/10.1111/jen.12042

Chapter 10
Conditional Expression Systems for *Drosophila suzukii* Pest Control

Syeda A. Jaffri, Ying Yan, Maxwell J. Scott, and Marc F. Schetelig

Abstract The sterile insect technique (SIT) is an environmentally friendly pest control method involving sterilization of mass reared insects and their release into the field to suppress insect populations. It has been a valuable tool for insect pest management for several decades. Besides the classical genetic approaches in use, transgenic systems have been established that have or could be transferred to the agricultural pest species *Drosophila suzukii* to improve the efficiency of population suppression such as through the production and release of only male flies. For male-only strains, conditional gene expression systems are required to inhibit female lethality due to expression of a lethal gene. Practically, such a strain with an "off-system" can be reared in a bisexual way in the mass rearing facility with food supplement and create male-only populations for field release by activating female lethality through removal of the supplement. In this chapter, we discuss conditional expression systems that have been developed in the past and their potential for *D. suzukii* control. Methods include the Tet-Off and Tet-On, Erythromycin-Off, Biotin-On, Vanillic acid regulated, Phloretin-Off, Bile acid-Off, and the Quinic acid systems for expression control. In addition, systems that work on stimuli based on light and temperature are discussed.

Keywords Binary expression · Conditional lethality · Insect control · Tet-Off · E-Off · Q system · BA-Off · BA-On · Biotin system · VA-Off · VA-On systems

S. A. Jaffri · Y. Yan (✉) · M. F. Schetelig (✉)
Department of Insect Biotechnology in Plant Protection, Justus-Liebig-University Giessen, Institute for Insect Biotechnology, Giessen, Germany
e-mail: ying.yan@agrar.uni-giessen.de; marc.schetelig@agrar.uni-giessen.de

M. J. Scott
Department of Entomology and Plant Pathology, North Carolina State University, Raleigh, NC, USA

F. R. M. Garcia (ed.), *Drosophila suzukii Management*,
https://doi.org/10.1007/978-3-030-62692-1_10

10.1 Introduction

Agricultural pests are a global challenge as they contribute to damage the crops. *Drosophila suzukii* Matsumura is now one global player of insect pests that has established around the world and is a devastating pest for the soft fruit industry (Mazzi et al. 2017; Yeh et al. 2020; Panel et al. 2018). The damage caused by *D. suzukii* females is due to oviposition into the soft fruits and their larvae feeding and destroying the fruit flesh. The negative impact of *D. suzukii* is further intensified due to the short generation time of *D. suzukii* that can complete several generations in a single season and is overwintering in many countries (Panel et al. 2018; Rendon et al. 2018). The presence of the alive *D. suzukii* larvae in fruit can cause an entire shipment to be rejected.

To control *D. suzukii* and its burden to economy and environment, several methods have been developed. A traditional control method for *D. suzukii* is the use of one or more chemical pesticides such as spinosyns, pyrethroids, or organophosphates (Cowles et al. 2015; Beers et al. 2011; Timmeren and Isaacs 2013). However, it should be highlighted that control methods based on chemical pesticides are weather-dependent and broad spectrum (i.e., not species-specific) and may cause negative effects on biodiversity (Geiger et al. 2010). It is anticipated that *D. suzukii* will develop resistance to one or more of the commonly used insecticides (Haviland and Beers 2012). However, the development of new chemical insecticides is costly and time-consuming (Borovok et al. 2008; Osei et al. 2003). Although insecticide resistance management helps to avoid the development of resistances, it is often overlooked or misused and can become infective under certain levels of resistance. Thus, there is an increased need to develop environmentally friendly strategies for pest control for *D. suzukii* that could offer a species-specific, sustainable, and cost-effective alternative for its control and is applicable in the framework of area-wide integrated pest management. The alternative strategies could be a supplement to current insecticide applications, but if so, it must be highly effective in the time period between the last (allowed) spray and the harvest. Common waiting periods are between 1 and 2 weeks with varying length depending on crops, formulation, dose, and chemicals used (Vijayasree et al. 2014).

The sterile insect technique (SIT) (Knipling 1955) is an efficient method to control insect pest populations. Classical SIT relies on the mass rearing of insect population, exposing them to irradiation to induce sterility and the mass release of the sterile insects into the field (Klassen 2005). Sterilized males mate with wild females in the field, which as a consequence do not produce any viable offspring that eventually cause the intended population reduction (Fig. 10.1) (Vreysen et al. 2007). The SIT has been implemented to control various insect species in large control programs including the Mediterranean fruit fly, *Ceratitis capitata* (Wiedemann) (Diptera: Tephritidae), the New World screwworm, *Cochliomyia hominivorax* (Coquerel) (Diptera: Calliphoridae), several other tephritids, tsetse flies, and Lepidopteran pests (Ingaramo and Beckett 2009; Robinson 2002b; Scott et al. 2017; Klassen and Curtis 2005; Dupont-Filliard et al. 2001). In the early days of the SIT

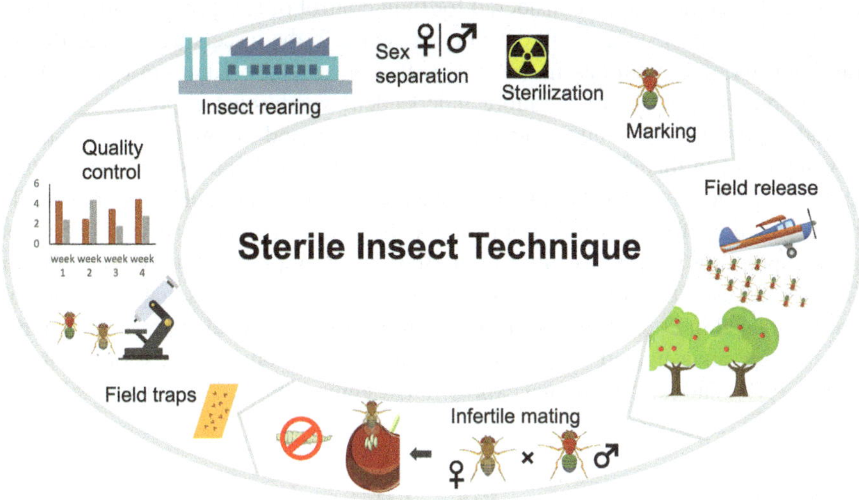

Fig. 10.1 The sterile insect technique (SIT). The SIT consists of mass rearing of insects in a rearing facility (sex sorting, elimination of females, marking and sterilization of males). The sterilized male-only populations are then released to the field area. Wild females that mate with sterile males produce no progeny. This technique helps to suppress the insect population. Targeted field area is continuously monitored by insect collection and analysis

and in the current *C. hominivorax* program (Concha et al. 2016), the release of insects from both sexes was used and resulted in a population reduction, but the release of male-only populations has proven to be most effective in the years and decades thereafter (Rendon et al. 2004). The release of only male fruitflies reduces the rearing costs and avoids additional fruit damage caused by egg deposition from released females. Therefore, to achieve maximum efficiency of the technique, it is important to consider male-only release and remove females from the population during early development within the rearing process. Consequently, for multiple invasive insects, the so-called genetic sexing strains have been established, which produce only males under certain rearing conditions (Franz 2005).

Strategies to enable SIT for novel invasive pest species include the generation of transgenic embryonic sexing strains (TESS) or unisex sterility that, like classical sexing or sterilization, induce lethality in the insects at the embryonic level. This can be achieved by expressing lethal factors in insect pests with the help of conditional expression systems (Schetelig and Handler 2012b; Yan and Scott 2015; Gong et al. 2005; Schetelig et al. 2009a; Ogaugwu et al. 2013; Lewandoski 2001; Meza et al. 2018; Schetelig et al. 2016). Functional systems in drosophilids include lethality and sexing system in *Drosophila melanogaster* (Meigen) (Heinrich and Scott 2000; Thomas et al. 2000; Horn and Wimmer 2003), while the establishment of sexing strains in *D. suzukii* has several constraints. The commonly used expression system for TESS is the Tet-Off system but to date this system has not been established in *D. suzukii*. Studies in *D. melanogaster* also suggest there are resistance constraints that need to be considered for tight control of any genetic trait. A single conditional control system will develop resistance in the insects based on different mechanisms and mutations in the flies (Zhao et al. 2020; Knudsen et al. 2020). Therefore, the use of two or more independent conditional expression systems would benefit a tight control and serve as a backup and mitigation for resistance development. In this chapter, we have compared potential expression systems that can be considered to develop effective stacked systems that could serve the sterile insect technique for *D. suzukii*. For practical use, any system needs to tightly control gene expression with a flexibility in the time window to switch expression On or Off. The focus of this article is on conditional expression systems that have a potential to be used in *D. suzukii*. Two broader groups of conditional expression systems have been investigated: (1) drug-inducible conditional expression systems (Table 10.1), and (2) conditional expression systems regulated by external stimuli.

10.2 Drug-Inducible Conditional Expression Systems

10.2.1 Gene Expression Systems Inducible by Antibiotics

10.2.1.1 Tetracycline-Controlled Gene Expression Systems

The tetracycline-based binary expression systems (Tet systems) originated from *Escherichia coli* and were developed by Gossen and Bujard in 1992. The Tet systems are comprised of three components: first, a tetracycline responsive element (TRE) that has a target gene under the control of a minimal or core promoter and multiple copies of the binding site of the tetracycline repressor (TetR) from the tet operon (tetO). The core promoter contains transcription initiation site and a polymerase binding site. Second, a tetracycline controllable transactivator (tTA) has been composed by fusing the TetR (Gossen and Bujard 1992) to the transcription activation domain from the herpes simplex virus protein VP16 (Triezenberg et al. 1988). Third, the antibiotics of the tetracycline family can be used as control molecules as binding of tetracycline to TetR inhibits binding to tetO (Fig. 10.2a). In the Tet-Off system, in the absence of tetracycline or its derivates, i.e., doxycycline, tTA binds to TRE and activates the minimal promotor that initiates transcription of an effector gene. When tetracycline is present and binds to tTA, the complex cannot bind to the TRE, and transcription of the targeted gene is terminated (Gossen and Bujard 1992). Another variant of a tetracycline-based system is the Tet-On (rtTA) system (Fig. 10.2b). Here, an engineered form of tTA, the rtTA has been created that can only bind to the TRE in the presence of tetracycline or doxycycline (Gossen et al. 1995).

Tet-Off systems have been established in *D. melanogaster* (Handler 2016; McGuire et al. 2004; Heinrich and Scott 2000; Thomas et al. 2000; Horn and Wimmer 2003). Combinations of promoter and reporter genes have been studied to develop female-specific lethal strains. For example, the female-specific yolk protein

Table 10.1 Overview of possible drug-inducible gene expression systems. Conditional gene expression systems are listed with their driver, operator, and repressor components as well as other known parameters

System	Origin	Regulator	Operator (Effector)	Complexity	Drug element	Drug source	Potential effect on insects	Stability of drug (Half-life)	Drug costs[a]	Conc for cells test	Tested in	In vivo test
Tet-On	*Escherichia coli*	rtTA rtetR: VP16	TRE (7x tetO)	Two genetic elements	Tetracycline/ Doxycycline	Antibiotic	Might affect microbiota	11 h	5 g Tet 27 € 5 g Dox 67 €	Tet: 1 μg/ mL Dox: 0.1 μg/ mL	Mammalian cells, insect cells	Mice, rats, mosquitoes, *L. cuprina, A. suspensa,* C. capitata, A. ludens, *D. melanogaster*
Tet-Off	*Escherichia coli*	tetR: VP16										
E-On	*Escherichia coli*	E:KRAB	ETR (8x operator 35 bp each)	Two genetic elements	Erythromycin	Antibiotic	Might affect microbiota	2 h	25 g 172 €	1 μg/mL	Mammalian cells	Mice
E-Off	*Escherichia coli*	E:VP16										
VA-On	*Caulobacter crescentus*	VanR: KRAB	VanO (8x operator, 12 bp each)	Two genetic elements	Vanillic acid	Plants, vanilla beans	No harmful effect known	N/A	25 g 27 €	16.8 μg/mL	Mammalian cells	Mice
VA-oOff	*Caulobacter crescentus*	VanR: VP16	VanO (2x operators)								Mammalian cells	
Ph-oOff	*Pseudomonas putida*	PTtgA-VP16	O_{rTgR} (2x operators)	Two genetic elements	Phloretine	Flavonoids in apples	No harmful effect known	70 h	25 mg 28 €	0.13 μg/mL	Mammalian cells	Mice

(continued)

Table 10.1 (continued)

System	Origin	Regulator	Operator (Effector)	Complexity	Drug element	Drug source	Potential effect on insects	Stability of drug (Half-life)	Drug costs[a]	Conc for cells test	Tested in	In vivo test
BA-On	*Campylo - bacter jejuni*	CmeR-KRAB	Ocme (14x operators)	Two genetic elements	Cholic acid/ bile acid	Cholesterol in liver	No harmful effect known	38.5 h	25 g 20 €	102 μg/mL	Mammalian cells	Mice
BA-Off		CmeR-VP16	Ocme (8x operators)			Cholesterol in liver					Mammalian cells	
Biotin-On	*Escherichia coli*	BirA:VP16	OBirA (3x operators)	Two genetic elements	Biotin	Vitamin H	No harmful effect known	2 h	5 g 170 €	0.002 μg/ mL	Mammalian cells and mice	Mice
Q system	*Neurospora crassa*	QF and QS 816 and 918 aa	QUAS (5x UAS units of the qa cluster)	Three genetic elements	Quinic acid	Soft fruits	No harmful effect known	N/A	5 g 14 €	5 mg/mL	Mammalian cells, plants, and insects	*Drosophila melanogaster, C. elegans,* mosquitoes and plants

[a]Drug costs (2020) are taken from www.alfa.com

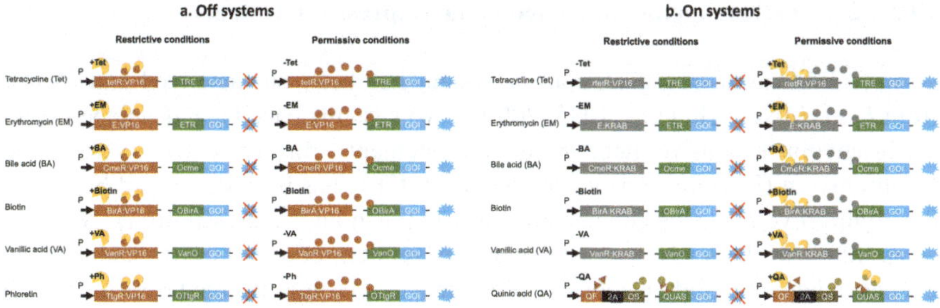

Fig. 10.2 Drug-inducible conditional expression systems. Mechanism of conditional gene expression systems controllable by tetracycline (Tet), erythromycin (E), bile acid (BA), biotin, vanillic acid (VA), phloretin, and quinic acid (QA) are displayed. (**a**) Off systems: all systems harbor a promoter (P) that drives the expression of a transcriptional factor consisting of a DNA binding domain fused to the VP16 transcription activation domain. The control molecules bind to the transcriptional factor and repress binding to its operator, resulting in no expression of the gene of interest (GOI). Gene expression is activated in the absence of the respective control molecule. (**b**) On systems: all systems harbor a promoter (P) that drives the expression of a transcriptional repressor consisting of a DNA binding domain fused to the KRAB transcription repression domain. In the absence of control molecules, the transcriptional repressor does not allow the expression of the gene of interest (GOI). The control molecules bind to the transcriptional repressor and inhibit binding to its operator and thus gene expression is activated. For the quinic acid (QA) system, the activation factor (QF) and suppressor (QS) can be combined with 2A peptides in a single construct. In the absence of quinic acid, QS binds to QF inhibiting gene expression. In presence of quinic acid (QA), QA inhibits QS and QF is released to activate the expression of the GOI

enhancer *yp3* has been used to drive tTA and activate the oncogene *Ras64B* resulting in female elimination in the absence of tetracycline (Thomas et al. 2000). Similarly, the *Yp1* promoter has been used to express tTA and activate the pro-apoptotic *hid* gene, inducing 100% female-specific lethality when reared on tetracycline-free food (Heinrich and Scott 2000). To induce embryo-specific lethality, Horn and Wimmer developed a transgene-based embryonic lethality system by using the cellularization gene promoters *sry-α* and *nullo*, the Tet-Off system, and the phosphor-mutated hid^{Ala5} gene (Horn and Wimmer 2003). For all systems, maternal contribution of tetracycline allowed the suppression of lethality by restricting the Tet-Off systems. A transfer of such systems from *D. melanogaster* to *D. suzukii* and the development of conditional genetic control strains and strategies should be possible. Tet systems have also been established in many other organisms including mice, rats (Zhu et al. 2002; Lewandoski 2001), mosquito species including *Aedes aegypti* (Linnaeus) (Fu et al. 2010) and *Anopheles stephensi* Liston (Diptera: Culicidae) (Lycett et al. 2004), and other insects, including *Lucilia cuprina* (Wiedemann) (Yan and Scott 2015), *C. hominivorax* (Concha et al. 2016) (Diptera: Calliphoridae), *Anastrepha suspensa* (Loew) (Schetelig and Handler 2012a), *C. capitata* (Ogaugwu et al. 2013), and *A. ludens* (Loew) (Diptera: Tephritidae) (Schetelig et al. 2016).

10.2.1.2 Erythromycin-Controlled Gene Expression Systems

A macrolide-based transgene control system has been characterized and cloned from *E. coli* (Noguchi et al. 1995, 2000). Two systems have been developed from erythromycin-responsive gene regulation elements. In the E-Off system, the Erythromycin (EM)-dependent transactivator (ET1) has been developed by fusing the erythromycin resistance gene repressor (E) also known as MphR(A) (Noguchi et al. 2000) to the VP16 transactivator (Triezenberg et al. 1988). In addition, an ET-dependent macrolide-responsive promoter (P_{ETR}) was assembled from the E. coli MphR(A)-specific operator (ETR) and a minimal promoter derived from the human cytomegalovirus promoter $P_{hCMVmin}$ (Fig. 10.2a). The system functions when ET1 binds and activates the transcription of P_{ETR} in the absence of EM. In the presence of EM, binding of ET1 to ETR is inhibited, P_{ETR} is not activated, and gene expression is stopped. In the E-On system, the repressor (E) has been fused to the trans-silencing domain (KRAB) to generate repressor transrepressor (ET4) (Moosmann et al. 1997) (Fig. 10.2b). ET4 represses P_{ETR}ON in the absence of EM. In the presence of EM, ET4 is released and gene expression is driven by the promoter.

Erythromycin-inducible expression systems have functional compatibility to tetracycline expression systems that make them highly efficient for generating a stacked lethality control system for *D. suzukii* pest control. Combining the E-Off and Tet-Off systems would enable further control of gene regulation (Weber et al. 2002).

10.2.1.3 Geneticin and Puromycin-Based System

A novel drug-inducible sex separation technique has been developed by Kandul et al. (2020) that is based on antibiotic-resistance genes rather than small-molecule-regulated DNA binding transcription factors. The two antibiotics geneticin and puromycin were lethal to *D. melanogaster* when added to the diet. Transgenic strains carrying constitutively expressed resistance genes, *NeoR* and *PuroR*, were viable. To create sex-specific systems, the *NeoR* and *PuroR* genes were combined with sex-specifically spliced introns from the *transformer* and *doublesex* genes. Here, males and females can be positively selected by rearing populations on either geneticin or puromycin (Kandul et al. 2020).

10.2.1.4 Disadvantages of Antibiotic-Based Systems

The addition of high concentration of antibiotics to the mass reared insect diet could reduce the fitness of male insects. It has been reported that tetracycline can impair fertility and male courtship, possibly through disruption of mitochondrial function (Zeh et al. 2012; Ballard and Melvin 2007; Moullan et al. 2015). A further problem is the constant use of antibiotics in insect diets, which could lead to antibiotic-resistant strains of gut microbiota or the shift of the insect microbiome due to the antibiotic pressure. For a mass rearing operation, the cost of the antibiotic is also a consideration. Geneticin and puromycin are in general more expensive than tetracycline.

10.2.2 Gene Expression Systems Inducible by Non-antibiotic Molecules

10.2.2.1 Quinic Acid-Controlled Gene Expression System

The Q system is a repressible binary expression system based on qa-gene cluster from the bread fungus, *Neurospora crassa.* The regulatory genes allow the fungus to use quinic acid as a carbon source (Giles et al. 1985). The Q system offers appealing features for a transgene expression as it can provide temporal control of gene expression. The synthetic Q system consists of four components: (1) a gene-regulator or effector QUAS, (2) a driver QF, (3) a suppressor QS, and (4) a food element quinic acid (QA). Quinic acid is a naturally occurring nontoxic compound, with antioxidative properties.

The driver component QF (also known as activation factor) of the Q system binds to the upstream region of effector component QUAS. The QUAS consists of five structural and enzymatic genes (Patel et al. 1981; Baum et al. 1987). In absence of quinic acid QA, the QS suppressor binds to QF (driver) and prevents activation of gene expression. In the presence of QA, QA binds to QS and releases QF, which can bind to QUAS and activate the expression of downstream genes (Giles et al. 1991) (Fig. 10.2b). The original QF consists of three structural domains, DBD (DNA binding and dimerization domain), MD (middle domain), and the transcriptional activation domain (AD). However, QF was found toxic in the *Drosophila* system (Riabinina et al. 2015). Two variants of QF have been designed to avoid toxicity and maintain the functional activity of QF—the QF2 and QF2w. QF2 was designed by deleting the middle domain (MD) and was still fully capable of driving gene expression in *D. melanogaster* (Riabinina et al. 2015). QF2w was further designed by changing the last two amino acids (glutamic acid and glutamine) of QF2 to four lysine(s) that change the charge on the C-terminus from negative to positive. This makes QF2w a weaker transcriptional activator but also less toxic. In addition, it can be more efficiently suppressed by QS than QF2. The Q system has been successfully used in *D. melanogaster* (Potter and Luo 2011), mammalian cells (Potter et al. 2010), *C. elegans* (Maupas) (Wei et al. 2012), *Danio rerio* (Hamilton) (Subedi et al. 2014), mosquitoes (Riabinina et al. 2016), and plants (Persad et al. 2020). Like any other gene control systems, QF expression can also be manipulated with strong or weak promoters.

QA temporal gene control can be achieved by the amount of QA fed to the flies, and duration of exposure. *Drosophila* larvae are more receptive to QA in the food than adult flies. That makes it a better control agent for embryonic lethality control (Riabinina et al. 2015). The Q system can be combined with other expression systems to induce tightly controlled, specific and multi-gene expression. For example, it can be used together with the widely established tTA (Mao et al. 2019; Eckermann et al. 2014) and GAL4 systems (Potter et al. 2010; Li and Stavropoulos 2016).

10.2.2.2 Biotin-Controlled Gene Expression System

The novel biotin-inducible gene expression system can be considered as an ideal control strategy for transgene expression due to nontoxic characters of biotin as a naturally occurring Vitamin H. Weber et al. in 2007 developed a synthetic model on *E. coli* biotin BirA (Chapman-Smith et al. 2001) (a bifunctional protein) that activates biotin by coupling to AMP (biotinyl-5′-AMP) and a transfer of the biotin group to the biotin carboxyl carrier protein subunit of acetyl-CoA-carboxylase which represses the biotin biosynthesis operon. The synthetic system consists of a biotin-dependent transactivator. BIT (BirA fused to the herpes simplex transactivation domain VP16 (Triezenberg et al. 1988) (Fig. 10.2a) that binds to a synthetic target promoter. BIT–Promoter interaction enables adjustable and reversible transgene expression. An initial test in mammalian cell lines with the biotin-inducible control system suggested that 10 nM (0.002 μg/mL) biotin is sufficient to activate gene expression. A downside of biotin-inducible expression systems can be the natural presence of biotin in insect diets that could lead to unexpected gene expression.

10.2.2.3 Vanillic Acid-Controlled Gene Expression System

Vanillic acid (VA)-controlled gene expression system is based on gene regulation elements from *Caulobacter crescentus* which is a freshwater bacterium. It can utilize VA as carbon source to convert metabolic energy in the citric acid cycle (Thanbichler et al. 2007; Harwood and Parales 1996). The system consists of a transcriptional repressor (Van R), an operator VanO, and a gene that is expressed with operator and repressor in VanAB. In the absence of VA, transcriptional repressor (VanR) binds to operator (VanO) upstream of the promoter region of VanAB gene cluster and inhibits VanAB gene expression. In the presence of VA, VanR binds to VA instead of VanO which derepresses the metabolic pathway (Thanbichler et al. 2007). VA is the oxidized form of vanillin and found at high concentrations in vanilla beans and has been used as a food additive (Sinha et al. 2008). VA was reported to be a suppressor of apoptosis in Neuro-2A cells (Huang et al. 2008) and also acts against snake venom (Dhananjaya et al. 2006) and cell carcinogenesis (Vetrano et al. 2005). Due to its nontoxic nature, VA can be used as a safe inducer molecule for controlling gene expression.

The synthetic VA-based systems have been designed as VA-Off and VA-On systems, respectively, which respond exclusively to the food additive VA. The VA-Off system (Fig. 10.2a) was designed by fusing the VanR DNA binding domain to the domain VP16 activation domain (Triezenberg et al. 1988) to generate a transcription factor (VanA1) that binds to VanO-operator sequences upstream of a core promoter. VA triggers the release of VanA1, and thus switches off gene expression. For the VAN_{ON} system (Fig. 10.2b), VanR was fused to the KRAB domain (Moosmann et al. 1997) to generate a trans-silencer (VanA4). Multiple copies of the VanO-operator sequence were placed between the gene of interest (GOI) and a constitutive promoter. Binding of VanA4 to Van-O sequences inhibits transcription of the GOI. VA triggers the release of VanA4, which derepresses expression and switches on gene expression. The VA-Off system has been tested in mammalian cells and mice, and suggested to be more efficient than VA-On system as it shows maximum gene regulation without epigenetic imprinting compared to the KRAB-containing VAN_{ON} design (Ayyanathan et al. 2003; Peng et al. 2009). In cell culture tests using VA-Off and VA-On systems, 100 μM (16.8 μg/mL) VA turned out to be sufficient for gene expression induction in VA-On and gene suppression in VA-Off system. However, use of Vanillic acid system in *D. suzukii* could be compromised by the presence of vanillic acid 2.8–16.1 mg/100 g in unripe to ripe strawberries (Mahmood et al. 2012), approx. 110 mg/kg in blue berries, and 45 mg/kg in black berries (Zadernowski et al. 2005). This could be problematic for future field releases of fertile transgenic males carrying a vanilla-regulated female lethal system.

10.2.2.4 Phloretin-Controlled Gene Expression System

Phloretin-controlled gene expression system is based on gene regulation elements from *Pseudomonas putida*—a soil bacterium from the habitat of plant rhizosphere. The bacterium has an evolved RND (resistance/nodulation/division) family transporter T_{tgABC} which is controlled by its repressor TtgR, binding to the specific operator O_{TtgR} in the T_{tgR} promoter (P_{TtgR}). Phloretin binds to O_{TtgR} operator to release TtgR repressor that results in the production of T_{tgABC}. (Teran et al. 2003). Phloretin is mainly found in apples and the root barks of the apple trees, and acts as an antibacterial plant defense metabolite (Teran et al. 2006). It protects skin from UV light (Oresajo et al. 2008) and has been used as a chemopreventive agent for cancer treatment (Wu et al. 2009) or a penetration enhancer for skin-based drug delivery (Valenta et al. 2001). A synthetic mammalian phloretin-adjustable control element (PEACE) has been designed by fusing TtgR repressor to VP16 (Fig. 10.2a) to generate a mammalian transactivator TtgA1, which can bind to the O_{TtgR} operator and activates the expression of downstream gene. In the presence of phloretin, gene expression is suppressed due to the intercept of binding between TtgA1 and O_{TtgR}. The phloretin system has been tested in mammalian cell culture and other derivatives, suggesting phloretin as ideal inducer molecule for complete gene repression with maximum concentration of 50 μM (0.13 μg/mL) (Gitzinger et al. 2009).

10.2.2.5 Bile Acid-Controlled Gene Expression System

Bile acid (BA)-controlled gene expression system is from gene regulation elements of *Campylobacter jejuni*—a bacteria that causes food poisoning (Klancnik et al. 2012). *C. jejuni* has a three-gene operon CmeABC that encodes for an efflux system to promote resistance to antimicrobial compounds. In *C. jejuni*, the CmeABC expression is controlled by the repressor CmeR, which is a member of tetR family and predicted to be involved in recognizing inducer molecule (Routh et al. 2009). CmeR binds to the operator O_{cme} that is controlled by a promoter P_{cmeABC} and represses the transcription of CmeABC (Lin et al. 2005). Bile acids can bind to the CmeR repressor and inactivate gene expression in a dose-dependent manner. Bile acids are known to improve digestion of lipids and fat-soluble vitamins in mammalian intestines and are synthesized from cholesterol in the liver (Hofmann 2009).

The bile acid-controlled gene expression elements have been used to develop the BA-Off and BA-On systems. The BA-Off system (Fig. 10.2a) is comprised of a bile acid-dependent transactivator CmeA1, in which the CmeR repressor has been fused to VP16 transactivator domain of Herpes simplex virus (Rossger et al. 2014). In the absence of bile acid, the transactivator CmeA1 binds to the operator Ocme sequences that are upstream of a core promoter and activates gene expression. While in the presence of bile acid, CmeA1 is prevented from binding to Ocme, and gene expression is not activated. On the other hand, the transsuppressor CmeA2 has been designed for the BEAR-on system (Fig. 10.2b) by fusing the CmeR repressor to the transsilencer human KRAB (Moosmann et al. 1997). Specifically, CmeA2 binds to Ocme and represses the gene expression when bile acid is absent, and CmeA is released from Ocme, thus the gene expression is activated when bile acid is present. This system has been successfully tested in mammalian cells (Rossger et al. 2014). Due to the presence of bile salts in fatal calf serum (FCS) of cell growth media, the BEAR-On system showed a basal level of gene expression. In mammalian cells, both BEAR-Off and BEAR-On systems were responsive to bile acid derivatives with max concentration of 250 μM (102 μg/mL). Meanwhile, other cholic acid derivatives have also been tested for BEAR-On system, and the results suggested that 100 mg/kg cholic acid, 30 mg/kg deoxycholic acid, and 30 mg/kg chenodeoxycholic acid are sufficient to trigger gene expression in mice (Rossger et al. 2014).

10.3 External Stimuli-Inducible Conditional Expression Systems

Classical genetic sexing strains (GSS) in *C. capitata* use elevated temperature to achieve sex separation. GSS females are homozygous for a *temperature-sensitive lethal (tsl)* mutation, while males have the same *tsl* mutation but in addition carry a wild-type copy of the *tsl* gene (unknown) translocated to the Y chromosome (Franz 2005). Incubation of embryos at 33–36 °C causes female-specific lethality (Robinson

2002a). Since GSS have been successfully used to produce billions of male *C. capitata* for field release, it is attractive to consider temperature-regulated transgenic systems for controlling insect viability or fertility. Heat-shock promoters have been studied and used to induce heat-activated temporal and spatial gene expression in *Drosophila* (Monsma et al. 1988). Similarly, a variety of heat-shock promoters have been used to regulate conditional expression of genes in insect species. For example, the heat-shock gene promoters *hsp26* and *hsp70* in *D. melanogaster* (Hara et al. 2008; Thomas et al. 2000), *hsp23*, *hsp70*, and *hsp90* in the blow fly *Lucilia sericata* (Meigen) (Diptera: Calliphoridae) (Tachibana et al. 2005), and *hsp70* in *C. capitata* (Kalosaka et al. 2009). In addition to heat-shock promoters, temperature-sensitive proteins such as the β2 proteasome subunit gene (*Prosβ2'*) of *D. melanogaster* (Smyth and Belote 1999) have been explored as a means for achieving environmental control of insect viability. *Prosβ2'* causes pupal lethality at 29 °C, but allows survival to adulthood at 25 °C and has also been tested in the tephritid *A. suspensa* (Nirmala et al. 2009). An interesting alternative is to use temperature-sensitive versions of the conserved sex determination gene *transformer 2*, which is essential for female development. Early studies in *D. melanogaster* found that the mutant Tra2 proteins appear to function normally at 16 °C but not 26 °C (Belote and Baker 1982). *D. suzukii* strains carrying the same temperature-sensitive mutations in the *tra2* gene were made using CRISPR/Cas9 technology (Li and Handler 2017). While the results were promising, the full potential could not be evaluated due to the low survival of *D. suzukii* above 26 °C in the laboratory strains.

In addition to temperature, light-inducible systems offer an alternative conditional means for controlling insect viability. Ramos et al. have generated a laser-inducible heat-shock-mediated ectopic gene expression in the butterfly, *Bicyclus anynana* (Butler) (Lepidoptera: Nymphalidae), using the *Drosophila hsp70* gene promoter (Ramos et al. 2006). In *C. elegans*, single cell expression of genes in a variety of cell types of endodermal, mesodermal, or ectodermal origin have been achieved by using *hsp70*-induced gene expression after pulsing with a laser (Stringham and Candido 1993). In addition, optogenetic switches like Cry2-CIB1 have been integrated into existing strategies to develop a robust light-controllable Tet system and accurately manipulate gene expression by light stimulation (Yamada et al. 2018). Light induction has the advantage that it can be fine-dosed and together with varying Dox concentrations, a fine-tuned and tight gene expression can be achieved (Yamada et al. 2020).

Such systems are promising technologies because the use of heat and light can open possibilities to artificially control gene expression in every possible developmental stage.

10.4 Summary

10.4.1 Comparison of Different Systems

Complexity of conditional systems plays an important role in the generation of transgenic insect strains. Most of the conditional expression systems consists of two gene components. A driver element that regulates and acts on an effector element. In addition, a molecule that can be added for conditional regulation of the system is needed. Such systems can and have been generated by establishing either two independent transgenic lines carrying the driver and effector constructs or as so-called all-in-one system that implements both into one transgenic strain. These transgenic lines are further crossed or inbred to generate double homozygous lines carrying driver and effector components (Hara et al. 2008). However, other tools like the quinic acid-based system involve and require triple homozygous strains (Potter and Luo 2011). Those systems could be simplified by using a bicistronic expression cassette carrying driver QF and QS (Eckermann et al. 2014; Schwirz et al. 2020). In that context, all the abovementioned systems could be established also in *D. suzukii*. The gene constructs can be integrated into the *D. suzukii* genome using either transposon-mediated germline transformation for elements up to 15 kb (Schetelig and Handler 2013), recombinase-mediated cassette exchange transformation and P[acman] system for large genetic elements of more than 100 kb (Haecker et al. 2017; Schetelig et al. 2009b; Venken et al. 2006) or CRISPR-mediated HDR for short genetic elements (Li and Handler 2017).

Toxicity of driver elements and small molecules: In several studies the overexpression of tTA was found to be toxic (Knudsen et al. 2020), and the transcriptional activator (QF) of the quinic acid system is also toxic (Riabinina et al. 2015). The expression of such toxic drivers under constitutive promoters would make it difficult to keep TESS strains for sexing and could interfere with fitness parameters even at low expression levels. This would affect the efficiency of insect strains in SIT programs, in which fitness and especially competitiveness of the released males are of greatest importance to the success of the program. For *D. suzukii*, to date the potential toxic effects of the drivers have not been reported.

Drug and cost efficacy: The induction ability of a system is defined by the minimum drug quantity required to promote or switch off gene expression. For Tet-Off systems, doxycycline has demonstrated better induction ability compared to other antibiotics from the tetracycline family, though the costs of doxycycline are usually higher than tetracycline based on small-scale price comparisons. Bulk quantities of these chemicals could be cheaper. Comparing the induction ability of the other systems, biotin and cholic acid (BA systems) require small amount (10 nM) of drugs to induce gene expression in mammalian cells. The biotin system is also challenging to use as the half-life of biotin is only 2 h, and it is the most expensive drug available among all (see Table 10.1). Cholic acid in comparison has a half-life of 38.5 h and requires only 10 nM to regulate gene expression. In the case of the VA system, the stability of vanillic acid is not known but has a good induction ability at 100 μM concentrations. A maximum of 50 μM concentration for phloretin with a half-life of 70 h can also be a most economical system considering price and stability of the compound.

10.4.2 Building Stacked Systems

An ideal approach to tightly regulate a transgene expression can be by building a stacked system with two or more completely independent conditional systems to induce female lethality or reproductive sterility as independent factors or to build two independent female-specific lethal strains with different apoptotic genes as a backup mechanism. A combination of Q and Tet-Off systems, where the Tet-Off system confers sex-specific lethality and the Q system sperm-specific lethality has been proposed but not yet demonstrated (Eckermann et al. 2014). The VA-Off system has been tested in parallel to the Tet-Off (Gossen and Bujard 1992) and E-Off systems (Weber et al. 2002) in mammalian cells. They present an interference-free independent gene regulation, suggesting these three systems can be used parallel a relatively tight gene regulation control (Gitzinger et al. 2012; Weber et al. 2002). Furthermore, these expression systems can be improved by using a diverse number of operators, transactivation domains, different distances between operator site and minimal promoter (Gitzinger et al. 2012).

However, to generate lethality strains to be used for SIT, the use of "On" systems is an economically suitable option because the effector molecule drug is not needed for constant maintenance of the population. Lethality can be induced by adding the respective molecules when sex separation and sterility are needed before release. Nevertheless, some developmental non-feeding stages (embryos and pupae) cannot be treated, which limits the flexibility of on systems.

Resistance in a genetically modified pest population can occur due to the possibility of primary and secondary site mutations as described recently by Knudsen et al. (2020) and Zhao et al. (2020). These mutations can be avoided by the establishment of dual redundant lethality systems that do not share functional components and function through different pathways or simply harbor redundant systems. Combinations of two or multiple systems are less likely to accumulate resistant mutations. Gene regulation with independent systems can ensure effective sex separation and reduce the chances of a genetic breakdown of conditional lethal expression systems. However, while using any substance for the control of conditional systems, it is important to consider the negative effects of them. Constant exposure to the substances may cause unwanted fitness effects, physiological changes, or changes in microbiota of the insects (Chatzispyrou et al. 2015; Wang et al. 2015; Zeh et al. 2012). In addition, mass rearing facilities and large-scale productions must calculate the efficiency of all economic factors included in SIT programs, when conditional systems are used. This has to consider regular mass rearing parameters and the fitness and stability of molecularly engineered strains as well as the waste management of insect food that includes additional effector molecules. In this respect, some insect diets might need special treatment like heat to decontaminate foods before disposal. At the same time, other substances could directly be disposed of and do not create extra costs and handling efforts during the production process.

References

Ayyanathan K, Lechner MS, Bell P et al (2003) Regulated recruitment of HP1 to a euchromatic gene induces mitotically heritable, epigenetic gene silencing: a mammalian cell culture model of gene variegation. Genes Dev 17(15):1855–1869

Ballard JW, Melvin RG (2007) Tetracycline treatment influences mitochondrial metabolism and mtDNA density two generations after treatment in *Drosophila*. Insect Mol Biol 16(6):799–802

Baum JA, Geever R, Giles NH (1987) Expression of qa-1F activator protein: identification of upstream binding sites in the qa gene cluster and localization of the DNA-binding domain. Mol Cell Biol 7(3):1256–1266

Beers EH, Van Steenwyk RA, Shearer PW et al (2011) Developing *Drosophila suzukii* management programs for sweet cherry in the western United States. Pest Manag Sci 67(11):1386–1395

Belote JM, Baker BS (1982) Sex determination in *Drosophila melanogaster*: analysis of transformer-2, a sex-transforming locus. Proc Natl Acad Sci U S A 79(5):1568–1572

Borovok N, Iram N, Zikich DG et al (2008) Assembling of G-strands into novel tetra-molecular parallel G4-DNA nanostructures using avidin-biotin recognition. Nucleic Acids Res 36(15):5050–5060

Chapman-Smith A, Mulhern TD, Whelan F et al (2001) The C-terminal domain of biotin protein ligase from *E. coli* is required for catalytic activity. Protein Sci 10(12):2608–2617

Chatzispyrou IA, Held NM, Mouchiroud L et al (2015) Tetracycline antibiotics impair mitochondrial function and its experimental use confounds research. Cancer Res 75(21):4446–4449

Concha C, Palavesam A, Guerrero FD et al (2016) A transgenic male-only strain of the New World screwworm for an improved control program using the sterile insect technique. BMC Biol 14(1):72

Cowles RS, Rodriguez-Saona C, Holdcraft R et al (2015) Sucrose improves insecticide activity against *Drosophila suzukii* (Diptera: Drosophilidae). J Econ Entomol 108:640–653

Dhananjaya BL, Nataraju A, Rajesh R et al (2006) Anticoagulant effect of Naja naja venom 5′nucleotidase: demonstration through the use of novel specific inhibitor, vanillic acid. Toxicon 48(4):411–421

Dupont-Filliard A, Billon M, Guillerez S et al (2001) Comparison by QCM and photometric enzymatic test of the biotin-avidin recognition on a biotinylated polypyrrole. Talanta 55(5):981–992

Eckermann KN, Dippel S, KaramiNejadRanjbar M et al (2014) Perspective on the combined use of an independent transgenic sexing and a multifactorial reproductive sterility system to avoid resistance development against transgenic sterile insect technique approaches. BMC genetics 15 Suppl 2(Suppl 2):S17

Franz G (2005) Genetic sexing strains in mediterranean fruit fly, an example for other species amenable to large-scale rearing for the sterile insect technique. In: Sterile insect technique: principles and practice in area-wide integrated Pest management. Springer, Dordrecht, pp 427–451. https://doi.org/10.1007/1-4020-4051-2_16

Fu G, Lees RS, Nimmo D et al (2010) Female-specific flightless phenotype for mosquito control. Proc Natl Acad Sci U S A 107(10):4550–4554

Geiger F, Bengtsson J, Berendse F et al (2010) Persistent negative effects of pesticides on biodiversity and biological control potential on European farmland. Basic Appl Ecol 11(2):97–105

Giles NH, Case ME, Baum J et al (1985) Gene organization and regulation in the qa (quinic acid) gene cluster of *Neurospora crassa*. Microbiol Rev 49(3):338–358
Giles NH, Geever RF, Asch DK et al (1991) The Wilhelmine E. Key 1989 invitational lecture. Organization and regulation of the qa (quinic acid) genes in Neurospora crassa and other fungi. J Hered 82(1):1–7
Gitzinger M, Kemmer C, El-Baba MD et al (2009) Controlling transgene expression in subcutaneous implants using a skin lotion containing the apple metabolite phloretin. Proc Natl Acad Sci U S A 106(26):10638–10643
Gitzinger M, Kemmer C, Fluri DA et al (2012) The food additive vanillic acid controls transgene expression in mammalian cells and mice. Nucleic Acids Res 40(5):e37
Gong P, Epton MJ, Fu G et al (2005) A dominant lethal genetic system for autocidal control of the Mediterranean fruitfly. Nat Biotechnol 23(4):453–456
Gossen M, Bujard H (1992) Tight control of gene expression in mammalian cells by tetracycline-responsive promoters. Proc Natl Acad Sci U S A 89(12):5547–5551
Gossen M, Freundlieb S, Bender G et al (1995) Transcriptional activation by tetracyclines in mammalian cells. Science 268(5218):1766–1769
Haecker I, Harrell Ii RA, Eichner G et al (2017) Cre/lox-Recombinase-mediated cassette exchange for reversible site-specific genomic targeting of the disease vector, *Aedes aegypti*. Sci Rep 7:43883
Handler AM (2016) Enhancing the stability and ecological safety of mass-reared transgenic strains for field release by redundant conditional lethality systems. Insect Sci 23(2):225–234
Hara K, Kuwayama H, Yaginuma T et al (2008) Establishment of a tetracycline-off system using a *piggyBac*-based vector as a gene functional analysis tool for the temporal targeting of gene expression. J Insect Biotechnol Sericol 77:159–166
Harwood CS, Parales RE (1996) The beta-ketoadipate pathway and the biology of self-identity. Annu Rev Microbiol 50:553–590
Haviland DR, Beers EH (2012) Chemical control programs for *Drosophila suzukii* that comply with international limitations on pesticide residues for exported sweet cherries. J Integr Pest Manag 3(2):F1–F6
Heinrich JC, Scott MJ (2000) A repressible female-specific lethal genetic system for making transgenic insect strains suitable for a sterile-release program. Proc Natl Acad Sci U S A 97(15):8229–8232
Hendrichs J, Kenmore P, Robinson AS et al (2007) Area-wide integrated pest management (AW-IPM): principles, practice and prospects. In: Vreysen MJB,Robinson AS and Hendrichs J (eds) Area-wide control of insect pests, 1st edn. Springer, Dordrecht, pp 3–33. https://doi.org/10.1007/978-1-4020-6059-5_1
Hofmann AF (2009) Bile acids: trying to understand their chemistry and biology with the hope of helping patients. Hepatology 49(5):1403–1418
Horn C, Wimmer EA (2003) A transgene-based, embryo-specific lethality system for insect pest management. Nat Biotechnol 21(1):64–70
Huang SM, Chuang HC, Wu CH et al (2008) Cytoprotective effects of phenolic acids on methylglyoxal-induced apoptosis in neuro-2A cells. Mol Nutr Food Res 52(8):940–949
Ingaramo M, Beckett D (2009) Distinct amino termini of two human HCS isoforms influence biotin acceptor substrate recognition. J Biol Chem 284(45):30862–30870
Kalosaka K, Soumaka E, Politis N et al (2009) Thermotolerance and HSP70 expression in the Mediterranean fruit fly *Ceratitis capitata*. J Insect Physiol 55(6):568–573
Kandul NP, Liu J, Hsu AD et al (2020) A drug-inducible sex-separation technique for insects. Nat Commun 11(1):2106
Klancnik A, Mozina SS, Zhang Q (2012) Anti-*Campylobacter* activities and resistance mechanisms of natural phenolic compounds in *Campylobacter*. PLoS One 7(12):e51800
Klassen W (2005) Area-wide integrated Pest management and the sterile insect technique. In: Dyck VA, Hendrichs J, Robinson AS (eds) Sterile insect technique principles and practice in area-wide integrated pest management. Springer, Dordrecht, pp 3–33

Klassen W, Curtis CF (2005) History of the sterile insect technique. In: Sterile insect technique. Springer, Dordrecht, pp 3–36

Knipling EF (1955) Possibilities of insect control or eradication through the use of sexually sterile males. J Econ Entomol 48:459–462

Knudsen KE, Reid WR, Barbour TM et al (2020) Genetic variation and potential for resistance development to the tTA overexpression lethal system in insects. G3 (Bethesda) 10(4):1271–1281

Lewandoski M (2001) Conditional control of gene expression in the mouse. Nat Rev Genet 2(10):743–755

Li J, Handler AM (2017) Temperature-dependent sex-reversal by a transformer-2 gene-edited mutation in the spotted wing drosophila, *Drosophila suzukii*. Sci Rep 7(1):12363

Li Q, Stavropoulos N (2016) Evaluation of ligand-inducible expression systems for conditional neuronal manipulations of sleep in *Drosophila*. G3 (Bethesda) 6(10):3351–3359

Lin JAM, Sahin O, Zhang Q (2005) CmeR functions as a transcriptional repressor for the multi-drug efflux pump CmeABC in *Campylobacter jejuni*. A A C 49(3):1067–1075

Lycett GJ, Kafatos FC, Loukeris TG (2004) Conditional expression in the malaria mosquito *Anopheles stephensi* with Tet-on and Tet-off systems. Genetics 167(4):1781–1790

Mahmood T, Anwar F, Abbas M et al (2012) Effect of maturity on phenolics (phenolic acids and flavonoids) profile of strawberry cultivars and mulberry species from Pakistan. Int J Mol Sci 13(4):4591–4607

Mao S, Qi Y, Zhu H et al (2019) A Tet/Q hybrid system for robust and versatile control of transgene expression in *C. elegans*. iScience 11:224–237

Mazzi D, Bravin E, Meraner M et al (2017) Economic impact of the introduction and establishment of *Drosophila suzukii* on sweet cherry production in Switzerland. Insects 8(1):18

McGuire SE, Roman G, Davis RL (2004) Gene expression systems in Drosophila: a synthesis of time and space. Trends Genet 20(8):384–391

Meza JS, Ul Haq I, Vreysen MJB et al (2018) Comparison of classical and transgenic genetic sexing strains of Mediterranean fruit fly (Diptera: Tephritidae) for application of the sterile insect technique. PLoS One 13(12):e0208880

Monsma SA, Ard R, Lis JT et al (1988) Localized heat-shock induction in *Drosophila melanogaster*. J Exp Zool 247(3):279–284

Moosmann P, Georgiev O, Thiesen HJ et al (1997) Silencing of RNA polymerases II and III-dependent transcription by the KRAB protein domain of KOX1, a Kruppel-type zinc finger factor. Biol Chem 378(7):669–677

Moullan N, Mouchiroud L, Wang X et al (2015) Tetracyclines disturb mitochondrial function across eukaryotic models: a call for caution in biomedical research. Cell Rep 10(10):1681–1691

Nirmala X, Zimowska Gj Fau-Handler AM et al (2009) Characterization of the proteasomebeta2 subunit gene and its mutant allele in the tephritid fruit fly pest, *Anastrepha suspensa*. Insect Mol Biol 18(3):330–334

Noguchi N, Emura A, Matsuyama H et al (1995) Nucleotide sequence and characterization of erythromycin resistance determinant that encodes macrolide 2′-phosphotransferase I in *Escherichia coli*. A A C 39(10):2359–2363

Noguchi N, Takada K, Katayama J et al (2000) Regulation of transcription of the mph(A) gene for macrolide 2′-phosphotransferase I in *Escherichia coli*: characterization of the regulatory gene mphR(A). J Bacteriol 182(18):5052–5058

Ogaugwu CE, Schetelig MF, Wimmer EA (2013) Transgenic sexing system for *Ceratitis capitata* (Diptera: *Tephritidae*) based on female-specific embryonic lethality. Insect Biochem Mol Biol 43(1):1–8

Oresajo C, Stephens T, Hino PD et al (2008) Protective effects of a topical antioxidant mixture containing vitamin C, ferulic acid, and phloretin against ultraviolet-induced photodamage in human skin. J Cosmet Dermatol 7(4):290–297

Osei E, Gassman PW, Hauck LM et al (2003) Environmental benefits and economic costs of manure incorporation on dairy waste application fields. J Environ Manag 68(1):1–11

Panel ADC, Zeeman L, van der Sluis BJ et al (2018) Overwintered *Drosophila suzukii* are the main source for infestations of the first fruit crops of the season. Insects 9(4):145

Patel VB, Schweizer M, Dykstra CC et al (1981) Genetic organization and transcriptional regulation in the qa gene cluster of *Neurospora crassa*. Proc Natl Acad Sci U S A 78(9):5783–5787

Peng H, Ivanov AV, Oh HJ et al (2009) Epigenetic gene silencing by the SRY protein is mediated by a KRAB-O protein that recruits the KAP1 co-repressor machinery. J Biol Chem 284(51):35670–35680

Persad R, Reuter DN, Dice LT et al (2020) The Q-system as a synthetic transcriptional regulator in plants. Front Plant Sci 11:245

Potter CJ, Luo L (2011) Using the Q system in *Drosophila melanogaster*. Nat Protoc 6(8):1105–1120

Potter CJ, Tasic B, Russler EV et al (2010) The Q system: a repressible binary system for transgene expression, lineage tracing, and mosaic analysis. Cell 141(3):536–548

Ramos D, Kamal F, Wimmer E et al (2006) Temporal and spatial control of transgene expression using laser induction of the Hsp70 promoter. BMC Dev Biol 6:55

Rendon P, McInnis D, Lance D et al (2004) Medfly (Diptera: Tephritidae) genetic sexing: large-scale field comparison of males-only and bisexual sterile fly releases in Guatemala. J Econ Entomol 97(5):1547–1553

Rendon D, Buser J, Tait G et al (2018) Survival and fecundity parameters of two *Drosophila suzukii* (Diptera: Drosophilidae) morphs on variable diet under suboptimal temperatures. J Insect Sci 18(6):8

Riabinina O, Luginbuhl D, Marr E et al (2015) Improved and expanded Q-system reagents for genetic manipulations. Nat Methods 12(3):219–225

Riabinina O, Task D, Marr E et al (2016) Organization of olfactory centres in the malaria mosquito *Anopheles gambiae*. Nat Commun 7:13010

Robinson AS (2002a) Genetic sexing strains in medfly, *Ceratitis capitata*, sterile insect technique programmes. Genetica 116(1):5–13

Robinson AS (2002b) Mutations and their use in insect control. Mutat Res 511(2):113–132

Rossger K, Charpin-El-Hamri G, Fussenegger M (2014) Bile acid-controlled transgene expression in mammalian cells and mice. Metab Eng 21:81–90

Routh MD, Su CC, Zhang Q et al (2009) Structures of AcrR and CmeR: insight into the mechanisms of transcriptional repression and multi-drug recognition in the TetR family of regulators. Biochim Biophys Acta 1794(5):844–851

Schetelig MF, Handler AM (2012a) Strategy for enhanced transgenic strain development for embryonic conditional lethality in *Anastrepha suspensa*. Proc Natl Acad Sci U S A 109(24):9348–9353

Schetelig MF, Handler AM (2012b) A transgenic embryonic sexing system for *Anastrepha suspensa* (Diptera: *Tephritidae*). Insect Biochem Mol Biol 42:790–795

Schetelig MF, Handler AM (2013) Germline transformation of the spotted wing drosophilid, *Drosophila suzukii,* with a *piggyBac* transposon vector. Genetica 141(4–6):189–193

Schetelig MF, Caceres C, Zacharopoulou A et al (2009a) Conditional embryonic lethality to improve the sterile insect technique in *Ceratitis capitata* (Diptera: Tephritidae). BMC Biol 7:4

Schetelig MF, Scolari F, Handler AM et al (2009b) Site-specific recombination for the modification of transgenic strains of the Mediterranean fruit fly *Ceratitis capitata*. Proc Natl Acad Sci U S A 106(43):18171–18176

Schetelig MF, Targovska A, Meza JS et al (2016) Tetracycline-suppressible female lethality and sterility in the Mexican fruit fly, *Anastrepha ludens*. Insect Mol Biol 25(4):500–508

Schwirz J, Yan Y, Franta Z et al (2020) Bicistronic expression and differential localization of proteins in insect cells and *Drosophila suzukii* using picornaviral 2A peptides. Insect Biochem Mol Biol 119:103324

Scott MJ, Concha C, Welch JB et al (2017) Review of research advances in the screwworm eradication program over the past 25 years. Entomol Exp Appl 164(3):226–236

Sinha AK, Sharma UK, Sharma N (2008) A comprehensive review on vanilla flavor: extraction, isolation and quantification of vanillin and others constituents. Int J Food Sci Nutr 59(4):299–326

Smyth KA, Belote JM (1999) The dominant temperature-sensitive lethal DTS7 of *Drosophila melanogaster* encodes an altered 20S proteasome beta-type subunit. Genetics 151(1):211–220
Stringham EG, Candido EP (1993) Targeted single-cell induction of gene products in *Caenorhabditis elegans*: a new tool for developmental studies. J Exp Zool 266(3):227–233
Subedi A, Macurak M, Gee ST et al (2014) Adoption of the Q transcriptional regulatory system for zebrafish transgenesis. Methods 66(3):433–440
Tachibana S, Numata H, Goto SG (2005) Gene expression of heat-shock proteins (Hsp23, Hsp70 and Hsp90) during and after larval diapause in the blow fly *Lucilia sericata*. J Insect Physiol 51(6):641–647
Teran W, Felipe A, Segura A et al (2003) Antibiotic-dependent induction of *Pseudomonas putida* DOT-T1E TtgABC efflux pump is mediated by the drug binding repressor TtgR. Antimicrob Agents Chemother 47(10):3067–3072
Teran W, Krell T, Ramos JL, Gallegos MT (2006) Effector-repressor interactions, binding of a single effector molecule to the operator-bound TtgR homodimer mediates derepression. J Biol Chem 281(11):7102–7109
Thanbichler M, Iniesta AA, Shapiro L (2007) A comprehensive set of plasmids for vanillate- and xylose-inducible gene expression in *Caulobacter crescentus*. Nucleic Acids Res 35(20):473–502
Thomas DD, Donnelly CA, Wood RJ et al (2000) Insect population control using a dominant, repressible, lethal genetic system. Science 287(5462):2474–2476
Timmeren SV, Isaacs R (2013) Control of spotted wing drosophila, *Drosophila suzukii,* by specific insecticides and by conventional and organic crop protection programs. Crop Prot 54:126–133
Triezenberg SJ, Kingsbury RC, McKnight SL (1988) Functional dissection of VP16, the trans-activator of herpes simplex virus immediate early gene expression. Genes Dev 2(6):718–729
Valenta C, Nowak M, Hadgraft J (2001) Influence of phloretin and 6-ketocholestanol on the permeation of progesterone through porcine skin. Int J Pharm 217(1–2):79–86
Venken KJ, He Y, Hoskins RA, Bellen HJ (2006) P[acman]: a BAC transgenic platform for targeted insertion of large DNA fragments in *D. melanogaster*. Science 314(5806):1747–1751
Vetrano AM, Heck DE, Mariano TM et al (2005) Characterization of the oxidase activity in mammalian catalase. J Biol Chem 280(42):35372–35381
Vijayasree V, Bai H, Mathew TB et al (2014) Dissipation kinetics and effect of different decontamination techniques on the residues of emamectin benzoate and spinosad in cowpea pods. Environ Monit Assess 186(7):4499–4506
Wang X, Ryu D, Houtkooper RH, Auwerx J (2015) Antibiotic use and abuse: a threat to mitochondria and chloroplasts with impact on research, health, and environment. Bioessays 37(10):1045–1053
Weber W, Fux C, Daoud-el Baba M, Keller B, Weber CC, Kramer BP, Heinzen C, Aubel D, Bailey JE, Fussenegger M (2002) Macrolide-based transgene control in mammalian cells and mice. Nat Biotechnol 20(9):901–907
Wei X, Potter CJ, Luo L, Shen K (2012) Controlling gene expression with the Q repressible binary expression system in *Caenorhabditis elegans*. Nat Methods 9(4):391–395
Wu CH, Ho YS, Tsai CY, Wang YJ, Tseng H, Wei PL, Lee CH, Liu RS, Lin SY (2009) In vitro and in vivo study of phloretin-induced apoptosis in human liver cancer cells involving inhibition of type II glucose transporter. Int J Cancer 124(9):2210–2219
Yamada M, Suzuki Y, Nagasaki SC, Okuno H, Imayoshi I (2018) Light control of the Tet gene expression system in mammalian cells. Cell Rep 25(2):487–500. e486
Yamada M, Nagasaki SC, Ozawa T, Imayoshi I (2020) Light-mediated control of gene expression in mammalian cells. Neurosci Res 152:66–77
Yan Y, Scott MJ (2015) A transgenic embryonic sexing system for the Australian sheep blow fly *Lucilia cuprina*. Sci Rep 5:16090
Yeh DA, Drummond FA, Gomez MI, Fan X (2020) The economic impacts and management of spotted wing Drosophila (*Drosophila suzukii*): the case of wild blueberries in Maine. J Econ Entomol 113(3):1262–1269

Zadernowski R, Naczk M, Nesterowicz J (2005) Phenolic acid profiles in some small berries. J Agric Food Chem 53(6):2118–2124
Zeh JA, Bonilla MM, Adrian AJ, Mesfin S, Zeh DW (2012) From father to son: transgenerational effect of tetracycline on sperm viability. Sci Rep 2:375
Zhao Y, Schetelig MF, Handler AM (2020) Genetic breakdown of a Tet-off conditional lethality system for insect population control. Nat Commun 11(1):3095
Zhu Z, Zheng T, Lee CG, Homer RJ, Elias JA (2002) Tetracycline-controlled transcriptional regulation systems: advances and application in transgenic animal modeling. Semin Cell Dev Biol 13(2):121–128

Chapter 11
Fine-Mesh Exclusion Netting for Control of *Drosophila suzukii*

Ryan Kuesel and David Gonthier

Abstract The exclusion of agricultural pests through physical barriers is a growing practice expanded from its initial use for climate control and season extension. Today, a diversity of approaches to physical barriers allow growers to suppress damage from birds, large arthropod pests, and the spotted wing Drosophila (*Drosophila suzukii*). Here, we review the history of physical barriers, the diversity of approaches, the efficacy for suppression of *D. suzukii*, and potential win-wins and trade-offs with economic and agricultural management. For long-lived perennial fruit systems, inexpensive spunbond materials have been eschewed in favor of stronger, woven, polyamide mesh nets. A flurry of 12 publications show that fine-mesh netting excludes *D. suzukii* in blueberry, blackberry, raspberry, and grapes compared to non-insecticidal and insecticidal controls. These nets also protect against insect and avian fruit pests, increase yield, have little to no impact on fruit quality, and minor impacts on temperature and humidity. However, pollination management challenges their implementation. Further, the high cost of both application and purchase of materials are potential barriers to adoption. Nonetheless, with increased longevity and decreasing costs, we conclude that netting physical barriers are effective and show great potential as an alternative or companion to insecticide use for the future.

Keywords *Drosophila suzukii* · Small fruits · Pest control · Exclusion barrier · Netting

R. Kuesel (✉) · D. Gonthier
Entomology Department, S-225 Agriculture Science Center, University of Kentucky, Lexington, KY, USA
e-mail: r.kuesel@uky.edu

F. R. M. Garcia (ed.), *Drosophila suzukii Management*,
https://doi.org/10.1007/978-3-030-62692-1_11

11.1 Introduction

Drosophila suzukii (Matsumura) threatens the economic feasibility of growing soft fruits such as raspberry, blackberry, strawberry, blueberry, cherry, and grape throughout the world in regions where it has invaded (Bolda et al. 2010). The presence of a hardened and serrated ovipositor makes *D. suzukii* uniquely adapted to damage fruit crops, as it can lay eggs inside of healthy fruits as they ripen. Once infested, fruits may never be harvested due to an increased speed of decomposition, or fruits may be rejected at the time of sale. While *D. suzukii* is a highly fecund pest species which reproduces within fruits from a wide variety of plants, some aspects of its life cycle can be exploited in order to protect fruit crops from its oviposition.

The protection of fruits from *D. suzukii* is most commonly achieved with chemical insecticides. However, effective cultural controls are also beginning to emerge as researchers grow to understand the life history of this pest. One promising cultural control for *D. suzukii* is the use of exclusion barriers. This technique may reduce insecticide use, but it is currently economically challenging to implement. Inquiry into this topic has been driven in part by concerns that the repeated application of insecticides could cause *D. suzukii* to rapidly build pesticide resistance. Indeed, some *D. suzukii* populations are reported to already have resistance to some insecticides (Gress and Zalom 2019). This calls for the increased importance of finding effective cultural controls that can reduce the needed frequency of pesticide sprays. Additionally, for many fresh fruit markets, a zero-tolerance policy challenges growers to prevent even a single fruit from becoming infested with *D. suzukii* (Bruck et al. 2011; Van Timmeren and Isaacs 2013). If *D. suzukii* larva emerge from a producer's berries, a market may have cause to reject the current fruit shipment. To meet these challenges, a multi-faceted approach may be needed that includes chemical and cultural controls. This is especially important for producers who practice organic compliant production, as the only effective, organically certified insecticides for *D. suzukii* control are spinosads and pyrethrins (Van Timmeren and Isaacs 2013).

In this chapter, we review the history of exclusion barrier usage for pest control and specifically for the control of *D. suzukii*. In Sect. 11.2, we begin by briefly summarizing the history of physical barriers in agriculture; the concept of exclusion nets for *D. suzukii* control did not arrive de novo, but instead arose as a result of the ingenuity and observations of producers and agricultural researchers as they used physical barriers to protect crops from inclement weather. In Sect. 11.3, we describe differences between the types of netting used for *D. suzukii* management and explore the current most promising uses for on-farm suppression of *D. suzukii*. In Sect. 11.4, we describe the methods researchers use to evaluate netting effectiveness. In Sect. 11.5, we summarize the results of peer-reviewed studies on the efficacy of netting barriers for the control of *D. suzukii* before moving into deeper discussion of the experimental findings for blueberry, cherry, raspberry, blackberry, and grape in Sects. 11.6–11.10, respectively. Fine-mesh netting is effective at reducing *D. suzukii* presence in fruit crops as supported by 11 of 12 studies. In Sect. 11.11, we discuss the additional effects of managing pests with mesh net; we see reduced incidence of other crop pests, reduced presence of pollinators, changes in peak temperatures, increased humidity, and a chance for increased disease incidence. Section 11.12 will look at the effects of mesh netting on crop yield. Some studies find slight increases in fruit crop yield under fine mesh but more often find statistically equivalent yields compared to other practical management strategies. Finally, in Sect. 11.13, we will discuss the need for economic analyses of fine mesh usage before concluding in Sect. 11.14.

Ultimately exclusion netting is an option that producers may turn to now and in the future for the control of *D. suzukii*. Netting barriers are a promising alternative or companion to pesticides for conventional and organic fruit producers, but while it is under investigation by researchers, it is used only by a niche group of fruit producers.

11.2 History of Exclusion Barriers in Agriculture

11.2.1 Row Covers for Weather Protection

Some of the earliest uses of physical barriers provided support that netting as a cultural control can be effective for the protection of crops against damage from birds, insects, and inclement weather. Before the turn of the century, parts of Israel widely adopted barriers to exclude whiteflies from greenhouse-grown tomatoes to control the tomato yellow leaf curl virus (Berlinger et al. 2002). Around the same era barriers began seeing use to protect crops from inclement weather and to prolong growing seasons by allowing cropping earlier in spring or later into the fall. For over four decades, growers in Europe, the Asian Pacific (especially in China), the Americas, Africa, and occasionally in the Middle East used thin, solid films of ethylene-vinyl acetate or ethylene-butyl acrylate (Fig. 11.1a) to create low-lying tunnels over crops or to act as groundcover (Espi et al. 2006). Tunnels served the purpose of stabilizing variable temperatures, while tunnels and groundcover both served to reduce water evaporation.

Spunbond fabrics are a relatively modern, light-weight plastic barrier (Fig. 11.1b). This fabric is created through a highly engineered process where threads of melted polyester plastic are blown into an overlapping pattern. The resulting flat, flexible, and slightly permeable sheets of spunbond fabrics gained agricultural usage as materials for low tunnels; uniquely, this fabric was light enough to see use as floating row covers where fabric is laid over immature crops without a support structure. For temperate parts of the world, early months of the growing season experience unexpected night-time frosts, and many vegetable crops need insulation to survive these weather events. Spunbond fabrics can be laid over crop fields when weather predictions estimate freezing temperatures (Poling et al. 1991; Rumpel 1993; Rekika et al. 2008). In these floating row cover systems, fabric is secured to the ground using stakes or heavy objects so that wind does not remove the cover overnight.

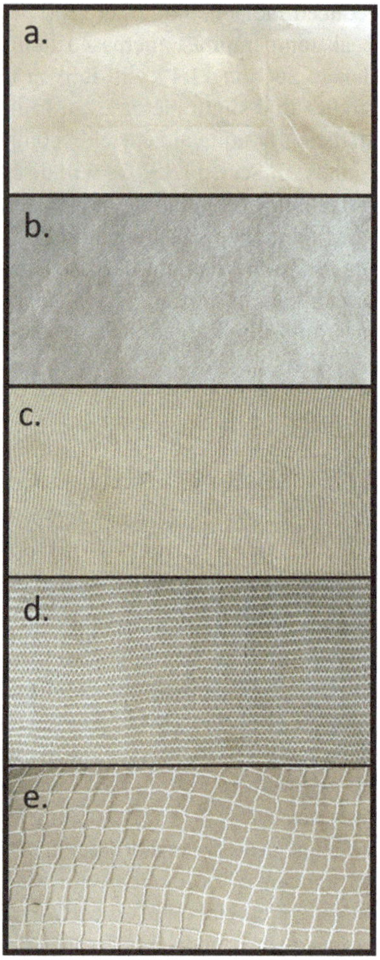

Fig. 11.1 Five types of agricultural barriers. (**a**) Solid plastic film, (**b**) spunbond fabric, (**c**) fine-mesh exclusion netting, (**d**) course-mesh hail netting, and (**e**) bird netting

11.2.2 The Evolution of Row Covers for Usage in Pest Exclusion

Many of the same protection barriers that guard against the weather also simultaneously protect crops against pests. The simple action of covering crops with a protective net or fabric can keep pests away while crops mature to harvest. Now, spunbond polyester fabrics are used for pest control purposes (Perring et al. 1989; Adams et al. 1990; Natwick and Laemmlen 1993; Andersen et al. 2006; Rojas et al. 2011; Skidmore et al. 2017).

In North America, this material is used for high-value crops that are extremely susceptible to damage from pests or diseases. As such, it has been used to protect melons, squash, and other cucurbitaceous crops from insects that vector diseases including many viruses and the bacterial wilt disease *Erwinia tracheiphila* (Perring et al. 1989; Natwick and Laemmlen 1993; Skidmore et al. 2017; Nelson and Gleason 2017, 2018; Rojas et al. 2011). The efficacy of these nets across this literature showed promising results. Perring et al. (1989) found this fabric decreased attack of sweet potato whiteflies, *Bemisia tabaci* (Gennadius) (Hemiptera, Aleyrodidae), on muskmelon (*Cucumis melo*), delayed the onset of lettuce infectious yellows virus, but only improved fruit yield in one of two repeated trials. Rojas et al. (2011) found that spunbound netting could increase yields in muskmelon by effectively excluding cucumber beetles *Acalymma vittatum* (F.) and *Diabrotica undecimpunctata* (Mannerheim) (Coleoptera: Chrysomelidae) as well as reducing bacterial wilt incidence. For best disease and insect prevention, the netting had to stay in place for 10 days after melon flowers opened, so commercial bumblebee hives had to be placed under the netting to allow for pollination. The outcomes of Skidmore et al. (2017) echoed the importance of pollinators under spunbond fabric in muskmelon and summer squash (*Cucurbita pepo* L.) as well. Spunbond row covers were used to protect a brassica crop, komatsuna (*Brassica rapa perviridis*), against flea beetles which eat the marketable crop leaves of komatsuna (Andersen et al. 2006). The row cover was more effective at limiting flea beetle damage than traditional sprays with carbaryl and spinosads. Whether crop damage is vectored to plants by insect attack or directly caused by insect feeding, row covers are supported by the literature as an effective pest control strategy.

Some issues exist with spunbond fabrics that preclude them from usage in large-scale pest exclusions for *D. suzukii* in fruit crops. For example, there is a striking delay in blueberry ripening due to the shading qualities of spunbond fabrics as noted in a 2014 experiment in Maine, USA (Alnajjar et al. 2017). To partially alleviate the climatic issues caused by spunbond fabrics and solid films, some folks have looked to use plastic mesh nettings. Though mesh nets do not perfectly permit sunlight, rain, and air (Castellano et al. 2007, 2008), their improvement in these areas allowed experimentation in the use of barriers for pest management. Perhaps the most wide-scale adoption of mesh netting is seen in fruit crop systems where birds are a major pest.

11.2.3 Physical Barriers for Fruit-Eating Bird Exclusion

Some fruit producers choose to invest in elaborate structures to support a woven netting (Fig. 11.1e) for the exclusion of birds (Taber 2002; Anderson et al. 2013). A survey of fruit growers from major fruit-producing regions of the USA (California, Michigan, New York, Oregon, and Washington) found that around 45% of wine grapes are protected by bird net. Around 10% of sweet and tart cherry growers and 25% of blueberry growers protect their fruits with bird exclosures (Anderson et al. 2013). One of the earliest mentions of a plastic mesh net for bird exclusion came from Minnesota, USA, in 1973 when the production of a plastic mesh was commissioned for research to protect grapes, blueberries, and cherries from birds (Stucky 1973). This mesh permits the entrance of insects, and blocks nearly no light, rain, nor airflow. However, difficulties in installing and removing these nets made their usage rare, until the early 2000s when creative advances in mechanisms to deploy nets increased the rates of their usage (Taber 2002). Notably, Stucky (1973) explained that very different support structures were needed to effectively exclude birds between these three different crops. Two methods of applying bird netting existed at the time, and both persist as common methods today. Individual rows of fruit crops can be covered with a single sheet of netting, meaning that across a field, many sealed tunnels of netting exist. Alternatively, multiple sheets of netting can be tied together to encapsulate an entire crop field as a full canopy exclosure. In both forms, the netting is stretched to reach the ground, where it is weighted or staked into place to exclude birds from entering at ground level.

11.3 Physical Barriers for *Drosophila suzukii* Exclusion

The exclusion of insects from fruit crops can be achieved by replacing wide-meshed bird net with smaller-meshed netting, and some plastic manufacturers now offer mesh marketed specifically for agricultural insect control (Fig. 11.1c). Like bird nets (Fig. 11.1e), nets of this type are woven or extruded and are most often composed of high-density polyethylene plastics (Castellano et al. 2008). For the purposes of insect exclusion, the size of net pores is an important consideration because if the common pests for a crop are, for instance, large-bodied beetles, a less expensive, larger mesh size can be used (Fig. 11.1d) (Archer et al. unpublished data). However, as interest grows in the use of exclusion barriers to prevent attack from *D. suzukii*, it is clear that net pores must be smaller than the body size of *D. suzukii*. The widths of male *D. suzukii* range from 0.70 to 0.94 mm, whereas female widths range from 0.85 to 1.24 mm (Kawase and Uchino 2005). A common guideline, and one supported by Kawase and Uchino (2005) is that net pores must be equal to or smaller than 1 mm by 1 mm to exclude these flies. More conservative suggestions also state that net pores should be as small as 0.5 mm by 0.8 mm (Grassi and Pallaoro 2012).

Due to the clandestine nature of these flies, maintaining a sealed exclosure at all times is important to prevent fruit infestation. In practice, we see that researchers and producers tend to carefully secure the netting to the ground with weights (such as bags filled with stones) more carefully than in exclosures for birds. Alternatively, the ends of the mesh net can be buried underground as in Rogers et al. (2016) or attached to permanent solid beams with plastic clips as in Riggs et al. (2016). A professor emeritus at Iowa State University and current blueberry producer, Dr. Lois Wright Morton, also secures her fine mesh net in this way (Fig. 11.2). However, fruit crops must be harvested over time, as not all fruits will mature to harvest simultaneously. We see a variety of measures to allow fruit harvest without compromising the exclosure (Fig. 11.3). The simplest method is that harvesters can briefly lift the netting in order to enter the exclosure before sealing the net behind them (Fig. 11.3a). Harvest can occur along the entire length of the exclosure before harvesters lift the netting once again to exit. Occasionally, some fruit crops are grown under existing

Fig. 11.2 One producer of blueberries established a half-acre plot covered with fine-mesh insect netting. The netting is supported on metal hoops and attached to horizontal supports at ground level. (Image Credit to Dr. Lois Wright Morton, Iowa State University)

high tunnels topped with permanent, impermeable plastic sheets. In these instances, researchers may choose to seal the open ends of their high tunnels with fine mesh, then install a netted swinging door on both ends to permit harvesters and tractors (Fig. 11.3b) Another highly engineered solution has been used by still other producers. By covering a large portion of a fruit crop field with one single net exclosure, a vestibule created by two zippered fabric doors allows entrance and exit to the plot (Fig. 11.3c). Harvesters open the first door to enter the vestibule before zipping the door shut and entering the plot.

11.4 Effectiveness of Mesh Exclusion Barriers for *Drosophila suzukii* Management

11.4.1 Literature Search

In the current collection of peer-reviewed articles and cooperative extension publications, we identified 12 papers that experimentally tested the ability of mesh nets for *D. suzukii* management. We identified seven studies in the production of

Fig. 11.3 Methods to access netted exclosures. (**a**) Ryan Kuesel and David Gonthier created no entrance to their blackberry exclosures. (**b**) Hannah Lee and colleagues created a netted swinging door to permit tractor access to the exclosure. Image Credit to Heather Leach, Penn State University. (**c**) Dr. Lois Wright Morton made a netted vestibule entrance to a full canopy exclosure. (Image Credit to Dr. Lois Wright Morton, Iowa State University)

blueberries, one in blackberries, three in red raspberries, and one that experimented in both wine grapes and cherries. The studies utilized a number of methods to evaluate the effectiveness of the netting barriers, and taken together they provide evidence that fine-mesh exclusions are effective at controlling *D. suzukii* in fruit crops. This efficacy for fruit crop protection has been compared against management both with and without pesticidal sprays, so we will identify the non-netting management strategy used for each study.

11.4.2 Methods to Evaluate Net Effectiveness

The 12 papers utilized a number of methods to evaluate the effectiveness of the netting barriers such as fruit dissection, fruit incubation, baited trapping, fruit dissections, and fruit float tests.

Baited trapping is performed by containing a liquid drowning solution and a liquid olfactory bait within a semi-permeable plastic container. A common trap design is a lidded plastic container perforated with holes just large enough to permit entrance to small flies. These containers are filled with a combination of fermenting yeasts and sugar, wine, or vinegar. Traps are then deployed into the field both inside and outside netted exclosures, and the difference in number of flies trapped acts as a metric for the effectiveness of the mesh net.

Fruit dissection has taken two forms in research: (1) removal of eggs, larva, or pupa from fruits under magnification and (2) float tests, whereby fruits are immersed in a solution of sugar and salt and crushed. This forces larval *D. suzukii* to leave the fruits and float to the top of the liquid allowing their numbers to be counted.

Finally, fruit incubation involves the storage of harvested fruits for a consistent number of days (up to 14 days). This allows maggots to mature, pupate, and emerge as adults. Through this process, the number of maggots within harvested berries are estimated, though potential mortality in the larval and pupal stages are overlooked. In this section, we will summarize this literature by looking into each *D. suzukii*-threatened crop individually.

11.5 General Findings

The literature contained 14 experiments from 12 publications using fine-mesh netting. We summarize their finding in Table 11.1. Five out of six blueberry studies found exclusion netting had lower levels of *D. suzukii* than did non-insecticide-treated controls. Further, both blueberry studies that tested against pesticide sprays found exclusion to be more effective than conventionally sprayed plants. The study in cherries found better effectiveness of netting than both an unsprayed

Table 11.1 Summary of findings on *D. suzukii* exclusion and secondary pest exclusion across the literature

				Compared netting barriers to non-insecticide control		Compared netting barriers to spray treatment		
Authors & Year	Crops	Journals	Other pests addressed	Nets reduced D. suzukii best	Nets reduced other pests best	Nets reduced D. suzukii best	Nets reduced other pests best	Organic or conventional
Grassi and Pallaoro (2012)	Cherry	Ecofruit	No	Yes	x	Yes	x	Conventional
Kuesel et al. (2019)	Blackberry	Insects	Yes	x	x	Yes	Yes	Organic
Archer et al. (2020)	Blueberry	Insects (In Review)	Yes	Yes	Yes	x	x	x
Kawase et al. (2008)	Blueberry	Bulletin of Chiba Prefectural	No	Yes	x	x	x	x
Grassi and Pallaoro (2012)	Blueberry	Ecofruit	No	Yes	x	Yes	x	Conventional
Riggs et al. (2016)	Blueberry	NY Fruit Quarrterly	No	x	x	x	x	Conventional
McDermott and Nickerson (2014)	Blueberry	NY Fruit Quarrterly	Yes	Equal	Yes	x	x	x
Cormier et al. (2015)	Blueberry	IOBC-WPRS Bulletin	Yes	Yes	Yes	x	x	x
Alnajjar et al. (2017)	Blueberry	Int J Entomology and Nematology	No	Yes	x	Yes/Equal	x	x
Ebbenga et al. (2019)	Grape	Horticultural Entomology	Yes	Yes	Yes	Yes	Yes	Conventional
Rogers et al. (2016)	Raspberry	J Pest Science	No	Yes	Yes	Equal	x	Conventional

Leach et al. (2016)	Raspberry	J Economic Entomology	Yes	Yes	x	x	x	Conventional
Leach et al. (2016)	Raspberry	J Economic Entomology	Yes	Yes	Yes	Yes	x	x
Stockton et al. (2020)	Raspberry	Crop Protection	No	x	x	x	x	Conventional

control and conventional sprays. All three raspberry studies that used non-insecticidal controls found fewer *D. suzukii* under nets and two of three found greater effectiveness compared to conventionally sprayed plants. The only study in blackberry production found netting to be more effective than organic pesticide sprays. The only study in wine grapes found netting to be more effective than an unsprayed control, and equally as effective as conventional pesticides. Importantly, the effectiveness of these exclusion barriers was rarely perfect and varied within season and across methodologies. This makes the findings of each study context-dependent and important considerations are needed for each crop, region, and method.

11.6 Studies in Blueberry

Seven studies evaluated the effectiveness of netting in blueberry. These studies ranged in age from experiments beginning in 2003 to one that concluded in 2019. This analysis in blueberries is represented by experiments from Japan, Italy, and the USA, and all of them showed that mesh nets can be more effective than either a pesticide spray regime or an unsprayed control.

The earliest publication examining fine meshes for *D. suzukii* control comes from the Chiba prefecture of Japan. Significantly fewer larva were trapped in berries under fine-mesh netting compared to management with only bird netting. By covering replicates of small blueberry plots with either a fine mesh (0.98 mm × 0.98 mm) or a bird net for three summers (2003, 2004, and 2005), Kawase was able to show a significant exclusion of *D. suzukii* (Kawase et al. 2008). The float test, performed on 6 separate days across the 3 years found no *D. suzukii* had entered into the fine-mesh exclosures, while an average of 33 flies were present per 100 blueberries under bird netting.

Grassi and Pallaoro (2012) examined the effectiveness of sealing plastic-covered high tunnels with a fine-mesh net (0.5 mm by 0.8 mm) in Trentino, Italy. Significantly fewer adult flies and larva were found in plots under fine-mesh-netted high tunnels compared to the non-insecticide control high tunnels. One hundred blueberries were searched weekly for 2 months from treatments of two netted tunnels and an unnetted control tunnel. Fruits collected from netted tunnels suffered no *D. suzukii* infestation until the final week of harvest, where 7.7% of fruits in one of the two tunnels showed infestation. The other netted tunnel remained uninfested. Fruits harvested from the non-insecticide tunnel peaked at 77% infestation. Weekly baited traps in each treatment replicate revealed zero or one *D. suzukii* adult was caught within netted tunnels, while an average of 33 flies were trapped each week in the unnetted tunnel. In the final week of harvest, concerning levels of infestation were found in one netted tunnel. This likely arose during harvest procedures. The opening of the net during blueberry harvest may have permitted the entrance of *D. suzukii* adults which could then breed inside the exclosure.

Cormier et al. (2015) found significantly fewer adult flies and larva under fine-mesh exclusions compared to non-insecticide treatment controls. The contents of baited traps from four netted and eight control replicates, each with five blueberry bushes, were examined for the effectiveness of this fine-mesh net (0.6 mm by 1 mm). They found no adult *D. suzukii* in traps under the net exclusions, while their control traps in non-insecticide bushes caught 16 flies. Using the incubation method, no *D. suzukii*-infested fruits were found under fine-mesh nets. In their control plots, high levels of *D. suzukii* infestation were found in the fourth, fifth, and sixth weeks of their experiment when an average of around 10, 50, and 72 adults emerged per 100 berries incubated.

A 2013 experiment in New York, USA, compared the effects of netting barriers in units of three blueberry bushes with a fine-mesh net (0.85 mm by 1 mm) (McDermott and Nickerson 2014). Only three *D. suzukii* were caught in non-insecticide-sprayed plots using mass bait trapping, while no flies were caught in fine-mesh treatments. However, there was no statistical difference between treatments. This was due to very low populations of *D. suzukii* at the study site.

In 2014, a two-year experiment was performed by researchers and a commercial blueberry grower (Riggs et al. 2016). This experiment looked to test the efficacy of a fine mesh (0.6 mm by 1 mm) and a course mesh (0.95 mm by 1.95 mm) against conventionally sprayed plants. They found significantly fewer larva under the fine-mesh exclusions compared to replicates managed with conventional pesticides and replicates managed with course mesh net. By examining harvested fruits from each of the three treatments through fruit incubation, the fine-mesh net significantly outperformed a sprayed control with a 10-week average of 0.53% compared to 15 percent of berries infested. The worst performer by far was the coarse-mesh net where 60% of fruits were found to be infested. Infestation in fruits under the fine mesh was also first detected 2 weeks later than in both other treatments. In the next year's experiment, only the fine mesh was compared against conventional pesticide sprays, and once again it significantly reduced *D. suzukii* infestation with a maximum weekly rate of 0.37% berries infested compared to 20%.

A 3-year study in Maine, USA, examined the effect of fine-mesh nets (0.72 mm by 0.97 mm) between 2014 and 2015, while spunbond fabric was examined in 2016 (Alnajjar et al. 2017). Significantly fewer *D. suzukii* larva were found under fine-mesh exclusions compared to unsprayed replicates. Here, 236 mL of blueberries were harvested from each replicate and examined using the flotation method. In both years where fine-mesh nets were used, the average number of *D. suzukii* larva per 236 mL sample were significantly lower in netted treatments (0.29 larva and 0.13 larva) compared to an untreated control (9.7 larva and 2.2 larva). The 2016 spunbond fabric trial also yielded success with 0.2 larva compared to 5.9 larva per 236 mL of berries.

A 2019 study carried out in Kentucky, USA, examined the effectiveness of three types of mesh netting for the control of *D. suzukii*, scarab beetle pests, and birds (Archer et al. 2020 unpublished data). Bird netting (19 mm by 19 mm), a course-mesh netting (1 mm by 2 mm), and a fine-mesh netting (1.4 mm by 0.85 mm) were trialed in order to tease apart the amounts of yield lost to these three classes of pests. There were significantly fewer *D. suzukii* adults in blueberry bushes under fine-mesh exclusions compared to course-mesh exclusions, bird net exclusions, and unsprayed management. Individual branches containing approximately 100 blueberries each were flagged off, and these berries were monitored until they achieved harvestable ripeness. Those berries were analyzed for pest damage and incubated to measure *D. suzukii* presence. A baited trap was paired under the netting of each plant and collected weekly for 11 weeks. The number of *D. suzukii* caught in these baited traps showed that the fine-mesh net outperformed all other treatments and an individual exclosure permitted less than one fly per week on average. Surprisingly the exclosures of course-mesh netting and bird netting were statistically equivalent (containing approximately 9 and 17 flies per week, respectively) but significantly effective compared to unnetted plants, where 36 flies were found on an average week. The analysis of incubated blueberries yielded few emerged adults, however. This emergence data showed no significant difference between exclosures composed of any net types nor the control.

11.7 Studies in Cherry

The only publication on fine-mesh nets for control of *D. suzukii* in sweet cherries found significantly fewer larva in cherries under fine-mesh netting compared to cherries managed both with conventional pesticides and as a non-insecticide control (Grassi and Pallaoro 2012). Here, they sewed large bags of fine-mesh net (0.5 mm by 0.8 mm) and sealed individual trees on two separate farms just before cherries began to color. At each farm they compared 250 fruits harvested equally from five replicates of netted trees, conventionally sprayed trees, and non-insecticide control trees at two time periods. By searching fruits for *D. suzukii* eggs, they found that fine-mesh nets nearly perfectly protected fruits from *D. suzukii* oviposition. One deposited egg was found inside bagged trees across the trial at the two farms. This is in comparison to high levels of fruit infestation in sprayed plots (70% and 94% of fruits infested) and a peak of 100% cherry infestation (at both farms) in one week's harvest of the non-insecticide control fruits.

11.8 Studies in Red Raspberry

Four experiments across three publications evaluated the effectiveness of netting in raspberry. These experiments were all performed between 2014 and 2017. Three of these experiments compared netting against non-insecticidal management, and all three found fine-mesh netting effective. Also, three experiments tested against conventional insecticide sprays. Two found nets more effective than the insecticide regime, while the third found nets equally as effective.

In 2014 and 2015 in Michigan, USA, two experiments were performed in regard to mesh nets (Leach et al. 2016). In one study, tall cages constructed of extremely fine fabric mesh were repurposed from commercial insect enclosures and were placed over less than 2 meters of raspberry row. In a factorial method, they applied conventional pesticides to raspberries in half of the plots under cage and open management so that four treatments existed: caged, caged and sprayed, uncaged, and uncaged and sprayed. Using the float method, unsprayed and uncaged plots tended to have the highest number of larval *D. suzukii* inside of fruits. Individual weekly numbers of larva varied quite significantly among the treatments, though, meaning that identifying the statistical significance of each treatment was difficult. When averaged across all harvest weeks, sprayed and netted raspberries did have the fewest larva inside while unsprayed nets and sprayed unnetted berries were equivalently effective.

In their second experiment, mesh nets (1.0 mm × 0.6 mm) were used to seal off plastic-film-topped high tunnels (Leach et al. 2016). These fine-mesh-netted high tunnels were paired with identical high tunnels as non-insecticide controls. The presence of *D. suzukii* was monitored in both marketable and over-ripe berries underneath sealed high tunnels versus unsealed high tunnels via the float test and baited traps. Under fine-mesh-netted high tunnels, 82% fewer eggs were found in raspberries, 74% fewer larvae were found in fruits, and 65% fewer adults were trapped compared to non-insecticide plots.

In 2015 in Minnesota, USA, mesh nets (1.0 mm × 0.6 mm) were used to net small, three-meter-long low tunnels (Rogers et al. 2016). An equal number of replicate tunnels were also sealed off with solid plastic films. They compared the *D. suzukii*-excluding effectiveness of fine-mesh nets and plastic films to non-insecticide management and conventional pesticide sprays. Though netting did not achieve perfect *D. suzukii* exclusion, significantly fewer larva were found in berries under netting and plastic compared to conventionally sprayed berries and non-insecticide berries. Using the incubation method, a lower percentage of berries were infested with larva in fine mesh and plastic-covered tunnels than in the non-insecticide and conventionally sprayed replicates. Over the course of the study, over 80% of fruits in the non-pesticide plot were infested and 60% of conventionally sprays fruits were infested. Over 34% of fruits under the fine mesh were infested, and only 2% under the plastic film were infested.

Researchers in New York, USA, performed a 2016 and 2017 experiment to test fine-mesh nets (1.0 mm × 0.6 mm) against conventionally sprayed red raspberries (Stockton et al. 2020). Two-meter-tall structures supported netting in 12-meter-long replicates of red raspberries. Early in the season, netting sufficiently controlled *D. suzukii* infestation, where 97.9% (year 1) and then 86.8% (year 2) fewer *D. suzukii* were found under nets than in conventionally sprayed plots. The effectiveness of nets fell later in the season and in some netted replicates, *D. suzukii* populations under fine-mesh netting were higher than in conventionally sprayed plots.

11.9 Studies in Blackberry

The only manuscript that examines fine-mesh nets for control of *D. suzukii* in blackberries comes from a 2018 study in Kentucky, USA (Kuesel et al. 2019). Significantly fewer adult flies and larva were found in fine-mesh netting treatments compared to organic insecticide un-netted treatments. Here, Kuesel et al. (2019) tested whether fine-mesh netting (0.85 mm by 1.4 mm) would control *D. suzukii* over 4 weeks. Three twenty-five-meter-long rows were netted and compared to three unnetted rows which received organic pesticide sprays. Incubation and baited traps were employed here. Few *D. suzukii* were trapped in the first 3 weeks of the study so no effect of treatment was found. However, in the fourth week, an average of six adults were trapped per control row, while an average of one adult was trapped per exclusion net row. From the incubation of blackberry fruit, adult *D. suzukii* emergence was significantly lower in exclusion treatments than in insecticide treatments. On average, blackberries in organic insecticide treatments contained nearly 31 times the number of *D. suzukii* flies per fruit than fruits under fine-mesh netting treatments.

11.10 Studies in Grape

The usage of fine-mesh nets for *D. suzukii* exclusion in grapes is represented in the literature by one study in 2017 and 2018 in Minnesota, USA. Here, the production of wine grapes under exclusion netting was tested against management with conventional pesticides and a non-pesticide control (Ebbenga et al. 2019). In order to fit the structure of wine grapes, the top canopy of the grapes was closed in a tube of net (0.6 mm by 1 mm), and the netting was sealed around each woody stem with metal wire. Here they also artificially inoculated half of their netted plots with 25 male and 25 female *D. suzukii* in order to observe how their populations may act if a breach in the exclosure permitted adult flies inside. The assessment of *D. suzukii* presence and oviposition was performed both with baited traps and by inspecting grapes for larva. The efficacy of their treatments varied across the 2 years of study and between sampling weeks, but taken in whole, netted grapes showed 95% lower grape infestation than the non-pesticide control. On some weeks, net treatments with and without artificial infestation showed statistically equivalent numbers of adult flies and larva compared to conventionally sprayed treatments. On other sampling weeks, both net treatments showed lower adult and larva infestation than the pesticide treatment, however. To explain the lack of significant oviposition within artificially infested plots, the authors discussed that *D. suzukii* females have trouble cutting through the skin of intact grapes during oviposition. Females instead tend to lay eggs in berries damaged by birds, yellow jacket wasps, or storms, but since wasps and birds were excluded by this netting, this damage was infrequent.

11.11 Additional Notable Effects

While studying the efficacy of fine-mesh nets for *D. suzukii*, it is clear that netting fruit crops can cause a variety of other changes including protection against bird and other arthropod pests, differences in the rates of flower pollination, struggles with weed management, changes in microclimate conditions within plots, and increases in disease incidence.

11.11.1 Fine-Mesh Netting Exclusion of Pests and Pollinators

Since fine-mesh exclusion netting is small enough to exclude *D. suzukii*, it also excludes larger insects and birds. Kuesel et al. (2019), Archer et al. (2020 unpublished data), and Ebbenga et al. (2019) all found fine-mesh netting excluded birds better than un-netted treatments. Further, fine-mesh nets reduced insect pest groups compared to controls: yellow jackets (Ebbenga et al. 2019), non-*Drosophila suzukii* flies (McDermott and Nickerson 2014; Kuesel et al. 2019), and all forms of unidentifiable insect damage Cormier et al. (2015). Leach et al. (2016) found pest insects overall were 44 percent less abundant under mesh-sealed tunnels than open tunnels; the presence of thrips, mites, leafhoppers, aphids, and raspberry beetles was all reduced. Archer et al. (2020, unpublished data) found that fine-mesh nets reduced Japanese and green June beetle abundance in blueberries. We summarize these findings in Table 11.1 under the two columns pertaining to "other pests."

However, fine-mesh netting also excludes natural enemies of pests and pollinators; an important consideration given that many fruit crops are pollinator-dependent crops. Importantly for raspberry crops, which are pollinator-dependent and flower multiple times across a given season, Leach et al. (2016) found 77% fewer pollinators under netted versus open high tunnels. Commercial bumblebee hives were stocked into each high tunnel to allow for pollination. Stockton et al. (2020) performed a similar procedure with bumblebees to ensure raspberry pollination. When they did not include these bumblebees, they saw a significant increase in "crumbly berry" which is a symptom of poor pollination.

11.11.1.1 Effects on Temperature, Humidity, and Disease

Some studies on fine-mesh net for *D. suzukii* control also analyzed how netting fruits may change the climate surrounding fruit crops. Some small changes in temperatures and humidity were found which could potentially alter the process of fruit ripening or cause higher rates of disease. This suite of literature, however, largely did not discuss disease incidence except for Stockton et al. (2020) who did not find different rates of botrytis gray mold incidence under fine-mesh exclusions. Within blueberries, two studies found no alteration in the temperature within the plots (Grassi and Pallaoro 2012; Cormier et al. 2015). Conversely, McDermott and Nickerson (2014) and Riggs et al. (2016) noted that temperatures under fine-mesh nets were slightly higher than unnetted controls. Slight increases in humidity were also found by two studies (Cormier et al. 2015; Riggs et al. 2016), and an increase in the splitting of blueberries was noted due to excess water content (Cormier et al. 2015). In raspberries grown under high tunnels, Stockton et al. (2020) and Leach et al. (2016) found no differences in temperature between netted and non-netted plots. Red raspberries grown in fine-mesh-covered low tunnels did see higher peaks in maximum and minimum temperatures as well as higher humidity (Rogers et al. 2016). Similarly, inside small, 1.8 by 1.8-m mesh-fabric cages maximum temperatures were slightly significantly higher than outside temperatures (Leach et al. 2016). When tubes of fine-mesh were used to protect wine grapes, Ebbenga et al. (2019) noted no differences in average temperatures. Differences in temperatures and humidity seem to be found in small-scale netting exclusions such as single row exclusions in McDermott and Nickerson (2014), low tunnels in Rogers et al. (2016) and small mesh cages in Leach et al. (2016). Evidence of climatic differences in large, multi-row canopy exclosures comes only from Riggs et al. (2016).

11.12 Effects on Total Yield

Through the reduction of damage from *D. suzukii* and other pests fine-mesh nets can increase total yields while maintaining the quality of harvestable fruit when compared to more widely used pest control strategies. However, still only seven studies have compared yields in netted trials. Overall, one study in blueberries and one in raspberries showed higher yields of fruit under management with fine-mesh nets as opposed to non-insecticidal management (Rogers et al. 2016 and Archer et al. 2020 unpublished data). Three experiments showed equivalent yields to non-insecticidal management (McDermott and Nickerson 2014; Cormier et al. 2015; Leach et al. 2016).

11.12.1 Yield in Blueberries

In blueberries, the effect of netting on yield and fruit quality were quite variable with no clear pattern. Kawase et al. (2008) found that while fine-mesh netting did not alter the amount of sugar present within blueberry fruits, it did increase their acidity and may cause some pale discoloration of berries (Kawase et al. 2008). McDermott and Nickerson (2014) found that blueberries under a fine-mesh net had slightly higher concentrations of sugar than berries under untreated control. They found that the total yields of harvested blueberries differed only very slightly between their factorial treatments, with year-long yields differing only by around 2.25 kg across all treatments; however, the average fruit was larger in the untreated control than under netting. Conversely, Cormier et al. (2015) found that blueberry fruits were significantly larger under fine mesh, but netting exclusion had no significant effects on total yield or sugar content. When Archer et al. (2020 unpublished data) examined the yields between a fine-mesh net, a course hail net, a bird net, and an unnetted control, unnetted bushes had 54.1% lower yield than the coarse-mesh netting treatment and 60.6% lower yield than the fine-mesh netting treatment. Bird netted bushes had 32.9 lower percent yield than the fine-mesh netted bushes. Here, between 13 and 15% of lost blueberries could be attributed to small pests including *D. suzukii*. Netting treatments did not affect the amount of sugar within each berry in this experiment.

11.12.2 Yield in Red Raspberries

In Rogers et al. (2016), two metrics of yield were examined for red raspberries grown in Minnesota, USA. The total weight of harvested berries and the percentage of undamaged, marketable berries was examined over 10 harvests. Here, the total yields across fine-mesh-covered high tunnels, plastic-film-covered high tunnels, conventionally sprayed plots, and unmanaged plots were variable but did not show significant differences between netted, conventionally sprayed, nor unmanaged plots. Notably, however, the percentage of marketable raspberries was greater under netting and plastic film-covered treatments compared to conventionally managed and unmanaged plots.

Raspberries grown by Leach et al. (2016) showed little differences in yields nor fruit quality. The average weights of raspberries were not significantly affected by netting (2.7 g under nets and 2.5 g without net). The average diameters of raspberries were not affected either (16.1 mm under nets and 15.7 mm without net). The sugar content of raspberries was not affected either (7.5°Brix under nets and 7.1°Brix without net).

Stockton et al. (2020) looked at marketable yields of red raspberries per meter of row and found that fine-mesh exclusion plots had significantly higher marketable yields than unmanaged control plots; however, when the overall yield across the season was examined, no difference was found between treatments.

11.12.3 Yield in Blackberries

In Kuesel et al. (2019), the yield of marketable blackberries was examined. Yields were on average 2.04 times higher for fine-mesh netting exclusion treatment (210 g per row each week) relative to organic insecticide treatment rows (103 g per row each week).

Table 11.2 Summary of findings on fruit crop yields under fine-mesh exclusion nets compared to organic, conventional, or unmanaged fruits

Authors & Year	Crop	Journal	Fruit yield greater than non-pesticide treatments	Fruit yield greater than pesticide treatments	Organic or conventional pesticides
Kuesel et al. (2019)	Blackberry	Insects	x	Yes	Organic
Archer et al. (2020)	Blueberry	Insects (In Review)	Yes	x	x
McDermott and Nickerson (2014)	Blueberry	New York Fruit Quarterly	Equal	x	x
Cormier et al. (2015)	Blueberry	IOBC-WPRS Bulletin	Equal	x	x
Rogers et al. (2016)	Raspberry	Journal of Pest Science	Yes	Yes	Conventional
Leach et al. (High Tunnel) (2016)	Raspberry	Journal of Economic Entomology	Equal	x	x
Stockton et al. (2020)	Raspberry	Crop Protection	x	Equal	Conventional

11.12.4 Yield Summary

We summarize fruit yield findings within the literature in Table 11.2. Due to the trend of positive evidence for the effectiveness of fine-mesh nets, it seems that fine-mesh exclusions are effective for the control of *D. suzukii*. Perhaps additional studies across a variety of crops and hardiness zones should be conducted to provide the scientific and farming communities with a better understanding of the possible function of fine-mesh exclusion netting in their region.

11.13 Future Research Needed to Get to Usage—Economic Analyses

A key challenge for the utility of fine-mesh netting is the high potential cost. Economic analyses on the subject are incomplete and exist only as material costs. One such cost estimate comes from Riggs et al. (2016). Here, covering half of an acre of blueberries with fine mesh (0.6 mm by 1 mm) costs 4600 USD, so one full acre of their system may cost 9200 USD. The authors suggested that a 5 year lifetime may be achieved for netting through proper care, such as patching torn holes. McDermott and Nickerson (2014) suggest that enclosing an acre of blueberries with 0.85 mm by 1 mm mesh net would cost between 7000 and 9000 USD. However, it is worth consideration that the installation of bird netting systems in highly susceptible crops such as blueberries is also expensive. In blueberry, cherry, and grape production where bird net is used, netting one acre of fruit crop can cost 2280 USD (Dellamano 2006).

Exclusionary netting is a leading option for non-insecticidal control of pests in small fruit crops. Its use in commercial agriculture will be determined by its cost effectiveness, which will require further comparisons of input costs, yield benefits, and longevity of netting materials compared to increases in fruit yield. Labor requirements for its installation may be costly at the start and end of production season as the netting is put up before fruit coloring and as it is removed at the end of fruit harvest. The durability of nets for on-farm use must also be analyzed in detail and may require individual farms to trial netting for their own independent scenarios. Statements from the netting manufacturer estimate the netting will last for 5 years. Studies published in the literature, however, place the longevity of similar fine-mesh nets from 7 to 10 years (Leach et al. 2016; Mazzi et al. 2017; McDermott and Nickerson 2014). Regardless, additional in-depth economic and agronomic analyses will be required to determine whether it is more profitable to switch from current organic pest control practices to exclusionary netting in blackberry production. Future studies should look to compare the costs of both the material inputs and labor required by both management schemes.

11.14 Conclusion

There is mounting evidence that fine-mesh netting can alleviate the effects of *D. suzukii* in a variety of fruit crops. Eleven out of twelve papers found that these barriers are effective at reducing *D. suzukii* presence and infestation of fruits, while three out of seven manuscripts found that netting can increase the yields of harvestable fruits. This suggests that netting can be a viable alternative or companion to insecticide treatment. In some parts of the world, absolutely no larval infestation in berries is the goal for producers because some markets practice a zero-tolerance policy for larval presence (Bruck et al. 2011; Van Timmeren and Isaacs 2013). For producers who aim to sell to zero-tolerance markets, exclusion netting may be combined with organic or conventional pesticide sprays. In less strict markets, growers who strive to help meet the demand for fresh, local, and organic fruit may make effective use of netting on its own to meet their pest control needs.

References

Adams RG, Ashley RA, Brennan MJ (1990) Row covers for excluding insect pests from broccoli and summer squash plantings. J Econ Entomol 83:948–954

Alnajjar G, Collins J, Drummond FA (2017) Behavioral and preventative management of *Drosophila suzukii* Matsumura (Diptera: Drosophilidae) in Maine wild blueberry (*Vaccinium angustifolium* Aiton) through attract and kill trapping and insect exclusion-netting. IJEN 3:51–61

Andersen CL, Hazzard R, VanDriesche R et al (2006) Alternative management tactics for control of *Phyllotreta cruciferae* and *Phyllotreta striolata* (Coleoptera: Chrysomelidae) on *Brassica rapa* in Massachusetts. J Econ Entomol 99:803–810

Anderson A, Lindell CA, Moxcey KM et al (2013) Bird damage to select fruit crops: the cost of damage and the benefits of control in five states. Crop Prot 52:103–109

Archer K, Kuesel R, Hicks DS, et al (2020) Effectiveness of insect and bird net barriers for control of pests in blueberry. Unpublished data

Berlinger MJ, Taylor RAJ, Lebiush-Mordech S et al (2002) Efficiency of insect exclusion screens for preventing whitefly transmission of tomato yellow leaf curl virus of tomatoes in Israel. Bull Entomol Res 92:367–373

Bolda MP, Goodhue R, Zalom FG (2010) Spotted wing drosophila: potential economic impact of a newly established pest. Agric Res Econ Update 13:5–8

Bruck DJ, Bolda M, Tanigoshi L et al (2011) Laboratory and field comparisons of insecticides to reduce infestation of *Drosophila suzukii* in berry crops. Pest Manag Sci 67:1375–1385

Castellano S, Candura A, Scarascia Mugnozza G (2007) Relationship between solidity ratio, colour and shading effect of agricultural nets. Paper presented at the international symposium on high technology for greenhouse system management: Greensys, 4 October 2007, vol. 801, pp 253–58

Castellano S, Scarascia Mugnozza G, Russo G et al (2008) Plastic nets in agriculture: a general review of types and applications. Appl Eng Agric Waaijenberg 24:799–808

Cormier D, Veilleux J, Firlej AJ (2015) Exclusion net to control spotted wing Drosophila in blueberry fields. IOBC-WPRS Bull 109:181–184

Dellamano F (2006) Controlling birds with netting: blueberries, cherries and grapes. J New York Fruit Quarterly 14:3–5

Ebbenga DN, Burkness EC, Hutchison WD (2019) Evaluation of exclusion netting for spotted-wing Drosophila (Diptera: Drosophilidae) management in Minnesota Wine Grapes. J Econ Entomol 112(5):2287–2294

Espi E, Salmeron A, Fontecha A et al (2006) Plastic films for agricultural applications. J Plastic Film Sheeting 22:85–102

Grassi A, Pallaoro M (2012) *Drosophila suzukii* (Matsumura), a revolution for soft fruits in Trentino. Paper presented at Ecofruit 15th international conference on organic fruit-growing. Proceedings for the conference, Hohenheim, 20 February 2012

Gress BE, Zalom FG (2019) Identification and risk assessment of spinosad resistance in a California population of *Drosophila suzukii*. Pest Manag Sci 75:1270–1276

Kawase S, Uchino K (2005) Effect of mesh size on *Drosophila suzukii* adults passing through the mesh. J Annual Report of the Kanto-Tosan Plant Protection Society Uchino 52:99–101

Kawase S, Uchino K, Yasuda M, et al (2008) Netting control of cherry drosophila *Drosophila suzukii* injurious to blueberry Vaccinium. J Bulletin of the Chiba Prefectural Agriculture Research Center Motoori

Kuesel R, Hicks DS, Archer K et al (2019) Effects of fine-mesh exclusion netting on pests of blackberry. Insects 10:249

Leach H, Van Timmeren S, Isaacs R (2016) Exclusion netting delays and reduces *Drosophila suzukii* (Diptera: Drosophilidae) infestation in raspberries. J Econ Entomol 109:2151–2158

Mazzi D, Bravin E, Meraner M et al (2017) Economic impact of the introduction and establishment of *Drosophila suzukii* on sweet cherry production in Switzerland. Insects 8:18

McDermott L, Nickerson L (2014) Evaluation of insect exclusion and mass trapping as cultural controls of spotted wing Drosophila in organic blueberry production. NY Fruit Q 22:25–28

Natwick ET, Laemmlen FF (1993) Protection from phytophagous insects and virus vectors in honeydew melons using row covers. Fla Entomol 76:120–126

Nelson H, Gleason M (2017) Improving row cover systems for organic management of Bacterial Wilt in muskmelon and squash-Year 1. Iowa State University Research and Demonstration Farms Progress Reports 4

Nelson H, Gleason M (2018) Improving row cover systems for organic management of Bacterial wilt in Muskmelon and Squash-Year 2. Iowa State University Research and Demonstration Farms Progress Reports 4
Perring TM, Royalty RN, Farrar CA (1989) Floating row covers for the exclusion of virus vectors and the effect on disease incidence and yield of cantaloupe. J Econ Entomol 82:1709–1715
Poling EB, Fuller HP, Perry KB (1991) Frost/freeze protection of strawberries grown on black plastic mulch. Hortscience 26:15–17
Rekika D, Stewart KA, Boivin G et al (2008) Reduction of insect damage in radish with floating row covers. Int J Veg Sci 14:177–193
Riggs DI, Loeb G, Hesler S et al (2016) Using insect netting on existing bird netting support systems to exclude spotted wing drosophila (SWD) from a small scale commercial highbush blueberry planting. NY Fruit Q 24:9–14
Rogers MA, Burkness EC, Hutchison WD (2016) Evaluation of high tunnels for management of *Drosophila suzukii* in fall-bearing red raspberries: potential for reducing insecticide use. J Pest Sci 89:815–821
Rojas ES, Gleason ML, Batzer JC et al (2011) Feasibility of delaying removal of row covers to suppress bacterial wilt of muskmelon (*Cucumis melo*). J Plant Dis Duffy 95:729–734
Rumpel J (1993) Plastic and agrotextile covers in pickling cucumber production. Paper presented at VII international symposium on timing field production of vegetable, 23 August 1993, vol. 371, pp 253–260
Skidmore A, Wilson N, Williams M et al (2017) The impact of tillage regime and row cover use on insect pests and yield in organic cucurbit production. Renew Agr Food Syst:1–11
Stockton DG, Hesler SP, Wallingford AK et al (2020) Factors affecting the implementation of exclusion netting to control *Drosophila suzukii* on primocane raspberry. Crop Prot 135:105191
Stucky JT (1973) Use of plastic netting bird control seminar proceedings, 121
Taber MR (2002) Netting applications for agricultural bird control. Proceedings of the vertebrate pest conference, vol. 20, p 20
Van Timmeren S, Isaacs R (2013) Control of spotted wing drosophila, *Drosophila suzukii* by specific insecticides and by conventional and organic crop protection programs. Crop Prot 54:126–133

Chapter 12
Management of *Drosophila suzukii* in Berry Crops

Oscar E. Liburd and Elena M. Rhodes

Abstract There has been significant progress in the development of management tools for *Drosophila suzukii* Matsumura populations. Initially, conventional growers relied almost exclusively on conventional insecticides, primarily synthetic pyrethroids, organophosphates, spinosyns, and neonicotinoids, for control of *D. suzukii*. Although these pesticides provided effective control, there has been an increase in secondary pest outbreaks due to the destruction of natural enemies that regulate these secondary pests. Recently, much emphasis has been placed in finding effective biorational pesticides as alternatives to these conventional pesticides. Organic growers had been limited to spinosad as the only effective organic option, which raised concerns over resistance development. Recently, other organic pesticides including azadirachtin + pyrethrins, *Chromobacterium subtsugae*, and sabadilla alkaloids have demonstrated some level of activity against *D. suzukii* and can be used in an organic rotation program. Cultural control tactics such as increasing harvest frequency, field sanitation, mulches, irrigation techniques, and exclusion netting have provided different levels of control for *D. suzukii* populations. The potential to use attract-and-kill techniques is currently being researched and has shown some efficacy. Finally, biological control for *D. suzukii* management has been studied intensively, and much information is available on predators, parasitoids, and pathogens that attack *D. suzukii*.

Keywords Spotted Wing Drosophila · Integrated Pest Management · Insecticide Rotation · Biological Insecticides · Cultural Control

---

O. E. Liburd (✉) · E. M. Rhodes
University of Florida, Gainesville, FL, USA
e-mail: oeliburd@ufl.edu

F. R. M. Garcia (ed.), *Drosophila suzukii Management*,
https://doi.org/10.1007/978-3-030-62692-1_12

12.1 Introduction

Spotted wing drosophila (SWD), *Drosophila suzukii* Matsumura, is an invasive pest seriously impacting the production of blackberries, blueberries, cherries, grapes, raspberries, strawberries, and other thin-skinned fruits around the world (Bellamy et al. 2013). *Drosophila suzukii* is an economic threat to fruit industries throughout North America (Hauser 2011; Walsh et al. 2011), Europe (Calabria et al. 2012), and South America (Deprá et al. 2014). Females possess a heavily sclerotized, serrated ovipositor that allows them to oviposit in ripening and ripe fruit (Hauser 2011). Male has a single spot on each of the wings. The spot location in *D. suzukii* differs from other indigenous European *Drosophila* species with black spots including *D. biarmipes* (Malloch) (Gompel et al. 2005) and *D. subpulchrella* (Takamori et al. 2006). *Drosophila suzukii* larvae develop inside the fruit and the presence of a single larva in a shipment of berries can cause that shipment to be rejected.

Current management practices for SWD in all impacted crops rely heavily on applications of insecticides (Bruck et al. 2011; Haviland and Beers 2012; Timmeren and Isaacs 2013). As this approach is not sustainable and can result in issues such as resistance development and secondary pest outbreaks, research into alternative tactics is ongoing. Tactics that have shown some efficacy include border sprays, attract-and-kill technologies, alternative oviposition sites in the form of a food-grade gum, various cultural control tactics, and biological control. An integrated pest management (IPM) strategy to manage SWD on different crops in growing regions throughout the world can be developed by incorporating various combinations of these tactics that will be discussed as appropriate for a given crop and region.

12.2 Insecticides

12.2.1 Conventional

Pesticidal tactics have been the primary tools used to manage high populations of *D. suzukii* on conventional farms since its arrival in the Americas in 2008 (Bruck et al. 2011; Timmeren and Isaacs 2013; Diepenbrock et al. 2016; Iglesias and Liburd 2017a). Pesticides from various classes including organophosphates, pyrethroids, spinosyns, ryanoids, and neonicotinoids have been effective but concerns about resistance development (Timmeren et al. 2019), maximum residue limits (MRLs) levels (Haviland and Beers 2012), and the negative effects on non-target organisms (Sarkar et al. 2020) have caused researchers to investigate other options for management. Different types of insecticide rotational programs have been investigated for management of *D. suzukii*. Several factors are considered when developing insecticide rotational programs including insecticide class, pre-harvest interval (PHI), and MRL levels.

Pesticides with short PHIs ranging from a few hours up to 24 h after application are usually in high demand by fruit growers who experience problems with *D. suzukii*. Once the fruit has reached maturity and ready to be harvested, growers want to be able to apply a pesticide (if needed) and harvest immediately without having to worry about pesticide residues on fruit. The MRL is the highest level of a pesticide residue that is allowed on a fruit. This level varies according to the country where the fruit is exported to (Haviland and Beers 2012). Therefore, depending on the target export market, growers take MRL into consideration when developing their insecticide rotational programs (Table 12.1).

12.2.2 Organic

The most effective organic insecticide against SWD is Entrust®, the organic formulation of spinosad (Fanning et al. 2018). Concerns over potential resistance development, which is already emerging in the Watsonville area of California (Gress and Zalom 2019), have produced research into both alternatives and compounds that can be used in rotation with Entrust®. Though used in some rotation programs, Pyganic (pyrethrins) has minimal impacts on SWD numbers and infestation (Timmeren and Isaacs 2013). Other organic insecticides that have shown some efficacy against SWD include *Chromobacterium subtsugae*, sabadilla alkaloids, and azadirachtin + pyrethrins (Fanning et al. 2018; Iglesias and Liburd 2017a). *Chromobacterium subtsugae* is a soil bacterium that acts as a repellent and antifeedant. Sabadilla alkaloids are made from sabadilla lily seeds and have a similar mode of action to pyrethroids. These products work best in rotation with spinosad.

12.2.3 Adjuvants and Phagostimulants

Two options for enhancing the efficacy of pesticides are the addition of adjuvants and phagostimulants. Adjuvants are added to pesticides to improve spray coverage (spreaders), allow more of a pesticide to adhere to the target crop (stickers), etc.

Table 12.1 Insecticide rotational program for *Drosophila suzukii* in southern highbush blueberries in North Central Florida

Program	Spray 1	Spray 2	Spray 3	Spray 4
1	Spinetoram	Malathion	Zeta-cypermethrin	Zeta-cypermethrin
2	Spinetoram	Phosmet	Malathion	Malathion
3	Fenpropathrin	Spinetoram	Malathion	Malathion
4	Malathion	Spinetoram	Zeta-cypermethrin + Bifenthrin	Zeta-cypermethrin + Bifenthrin
5	Spinetoram	Cyantraniliprole	Cyantraniliprole	Spinetoram
6[a]	Spinosad	Spinosad	Pyrethrins	Pyrethrins

aPesticides are labeled for organic use

(Foy 1996). The adjuvants poly-1-p-menthene, alcohol ethoxylate (Fig. 12.1), and polyether-polymethylsiloxane-copolymer did not improve the efficacy of azadirachtin, azadirachtin + pyrethrins, *Burkholderia* spp., *Chromobacterium subtsugae*, pyrethrins, sabadilla alkaloids, spinosad, or two hydrogen peroxide-based crop sanitizers against SWD even though both poly-1-p-menthene and alcohol ethoxylate caused some SWD mortality on their own (Roubos et al. 2019a).

Phagostimulants are food-based products added to insecticides to increase the target pest's exposure to the pesticide because the phagostimulants are attractive, both keeping the target pest in contact with the insecticide for longer and triggering an increase in feeding behaviors. Sucrose and yeasts have been examined as phagostimulants to improve SWD control by both conventional and organic insecticides. At 1.2 g/L, sucrose increased the efficacy of spinetoram, cyantraniliprole, and acetamiprid (Cowles et al. 2015). In laboratory studies, fermented strawberry juice, the yeast *Hanseniaspora uvarum*, and a combination of the two increased the efficacy of Spinosad, cyantraniliprole, and lambda-cyhalothrin against SWD (Noble et al. 2019). Unfortunately, neither the addition of sucrose nor the yeast *Saccharomyces cerevisiae* increased the efficacy of azadirachtin, azadirachtin + pyrethrins, *Burkholderia* spp., *Chromobacterium subtsugae*, pyrethrins, sabadilla alkaloids, or two hydrogen peroxide-based crop sanitizers against SWD (Roubos et al. 2019b).

Another line of research is focused on erythritol, a sucrose substitute that cannot be digested or converted to a digestible carbohydrate (Choi et al. 2017) that is toxic to SWD and other flies because it accumulates in the body causing an imbalance in osmotic pressure (Tang et al. 2017). Erythritol and some of its less expensive derivatives cause 80–100% mortality in all SWD life stages in laboratory experiments and reduced fruit infestation in a blueberry field by up to 93% (Sampson et al. 2019). The presence of sucrose sources and wounded berries reduces the efficacy of erythritol, however (Choi et al. 2019). The addition of erythritol increased the efficacy of

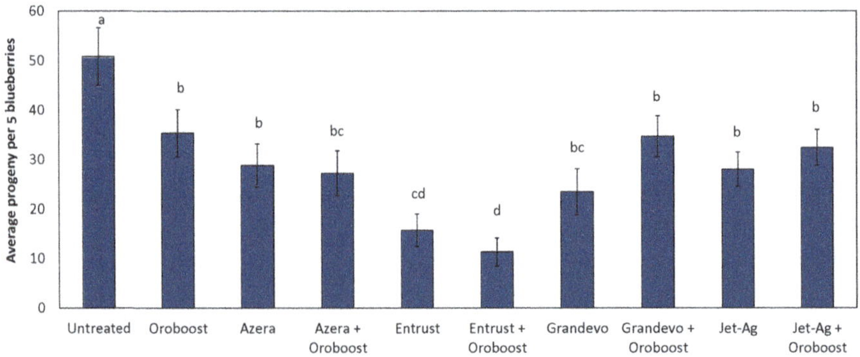

Fig. 12.1 Mean spotted wing Drosophila emerged per five blueberries in a semi-field bioassay treated with various organic insecticides with and without alcohol ethoxylate (Oroboost) in Florida. Azera is azadirachtin + pyrethrins, Entrust is spinosad, Grandevo is *C. subtsugae*, and Jet-Ag is a peroxide-based sanitizer

C. subtsugae and spinosad in laboratory trials (Gullickson et al. 2019). More field trials are needed before erythritol or its derivatives can be recommended as an effective control tactic alone or in combination with insecticides.

12.2.4 *Border Sprays*

It is well documented that SWD have many wild host plants (Lee et al. 2015; Little et al. 2017; Thistlewood et al. 2019), and numerous studies have shown that wild hosts near crop fields increase SWD numbers in the adjacent crops (Klick et al. 2016; Ballman and Drummond 2017; Santoiemma et al. 2018, 2019; WeiBinger et al. 2019). Border sprays, which are insecticide applications applied only to the border of a crop field, can successfully manage pests migrating into crop fields from outside including another fruit fly pest, the apple maggot, *Rhagoletis pomonella* (Trimble and Vickers 2000). Iglesias and Liburd (2017b) found that border sprays reduced numbers of SWD in organic blackberries in Florida (Fig. 12.2), and the sprays did not adversely impact natural enemy populations.

12.3 Behavioral-Based Tactics

12.3.1 *Attract-and-Kill*

One alternative to applying insecticides to entire fields is the attract-and-kill technique. An attractant, such as a food bait or pheromone, draws large numbers of a pest insect to a specific area where the pests encounter a killing agent, which is often an insecticide. Bait stations, liquid gels that partially solidify once applied (Vargas

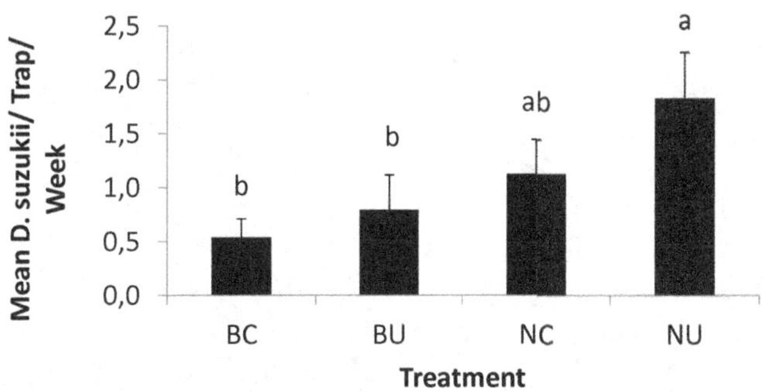

Fig. 12.2 Mean SWD per trap per week in plots treated with boarder sprays and cultivated (BC), treated with a border sprays only (BU), cultivated only (NC), and untreated (NU). Means with the same letter are not significantly different at $P \leq 0.05$. (Reprinted with permission from Iglesias and Liburd (2017b))

et al. 2008), and insecticide-treated spheres (Rice et al. 2017) are common ways of deploying attract-and-kill tactics. Both the insecticide-treated spheres and liquid gels have shown efficacy in SWD management.

Insecticide-treated spheres are used in the management of Tephritid fruit fly pests including the apple maggot, *Rhagoletis pomonella* Walsh (Wright et al. 2012; Morrison et al. 2016), and the blueberry maggot, *R. mendax* Curran (Liburd et al. 2003). The spheres are both colored and baited with sucrose to attract files. The spheres are impregnated with an insecticide that kills flies that land on the spheres and try to feed on them. Rice et al. (2017) developed and tested an insecticide-treated sphere for SWD management. The plastic spheres were painted red and included a cap made of wax, sucrose, and an insecticide. Both dinotefuran- and spinosad-treated spheres reduced SWD infestation of raspberries in the field. Reduction was increased by combining the treated spheres with insecticide applications.

The company ISCA Technologies Inc. has developed attract-and-kill tools for a variety of pests using their SPLAT gel matrix as a carrier. They have developed a prototype product for SWD management called HOOK SWD. The attractant components are the gel's red color and a food-based lure that is proprietary (Fig. 12.3). The insecticide typically employed in the product is spinosad. However, a variation of the product without insecticide is now available so that other insecticides can be mixed with it. Research is ongoing into which insecticides will be effective in combination with the SPLAT gel matrix. Klick et al. (2019) found that weekly

Fig. 12.3 HOOK SWD applications to (**a**) the lower leaves of a blackberry plant and (**b**) the stem of a blueberry bush. (Photo Credit: E. M. Rhodes, University of Florida)

applications of HOOK SWD in combination with a single application of spinetoram effectively reduced SWD infestation in highbush blueberries in New Jersey and red raspberry in California. Research on HOOK SWD is ongoing.

12.3.2 Alternate Oviposition Sites

Another behavioral-based tactic being developed for SWD management is a food-grade gum (Tait et al. 2018). The gum matrix has been developed as a solid formulation with a gel-like consistency and a liquid formulation with a cream-like consistency that can be applied with standard spray equipment. The matrix is entirely food-based and water-soluble. It attracts SWD females with its red color and a six-component synthetic blend of compounds like those found on the surface of fruit containing SWD eggs (Tait et al. 2020). Female SWD oviposit in the gum, where larvae are unable to complete development, instead of in fruit (Tait et al. 2018). Tait et al. (2018) found a mean reduction in fruit infestation of 48.3% in laboratory studies with blackberry, blueberry, cherry, raspberry, and strawberry fruits. In a highbush blueberry field in Oregon, both the solid and liquid matrixes caused a mean reduction in fruit infestation of 51.2%. These preliminary data indicate that the gum will likely be used in combination with other IPM tactics. Research on the food-based gum is ongoing, and it is not commercially available yet.

12.4 Cultural Control Tactics

12.4.1 Frequent Harvests

Harvesting fruit at more frequent intervals can reduce SWD infestations. For example, Leach et al. (2018) found that a 2-day harvest interval was ideal in Michigan raspberries as it reduced SWD infestation and increased fruit harvested per unit effort compared to the standard 3-day harvest interval. Organic blueberry growers in Florida utilize frequent harvests and report that it reduces SWD infestation (E. M. Rhodes, personal observation). Research is needed on optimized harvest intervals for other fruit crops in other areas.

12.4.2 Field Sanitation

Field sanitation is an important cultural control tactic for SWD management. Overripe and rotting fruit left in the field can serve as both oviposition sites and food sources for SWD females (Bal et al. 2017; Cai et al. 2019). Cai et al. (2019) found that SWD females mainly use rotting fruit as a food source and prefer to oviposit in ripe fruit when available. For these reasons, removing overripe and rotting fruit from the field can reduce SWD numbers. Leach et al. (2018) caused 99% mortality in waste berries by sealing the berries in plastic bags for 32 h. Fruit in clear plastic bags reached the highest temperatures.

12.4.3 Netting and Tunnels

Barriers physically prevent a pest from accessing a crop. Leach et al. (2016) found that exclusion netting reduced SWD infestation in red raspberries as well as insecticide applications to unnetted raspberries. Combining the two tactics reduced SWD infestation even further. Using exclusion cages in high tunnels also reduced SWD infestation though populations eventually increased (Leach et al. 2016). Plastic-covered high tunnels reduced SWD infestation to 2% compared with 35% in netted tunnels, 60% in insecticide-treated open plots, and 81% in open, untreated control plots (Rogers et al. 2016). Rogers et al. (2016) showed that the plastic covering caused the microclimate to become unfavorable for SWD reproduction and development. Barriers are most useful in cooler growing conditions. The biggest drawback is the cost (Rogers et al. 2016).

12.4.4 Irrigation

Using drip irrigation may reduce the population of SWD in fruit crops (Rendon and Walton 2019). Emergence of adult SWD from pupae was reduced in drip irrigated plots in blueberries because humidity was lower than in sprinkler irrigated blueberries, which caused pupae to desiccate. Research into the effects of irrigation on SWD in other fruit crops is ongoing.

12.4.5 Mulches

Rendon and Walton (2019) found higher temperatures and lower humidity above sawdust mulch compared with below the mulch, which caused fewer SWD above the mulch to survive. Rendon et al. (2020) found that SWD larvae can burrow through sawdust mulch to pupate underneath but not through weedmat (Fig. 12.4). The presence of mulch resulted in higher temperatures and lower SWD emergence. This effect was most pronounced at sites with young plants. Therefore, weedmat may reduce SWD numbers by preventing SWD larvae from burrowing into the soil, which can expose them to unfavorable temperatures and predators.

Fig. 12.4 Weedmat in blackberry planting

12.5 Biological Control

An excellent and comprehensive review of current and pending biological control tactics for SWD management has already been published by Lee et al. (2019). The main points will be summarized in this section. Predators observed consuming SWD in the field include earwigs, damsel bugs, spiders, ants, and minute pirate bugs. Sentinel pupal studies have indicated high predation in various crops, although this is likely an overestimation. Native parasitoids have been reared from SWD larvae and pupae in various locations though parasitism rates are generally low. These natural enemies will have a direct role in SWD suppression on organic farms and in wild hosts adjacent to crop areas. Research on classical biological control for SWD has been initiated and the most promising candidate is *Ganapsis brasiliensis* (Ihering), a larval parasitoid that is specific to SWD and closely-related drosophilids.

Several fungi, bacteria, nematodes, and viruses have been found to infect and kill SWD. Commercially available formulations of the fungus *Beauveria bassiana* have shown efficacy against SWD adults (Fig. 12.5) under laboratory conditions. Because fungi need high humidity and can be sensitive to UV degradation, field applications of *B. bassiana* have highly variable levels of success. Similarly, some strains of *Bacillus thuringiensis* bacteria and species of entomopathogenic nematodes have

Fig. 12.5 SWD infected with *Beauveria bassiana* after 1 week in an environmental chamber. (Photo Credit: E. M. Rhodes, University of Florida)

been shown to cause SWD mortality in the laboratory. The effects of naturally occurring bacteria and nematodes are unknown. In contrast, viral infections of SWD in the field have been documented, but none have been developed into commercial products at this time.

12.6 Conclusions

There are many chemical options for conventional growers to use to manage SWD infestations. Using these insecticides in rotation will delay the development of resistance. For organic growers, spinosad is still the most effective insecticide available. Other organic insecticides tested so far that could be used in rotation with the organic formulation of spinosad include *Chromobacterium subtsugae*, sabadilla alkaloids, and azadirachtin + pyrethrins. There are many alternative management tactics and cultural control practices that may reduce the number of insecticide applications needed to manage SWD populations. Alternative management tactics under development include border sprays, attract-and-kill technologies, and alternative oviposition sites in the form of a food-grade gum. In terms of cultural control, frequent harvests and field sanitation reduce the amount of SWD oviposition and feeding sites in crop fields. Weed mat mulch prevents SWD larvae from burrowing into the soil to pupate making them easier prey for generalist predators. The use of drip irrigation creates a microclimate unfavorable to SWD development compared with overhead irrigation. Netting and tunnels can provide a physical barrier that prevents SWD from accessing the crop. Progress is being made toward releasing a parasitoid from the native range of SWD that may, in time, reduce SWD populations in natural areas, which will, in turn, reduce pressure on crop fields. All these tactics can be used to develop a robust IPM strategy to manage SWD in a variety of fruit crops whether they are managed conventionally or organically.

References

Bal HK, Adams C, Grieshop M (2017) Evaluation of off-season potential breeding sources for spotted wing drosophila (*Drosophila suzukii* Matsumura) in Michigan. J Econ Entomol 110:2466–2470

Ballman ES, Drummond FA (2017) Infestation of wild fruit by *Drosophila suzukii* surrounding Maine wild blueberry fields. J Agric Urban Entomol 33:61–70

Bellamy DE, Sisterson MS, Walse SS (2013) Quantifying host potentials: indexing postharvest fresh fruits for spotted wing drosophila, Drosophila suzukii. PLoS One 8:e61227

Bruck DJ, Bolda M, Tanigoshi L et al (2011) Laboratory and field comparisons of insecticides to reduce infestation of *Drosophila suzukii* in berry crops. Pest Manag Sci 67:1375–1385

Cai P, Song Y, Yi C et al (2019) Potential host fruits for *Drosophila suzukii*: olfactory and oviposition preferences and suitability for development. Entomol Exp Appl 167:880–890

Calabria G, Máca J, Bächli G, Serra L, Pascual M (2012) First records of the potential pest species *Drosophila suzukii* (Diptera: Drosophilidae) in Europe. J Appl Entomol 36:139–147

Choi M, Tang YSB, Ahn SJ et al (2017) Effect of non-nutritive sugars to decrease survivorship of spotted wing drosophila, *Drosophila suzukii*. J Insect Physiol 99:86–94

Choi M-Y, Lucas H, Sagili R et al (2019) Effect of erythritol on *Drosophila suzukii* (Diptera: Drosophilidae) in the presence of naturally-occurring sugar sources, and on the survival of *Apis mellifera* (Hymenoptera: Apidae). J Econ Entomol 112:981–985

Cowles RS, Rodriguez-Saona C, Holdcraft R et al (2015) Sucrose improves insecticide activity against *Drosophila suzukii* (Diptera: Drosophilidae). J Econ Entomol 108:640–653

Deprá M, Poppe JL, Schmitz HJ et al (2014) The first records of the invasive pest *Drosophila suzukii* in the south American continent. J Pest Sci 87:379–383

Diepenbrock LM, Rosensteel DO, Hardin JA et al (2016) Season-long programs for control of *Drosophila suzukii* in southeastern U.S. blueberries. Crop Protect. 81:76–84

Fanning PD, Grieshop MJ, Isaacs R (2018) Efficacy of biopesticides on spotted wing drosophila, *Drosophila suzukii* Matsumura in fall red raspberries. J Appl Entomol 142:26–32. https://doi.org/10.1111/jen.12462

Foy CL (1996) Adjuvants—current trends and technology. In: Foy CL, Pritchard DW (eds) Pesticide formulation and adjuvant technology. CRC Press, Boca Raton, pp 323–352

Gompel N, Prud'homme B, Wittkopp PJ et al (2005) Chance caught on the wing: cis-regulatory evolution and the origin of pigment patterns in *Drosophila*. Nature 433:481–487

Gress BE, Zalom FG (2019) Identification and risk assessment of spinosad resistance in a California population of *Drosophila suzukii*. Pest Manag Sci 75:1270–1276

Gullickson MG, Rogers MA, Burkness MC et al (2019) Efficacy of organic and conventional insecticides for *Drosophila suzukii* when combined with erythritol, a non-nutritive feeding stimulant. Crop Protect 125:1–6

Hauser M (2011) A historic account of the invasion of *Drosophila suzukii* (Matsumura) (*Diptera*: *Drosophilidae*) in the continental United States, with remarks on their identification. Pest Manag Sci 67:1352–1357

Haviland DR, Beers EH (2012) Chemical control programs for *Drosophila suzukii* that comply with international limitations on pesticide residues for exported sweet cherries. J Integr Pest Manag 3:F1–F6

Iglesias LE, Liburd OE (2017a) Identification of biorational insecticides for managing spotted wing drosophila in organic blueberry production. Acta Hortic 1180:283–292

Iglesias LE, Liburd OE (2017b) The effect of border sprays and between-row soil tillage on *Drosophila suzukii* in organic blackberry production. J Appl Entomol 141:19–27

Klick JW, Lang Q, Walton VM et al (2016) Distribution and activity of *Drosophila suzukii* in cultivated raspberry and surrounding vegetation. J Appl Entomol 140:37–36

Klick J, Rodriguez-Saona CR, Cumplido JH, Holdcraft RJ, Urrutia WH, de Silva RO, Borges R, Mafra-Neto A and Seagraves MP (2019) Testing a novel attract-and-kill strategy for *Drosophila suzukii* (Diptera: Drosophilidae) management. J Insect Sci 19:3

Leach H, Timmeren SV, Isaacs R (2016) Exclusion netting delays and reduces *Drosophila suzukii* (Diptera: Drosophilidae) infestation in raspberries. J Econ Entomol 109:2151–2158

Leach H, Moses J, Hanson E et al (2018) Rapid harvest schedules and fruit removal as non-chemical approaches for managing spotted wing drosophila. J Pest Sci 91:219–226

Lee JC, Dreves AJ, Cave AM et al (2015) Infestation of wild and ornamental noncrop fruits by *Drosophila suzukii* (Diptera: Drosophilidae). An Entomol Soc Am 108:117–129

Lee JC, Wang X, Daane KM et al (2019) Biological control of spotted-wing drosophila (Diptera: Drosopilidae)—current and pending tactics. J Integr Pest Manag 10(13):1–9

Liburd OE, Finn EM, Pettit KL et al (2003) Response of blueberry maggot fly (Diptera: Tephritidae) to imidacloprid-treated spheres and selected insecticides. Can Entomol 135:427–438

Little CM, Chapman TW, Moreau DL et al (2017) Susceptibility of selected boreal fruits and berries to the invasive pest *Drosophila suzukii* (Diptera: Drosophilidae). Pest Manag Sci 73:160–166

Morrison W, Lee DH, Reissig D et al (2016) Inclusion of specialist and generalist stimuli in attract-and-kill programs: their relative efficacy in apple maggot fly (Diptera: Tephritidae) pest management. Environ Entomol 45:202–203

Noble R, Dobrovin-Pennington A et al (2019) Improved insecticidal control of spotted wing drosophila (*Drosophila suzukii*) using yeast and fermented strawberry juice baits. Crop Protect 125:1–9

Rendon D, Walton VM (2019) Drip and overhead sprinkler irrigation in blueberry as cultural control for *Drosophila suzukii* (Diptera: Drosophilidae) in northwestern United States. J Econ Entomol 112:745–752

Rendon D, Hamby KA, Arsenault-Benoit AL et al (2020) Mulching as a cultural control strategy for *Drosophila suzukii* in blueberry. Pest Manag Sci 76:55–66

Rice KB, Short BD, Leskey TC (2017) Development of an attract-and-kill strategy for *Drosophila suzukii* (Diptera: Drosophilidae): evaluation of attracticidal spheres under laboratory and field conditions. J Econ Entomol 110:535–542

Rogers M, Burkness EC, Hutchison WD (2016) Evaluation of high tunnels for management of *Drosophila suzukii* in fall-bearing red raspberries: potential for reducing insecticide use. J Pest Sci 89:815–821

Roubos CR, Gautam BK, Fanning PD et al (2019a) Evaluation of adjuvants to improve control of spotted-wing drosophila in organic fruit production. J Appl Entomol 143:706–720

Roubos CR, Gautam BK, Fanning PD et al (2019b) Impact of phagostimulants on effectiveness of OMRI-listed insecticides used for control of spotted-wing drosophila (*Drosophila suzukii* Matsumura). J Appl Entomol 143:609–625

Sampson BJ, Easson MW, Stringer SJ et al (2019) Laboratory and field assessments of erythritol derivatives on the survival, reproductive rate, and control of *Drosophila siuzukii* (Diptera: Drosophilidae). J Econ Entomol 112:173–180

Santoiemma G, Mori N, Tonina L et al (2018) Semi-natural habitats boost *Drosophila suzukii* populations and crop damage in sweet cherry. Agric Ecosyst Environ 257:152–158

Santoiemma G, Trivellato F, Caloi V et al (2019) Habitat preference of *Drosophila suzukii* across heterogenous landscapes. J Pest Sci 92:485–494

Sarkar N, Rhodes EM, Spies JM et al (2020) Evaluation of non-target effects of OMRI-listed insecticides for management of *Drosophila suzukii* Matsumura in berry crops. J Appl Entomol 144:12–25. https://doi.org/10.1111/jen.12713

Tait G, Kaiser C, Rossi-Stacconi MV et al (2018) A food-grade gum as a management tool for *Drosophila suzukii*. Bull Insectology 71:295–307

Tait G, Park K, Nieri R et al (2020) Reproductive site selection: evidence of an oviposition cue in a highly adaptive dipteran, *Drosophila suzukii* (Diptera: Drosophilidae). Environ Entomol 49:355–363

Takamori H, Watabe H, Fuyama Y et al (2006) *Drosophila subpulchrella*, a new species of the *Drosophila suzukii* species subgroup from Japan and China (Diptera: Drosophilidae). Entomol Sci 9:121–128

Tang SB, Lee JC, Jung JK, Choi MY (2017) Effect of erythritol formulation on the mortality, fecundity and physiological excretion in *Drosophila suzukii* (Diptera: Drosophilidae). J Insect Physiol 101:178–184

Thistlewood HMA, Rozema B, Acheampong S (2019) Infestation and timing of use of non-crop plants by *Drosophila suzukii* (Matsumura) (Diptera: Drosophilidae) in the Okanagan Basin, Canada. Can Entomol 151:34–48

Timmeren SV, Isaacs R (2013) Control of spotted wing drosophila, *Drosophila suzukii*, by specific insecticides and by conventional and organic crop protection programs. Crop Protect 54:126–133

Timmeren SV, Sial AA, Lanka SK, Spaulding NR, Isaacs R (2019) Development of a rapid assessment method for detecting insecticide resistance in spotted wing drosophila (*Drosophila suzukii* Matsumura). Pest Manag Sci 75:1782–1793

Trimble RM, Vickers PM (2000) Evaluation of border spays for managing the codling moth (Tortricidae: Lepidoptera) and the apple maggot (Tephritidae: Diptera) in Ontario apple orchards. J Econ Entomol 93:777–787

Vargas RJ, Stark JD, Hertlein M, Mafra Neto A, Coler R, Piñero JC (2008) Evaluation of SPLAT with spinosad and methyl eugenol or cue-lure for "attract-and-kill" of oriental and melon fruit flies (Diptera: Tephritidae) in Hawaii. J Econ Entomol 101:759–768

Walsh DB, Bolda MP, Goodhue RE, Dreves AJ, Lee J, Bruck DJ, Walton VM, O'Neal S, Zalom FG (2011) *Drosophila suzukii* (Diptera: Drosophilidae): invasive pest of ripening soft fruit expanding its geographic range and damage potential. J Integr Pest Manag 2:1–7

WeiBinger L, Schrieber K, Breuer M, Müller C (2019) Influences of blackberry margins on population dynamics of *Drosophila suzukii* and grape infestation in adjacent vineyards. J Appl Entomol 143:802–812

Wright SE, Leskey TC, Jacome I, Pinero JC, Prokopy RJ (2012) Integration of insecticidal, phagostimulatory, and visual elements of an attract and kill system for apple maggot fly (Diptera: Tephritidae). J Econ Entomol 105:1548–1556

Chapter 13
Postharvest Quarantine Treatments for *Drosophila suzukii* in Fresh Fruit

Spencer S. Walse, Dong H. Cha, Byung-Ho Lee, and Peter A. Follett

Abstract Certain countries have imposed restrictions on the trade of fresh fruit due to possible infestation by *Drosophila suzukii*. A stand-alone postharvest treatment is often the simplest means to provide quarantine security against *D. suzukii*. With respect to stand-alone postharvest fumigation, efficacious parameters for methyl bromide and phosphine have been published, while those for ethyl formate are ongoing. Cold treatments have been developed for several types of fruit, with durations lasting 12–14 days at <1 °C. *D. suzukii* in fruit subjected to irradiation at a dose of 80 Gy were unable to produce F1 adults. System approaches combining multiple postharvest treatments have also been implemented, primarily in the context of reducing time and dosage requirements of the stand-alone treatments as well as harnessing control measures intrinsic to the commercial marketing of fruit (e.g., packing-line sorting, fungicide application, and *in transit* cold storage). The phytosanitary approaches cited above must be compliant with the regulations at the location of treatment and, as is the case for any phytosanitary treatment proposed for international trade, the importing country issues approval based on risk assessments and other regulatory considerations (e.g., residue tolerances, consumer exposure, and environmental health).

Keywords Quarantine · Phytosanitary · Methyl bromide · Phosphine · Irradiation · Ethyl formate

---

S. S. Walse
USDA-ARS, San Joaquin Valley Agricultural Sciences Center, Parlier, CA, USA

D. H. Cha · P. A. Follett (✉)
USDA-ARS, U.S. Pacific Basin Agricultural Research Center, Hilo, HI, USA
e-mail: peter.follett@usda.gov

B.-H. Lee
Institute of Life Science, Gyeongsang National University, Jinju, Republic of Korea

F. R. M. Garcia (ed.), *Drosophila suzukii Management*,
https://doi.org/10.1007/978-3-030-62692-1_13

13.1 Introduction

Spotted wing drosophila, *Drosophila suzukii* Matsumura (Diptera: Drosophilidae) is native to Asia and had been recorded from China, Japan, Korea, and Thailand. In 2008, *D. suzukii* was trapped for the first time in California and has since spread to many other states in the USA (Lee et al. 2011a) as well as to Europe (Cini et al. 2012) and South America (dos Santos et al. 2017). *D. suzukii* mainly infests small fruits, and an index of host potential has been proposed (Bellamy et al. 2013). Damage is caused by larvae feeding internally on the fruit pulp, and by the introduction of rot-type pathogens at the site of oviposition (Lee et al. 2011b; Cini et al. 2012). Field control of *D. suzukii* currently relies on calendar- and trapping-synchronized sprays of various insecticides such as pyrethroids, organophosphates, spinosyns, and neonicotinoids (Bruck et al. 2011; Van Timmeren and Isaacs 2013). Field control cannot prevent low levels of infestation at harvest, and a postharvest treatment may be required for movement of fruit to areas where *D. suzukii* does not exist. For example, Australia as well as New Zealand have imposed restrictions and require a phytosanitary treatment to import certain fruits from the USA.

A single, stand-alone postharvest treatment is often the simplest means to overcome an insect-related trade barrier, and several treatment options are available that provide quarantine security against *D. suzukii*. Chemical treatment options, such as postharvest fumigation, and physical treatment options including irradiation and cold are briefly reviewed below. In addition, we present and discuss the potential for sequential treatments, such as combination treatments and/or systems approaches, as phytosanitary measures that result in an acceptable level of quarantine security for *D. suzukii*.

13.2 Methyl Bromide

One fumigant, methyl bromide, has dominated the postharvest treatment of horticultural crops over the last four decades, being essentially the only available option for quarantine and pre-shipment (QPS) disinfestations, where pest-free security must be "guaranteed." Methyl bromide use is regulated under the Montreal Protocol, where Decision XX/6 by the Methyl Bromide Technical Options Committee (MBTOC) recognizes "that methyl bromide use for quarantine and pre-shipment (QPS) purposes is an important remaining use of an ozone-depleting substance that is not controlled pursuant to paragraph 6 of Article 2H" (UNEP 2010). This use has resulted in a global fruit industry, producers, and port facilities alike, with logistics and infrastructure largely geared to conduct QPS methyl bromide chamber fumigations. Importantly, methyl bromide is registered for postharvest use on most fresh fruits and has food tolerances/maximum residue levels in most countries. Accordingly, postharvest fumigation with methyl bromide is often used to rapidly address requirements for insect pest control, as was the case when *D. suzukii* rapidly spread through key production regions of the USA, halting fresh fruit exports to Australia and New Zealand.

Postharvest methyl bromide fumigation schedules were rapidly developed for strawberries, sweet cherries, peaches, nectarines, Japanese plums, apricots, blueberries, raspberries, and table grapes. The specifics of each schedule, and those of every other schedule discussed below, are available in scientific journals, the federal registry of the import country, or in the National Plant Protection Organization (NPPO) export or import "work plans." Several over-arching features of the methyl bromide treatments are discussed below.

Different types of fruit and packaging, as well as the amount of fruit and packaging, critically affect the efficacy of a methyl bromide fumigation. Haber's rule relates concentration (C), time (t), and ultimately an exposure (Ct) to a level of lethality (ω), at least with respect to fumigation science (Bliss 1940; Miller et al. 2000; Winks 1984). In its most familiar expression, $C^z t = \omega$, ω (mg L^{-1} h) is an empirical level of lethal response for a given endpoint, in this case a proxy for mortality (a) specific to an exposure, and z (unitless) is the relative effect of C versus t toward the evoked response (i.e., mortality). Note that when $z = 1$, C and t contribute equally to ω (and a), such as the case for MB fumigations under kinetic control, i.e., those lasting <~4 h (Leesch et al. 1999; Maindonald et al. 2001; Walse et al. 2016, 2019). Importantly, a linear relationship exists between ω and (Ct) exposure (slope = unity; $\omega = Ct$), positively correlated with a, specific to controlling *D. suzukii* in a given life stage. In these cases, despite the fumigation of different types and/or amounts of commodity, the expectation of achieving a given % mortality (a) is justified if C and t are measured over the course of fumigation, and the resultant Ct exposure is determined to be adequate based on previous toxicological demonstration(s). A predictive kinetic model was developed by Walse et al. (2013, 2016) to quantitatively estimate the relationship between exposure (Ct), load, and load geometry. The model identifies how these parameters can be modulated (i.e., tuned) to ensure adequate toxicological efficacy (a) is attained when fumigating loads that vary in the amount and type of fruit and/or packaging. To develop these methyl bromide treatments, the rate of sorption (i.e., amount of methyl bromide sorbed per unit time) was measured for each fruit and packaging type, allowing one to predict a resultant Ct and its associated mortality (a) following a commercial fumigation. It should be noted that achieving mortality (a) >99.9% did not occur at pulp temperature (T) <8.3 °C with an applied dose <80 mg L^{-1} and a duration of 4 h, the maximum applied dose and time allowed in the USA, regardless of fruit type even with a much less than desirable chamber load of 30% (>50% is preferred). As T was increased, there was a concomitant decrease in the Ct exposure required for $a = 99.9\%$. Fruits tolerate postharvest temperature fluctuation differently. Certain fruits, such as sweet cherries, have little tolerance and therefore received the methyl bromide fumigation in field bins just after harvest, before cooling. Fruit types with greater tolerance are packed in cartons, palletized, refrigerated, and then warmed prior to the methyl bromide fumigation.

The most methyl bromide-tolerant life stage was second and third instar larvae, which tend to reside within the fruit, farthest away from the fruit periphery and the greatest relative *C*. To target the second and third instar larvae in studies, a life history table was developed for each fruit type and a predictive calculator was developed (Walse and Bellamy 2013). Even at the same rearing temperature, the speed of *D. suzukii* development depended on the fruit host. Greater homogeneity in development was observed for more suitable hosts, and heterogeneity for relatively poor hosts. The predictive understanding of *D. suzukii* development was incorporated into a model that allowed one to calculate the probability that a surviving adult was of a particular life stage at the time of treatment, particularly useful information when survivorship has to be traced through multiple treatments or through a system as described below.

13.3 Phosphine

Owing to the pioneering work of Dr. Fransiskus Horn in the late 1990s (Horn and Horn 2004), cylinderized phosphine is now registered in major fruit-producing nations, with South Africa a noted exception, where it is typically used to treat packed-palletized cartons/boxes at the cold-storage temperature ideal for the subject fruit, and maximum residue levels (MRLs) of 10 μg kg^{-1} (ppb) are established across the globe. Phosphine is a unique fumigant from a toxicological perspective (Winks 1984, 1985; Winks and Waterford 1986), as in the context of Haber's rule, z changes as a function of C (Waterford and Winks 1994). Winks operationally defined the "narcosis threshold" as the region where $z \cong 0$, whereby increases or decreases in C did not change the duration (t) required for the particular level of control (a). Levels of C below the "narcosis threshold" required a longer duration, while those above result in the narcotic effect and a longer duration required to cause a same efficacy. In this case of treating fresh fruit with phosphine, or any other commodity in which minimizing the duration required for efficacy is desired, an "optimal" treatment maintains C within the upper and lower limits of the narcosis threshold. If the level of C extends beyond the limits, a longer, "sub-optimal," treatment duration is required to achieve an equivalent level of efficacy.

In the case of *D. suzukii* infesting fruit a $T = 0.5$ °C, the shortest treatment time to achieve mortality (a) > 99.9% in eggs, the most phosphine-tolerant life stage, was 60 h over the range, $0.4 < C < 1.5$ g m^{-3} (i.e., $250 < C < 1000$ ppmv). The optimal range in C did not change as a function of life stage, or temperature. However, as would be expected, warming the fruit 5 °C did reduce the duration required for 99.9% mortality of eggs to ca. 36 h. Moreover, it should be noted that loads of fresh fruits that vary by amount (load factor) and type (variety, size, etc.) are known to only negligibly influence [PH_3] levels as equilibrium between headspace in the enclosure and the load is typically reached within 30 min of application.

13.4 Ethyl Formate

Ethyl formate is highly efficacious toward many types of insect pests (Haritos and Dojchinov 2003; Simmons and Fisher 1945), having significant potential to replace certain QPS uses of methyl bromide on fresh fruit. The registration of ethyl formate is forthcoming in most major fruit-producing countries. Ethyl formate will be commercially formulated in compressed cylinders as eFUME®, an ~17% by mass dilution of ethyl formate in carbon dioxide, or in a liquid formulation (99%, FUMATE®) to be dispensed with a vaporizer and nitrogen gas. The use pattern (i.e., labeled maximum dose and durations, re-entry protocols, frequencies of use) for ethyl formate treatment of fresh fruit has not been firmly established; however, it will likely be used to treat recently harvested fruit before cooling as opposed to an application that follows sorting, packing, and palletizing. With respect to Haber's rule, z is ≥ 1 for certain species (Kawagoe et al. 2020), indicating that evaluating ethyl formate sorption, which occurs quickly, is critical for establishing efficacy. Recent work to support registration in the USA has quantified the kinetics of ethyl formate sorption into different types of fresh fruit for two proposed scenarios, that involving applications prior to packing and cooling versus applications occurring afterward. Results provide evidence that ethyl formate is rapidly hydrolyzed into relatively nonvolatile and nontoxic residues (e.g., formic acid, formic acid adducts, and ethanol) (FDA 2019), and the off gassing of ethyl formate in cold storage poses minimal risk to workers and consumers. The effect of ethyl formate on fruit quality has been equivocal and depends on species and cultivars of fruit and the dose treated (Simpson et al. 2004; Zoffoli et al. 2013). For example, studies have reported no or negligible ethyl formate phytotoxicity on apples for *Gonipterus platensis* (Agrawal et al. 2015), apricots for *Thrips obscuratus* (Chhagan et al. 2013), bananas for Planococcus citri (Park et al. 2020), plums, nectarines, and pears for *Macchiademus diplopterus* (Smit et al. 2000), and strawberries for Frankliniella occidentalis (Simpson et al. 2004) at the dose necessary for the complete control of target insects, while ethyl formate showed phytotoxicity in some other cases on various fruit species (e.g., Zoffoli et al. 2013; Jamieson et al. 2014; Pidakala et al. 2018).

Research to document the toxicity of ethyl formate toward *D. suzukii* in fresh fruit is just underway, and a recent example is briefly described below. Kwon et al. (2021) evaluated the efficacy of ethyl formate against different life stages of *D. suzukii* and assessed phytotoxicity in blueberries. The lethal Ct exposures required for 50% control (i.e., a = 50%), LCt50%, and the LCt99% (a = 99%) of ethyl formate against *D. suzukii* eggs, the most ethyl formate-tolerant life stage, were 24.0 and 207.7 $g \cdot h \cdot m^{-3}$, respectively, at 5 °C, and 21.9 and 168.5 $g \cdot h \cdot ghm^{-3}$, respectively, at 21 °C. In scale-up (10 m^3) trials, treatment of *D. suzukii* eggs in blueberries using 70 $g \cdot m^{-3}$ ethyl formate for 4 h at 5 °C with 5% loading ratio (w/v) resulted on 0 survivors in 1332 treated eggs. When assessed at 14-day post-fumigation at 5 °C, no significant difference was observed between deterioration rate of ethyl formate-treated blueberries and the non-treated controls.

13.5 Irradiation

The United States Food and Drug Administration has approved radiation doses up to 1000 Gy (1 kGy) for preservation and disinfestation of fresh fruits and vegetables (USDA-APHIS 2006). Disinfestation means controlling any arthropod pests infesting the commodity, particularly insects. The 1 kGy limit has been widely adopted by other countries and the International Plant Protection Convention (IPPC). A generic dose of 150 Gy is approved by the United States Department of Agriculture (USDA-APHIS 2006) and the IPPC (FAO 2007) for control of tephritid fruit flies in fresh commodities (Follett 2009; Roberts and Follett 2018). The United States (USDA), and Australia and New Zealand (FSANZ) have also approved a 400-Gy dose for all other insects except pupae and adults of Lepidoptera, which may require a higher dose. Specific doses have been approved for many economically important quarantine pests as well (USDA-APHIS 2006; Barkai-Golan and Follett 2017). Insects generally can be controlled at radiation doses of 50–400 Gy.

The source of ionizing radiation can be gamma rays produced by radionuclides (^{60}Co or ^{137}Cs), or electrons or X-rays generated from machine sources operated within certain energy limits (Diehl 1995). Irradiation whether by isotopes or machine sources (e-beam, X-rays) has the same mode of action: the energy absorbed from gamma rays, X-rays, or electrons breaks chemical bonds within DNA and other biomolecules, thereby disrupting normal cellular function in the insect (Roberts and Follett 2018). The different sources of ionizing radiation are equally effective in controlling insects (Mastrangelo et al. 2010).

Radiotolerance in insects can vary among the different life stages and between different taxonomic groups (e.g., insect families and orders) (Follett 2014). For example, the adult and pupal stages are more tolerant of radiation than the larval and egg stages, and Lepidoptera (moths and butterflies) are generally more tolerant than Coleoptera (beetles) and Diptera (flies). Unlike other disinfestation techniques, irradiation does not need to kill the pest immediately to provide quarantine security, and therefore live (but sterile or not viable) insects may occur with the exported fruit. The goal of a disinfestation treatment normally is to prevent reproduction, and therefore the desired response for an irradiation treatment is to prevent development to the adult stage or induce sterility in adult insects. The approvals for phytosanitary irradiation to control quarantine insect pests of fresh horticultural commodities for export, and the rapidly expanding use of the technology for this purpose (Follett 2009; Roberts and Follett 2018), demonstrate that the presence of live but sterile or nonviable insects is acceptable to regulatory authorities, and facility certification and proper documentation can be used to ensure treatment at the required dose. Under current USDA and FSANZ regulations, fruit for export could be irradiated using the generic 400 Gy dose to control *Drosophila suzukii*. However, lowering the dose for *D. suzukii* would reduce treatment time and costs and avoid any negative effects on fruit quality (Follett 2009).

Follett et al. (2014) tested the effects of phytosanitary irradiation on larval and pupal development and adult reproduction in *D. suzukii* in sweet cherries and grapes. Larvae (first, second, and third instars) and pupae (1- to 2-day old, 3- to 5-day old, and 7- to 8-day old) on diet were irradiated at target doses of 20, 30, 40, and 50 Gy in replicated factorial experiments, and survival to the adult stage was recorded. Tolerance to radiation increased with increasing age and developmental stage. Males and females were equally susceptible. A radiation dose of 40 Gy applied to first- and second-instar larvae prevented adult emergence. The late-stage pupa was the most radiation-tolerant stage that occurs in fruit, and individuals irradiated at this stage readily emerged as adults; therefore, prevention of F1 adults was the desired treatment response for large-scale validation tests with naturally infested fruit. In large-scale tests, a radiation dose of 80 Gy applied to late-stage pupae in sweet cherries or grapes resulted in no production of F1 adults in >33,000 treated individuals, which meets the zero-tolerance requirement for market access. In sterile insect release research, Kruger et al. (2018) showed that a radiation dose of 75 Gy applied to late-stage *D. suzukii* pupae caused sterility in adult females, even when mated with unirradiated males, lending support for the efficacy of an 80 Gy phytosanitary treatment.

Modified atmosphere packaging (MAP) creates a low-oxygen (O_2) environment that can increase the shelf life of fresh produce by decreasing respiration and the growth of pathogens. Low oxygen may also increase insect tolerance to irradiation, and regulators fear that the use of MAP with products treated by irradiation to control quarantine pests before export may inadvertently compromise treatment efficacy. The IPPC and USDA have restricted the use of modified atmospheres during phytosanitary irradiation (Follett and Neven 2020). MAP is commonly used during export of perishable commodities such as sweet cherries and blueberries and other small fruits potentially attacked by *D. suzukii*. Follett et al. (2018) tested the effect of low oxygen generated by MAP at ambient temperatures on the radiation tolerance of *D. suzukii* infesting sweet cherries. Early pupal stage *D. suzukii* were inserted into ripe sweet cherries and treated by (1) MAP + irradiation, (2) irradiation alone, (3) MAP alone, or (4) no MAP and no irradiation and held for adult emergence. Three types of commercially available MAP products were tested that produced different oxygen concentrations between 3 and 15%, and a sublethal radiation dose (60 Gy) was used to allow comparisons between the treatments. Xtend PP61 bags (3.2–4.8% O_2), Xtend PP71 bags (5.4–8.6% O_2), and Xtend PP53 bags (13.6–15.4% O_2) did not enhance survivorship to the adult stage in *D. suzukii* pupae irradiated at 60 Gy in sweet cherries. MAP use should not compromise phytosanitary irradiation treatment against *D. suzukii* in exported sweet cherries or other fruit (Follett et al. 2018; Follett and Neven 2020).

13.6 Cold Treatment

Having been used for decades to control tephritid fruit flies, cold treatments have been adapted to control *D. suzukii*. Schedules have been developed for peaches, nectarines, citrus, and table grapes from California, USA, with durations lasting

12–14 days at *T* of ca. <1 °C. Experimental guidelines for validating a cold treatment are published by the Phytosanitary Measure Research Group, convened by the Food and Agricultural Organization (FAO) (PMRG 2019). Notably, testing occurs on each fruit type as well as each varietal thereof. Such factors, when combined with the practical and operation limitations of achieving temperature requirements for durations that are relatively long compared to fumigation and irradiation have limited the commercial implementation of cold treatments for *D. suzukii*. Nevertheless, cold treatments provide a chemical- and irradiation-free "organic" option for *D. suzukii*-infested fresh fruits.

13.7 Postharvest Combination Treatments

To achieve the required mortality of *D. suzukii*, ca. $a = 99.9\%$, stand-alone postharvest treatments may be combined as components of a "systems approach" as defined in ISPM 14 (FAO 2002). The combination is driven by many considerations including economics, regulations, operations, logistics, practicality, fruit quality, and decay control. In certain cases, the selected treatments are already components of the preexisting "system" used to produce, pack, and market that particular fruit, other times, one or both of the selected treatments are new, and must be added to the preexisting "system." Examples of combination treatments originating from different perspectives are briefly described below.

Researchers may identify that a stand-alone treatment developed for 99.9% efficacy may not be practical, as it damages the fruit, takes too long, or is restricted by regulations. If the treatment is adjusted to achieve <99.9% efficacy, however, it is potentially commercially practical. The addition of a sequential and complimentary treatment may provide a means to maintain practicality and increase efficacy. Such as the case for the ethyl formate research referenced above, which is not part of a preexisting system for blueberries, but rather a proposed incorporation. Up to 9-day cold treatment, a pulp temperature (*T*) = 5 °C controlled *D. suzukii* eggs and larvae but not pupae. However, in small-scale tests (4 replicates, 20 insects/replicate) when a LCt50%-level ethyl formate fumigation was followed by cold treatment at 5 °C for 5, 7, and 7 days, >99.9% mortality of SWD eggs, larvae, and pupae, respectively, was achieved (Fig. 13.1). (Large-scale tests are still needed to confirm these results). Together, the results suggest that a systems approach combining the lower exposure ethyl formate fumigation with a subsequent cold treatment is a better-suited postharvest treatment option for *D. suzukii* control in blueberries, at least relative to the stand-alone ethyl formate fumigation.

Although a methyl bromide schedule for *D. suzukii* control was developed for the export of table grapes from California, USA, to Australia and New Zealand, the fumigation added an additional "event" to the preexisting "system" which cost money and time as well as impacted the quality of fruit reaching the consumer. Most commercial table grape operations fumigate recently harvested berries with either sulfur dioxide, a sulfur dioxide-carbon dioxide mixture, or ozone, to control

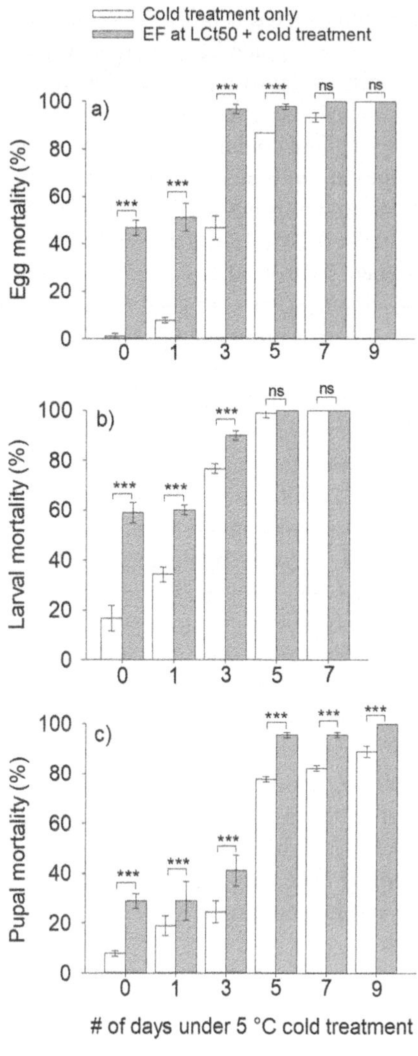

Fig. 13.1 Comparison of the effects of stand-alone cold treatment at 5 °C (white bars) and combination treatment of low dose ethyl formate ($LCt_{50\%}$ level) + the cold treatment (gray bars) on mortality of different life stages of *Drosophila suzukii*: (**a**) eggs, (**b**) larvae, and (**c**) pupae. Cold treatment was sequentially applied immediately after ethyl formate treatment (or blank ethyl formate treatment for stand-alone cold treatment) for up to 9 days. * indicates significant difference by *t*-test at $P < 0.05$. *ns* = no significant difference (Kwon et al. 2021 [submitted])

undesirable pests such as gray mold and spiders. After such fumigation, the packed grapes are refrigerated in storage at ca. 0 °C for a period lasting up to 8 weeks, although those intended for export are usually shipped within 1 week of harvest. Combination treatments were developed, using a fumigation step followed by a cold treatment step, for many permutations of the above "systems approach." The vast majority of shipments are exported following a fumigation at 15.6 ± 0.6 °C (x ± 2 *s*) for 30 min with sulfur dioxide–carbon dioxide mixture (1:6% vol.), and then refrigeration at −0.5 ± 0.7 °C (x ± *s*) for 6 days. With efficacy calculated by the method of Couey and Chew (1986), as later modeled in Liquido and Griffin (2010), the combination treatment resulted in 1 survivor out of 93,512 ± 2427 (*n* ± *s*) specimens (probit 9.03, 99.9972% mortality) if adult emergence from non-fumigated as well as non-refrigerated controls was used to determine the total number treated. Data on SWD development as well as adult emergence from fumigated berries were used to identify the most fumigant-tolerant timespan of development, estimate the fraction of treated specimens within the most fumigation-tolerant timespan that would be entering into cold treatment, and estimate the relative probability distribution of life stages at the time of fumigation that ultimately emerged during the most fumigation-tolerant timespan of development (eggs = 0.095, first instar = 0.153, second instar = 0.205, third instar = 0.514, and pupae = 0.032). Since 66.8% of the non-treated control adults emerged within the most fumigation-tolerant timespan of development, an estimated 62,434 ± 2427 ($n \pm s$) of the most fumigation-tolerant specimens were treated prior to entering cold treatment. Adult mortality following the described fumigation was 100%, effectively excluding any adult from entering cold treatment (Walse and Bellamy 2012, 2013).

In yet another example, California citrus exports to New Zealand are fumigated with phosphine for a duration $t > 12$ h at pulp temperature (T) > 8 °C to control bean thrips, *Caliothrips fasciatus* (Walse and Jimenez 2020). On very infrequent occasions, exports contain blemished fruits, and such fruits have a greater potential to host *D. suzukii*. While the stand-alone phosphine fumigation discussed above requires >36 h for 99.9% control, the 12-h fumigation already in the "system" for thrips control results in ca. 95% control of *D. suzukii*. When considering that the *in transit* container shipping takes ca. 15–20 days and typically occurs at T < 8 °C, as well as previous work on stand-alone and combination cold treatments, researchers and industry were curious to determine if any additional perturbation to the "system" was required to demonstrate control of *D. suzukii* at 99.9% efficacy. Indeed it was not, as when the 12-h fumigation was followed by 10 days of refrigeration at $T \cong 5$ °C, an applied dose of ca. 1.5 g^{-3} (1000 ppmv) resulted in 0 survivors from 50,560 ± 1313 ($n \pm s$) treated (probit 8.68, 95% CL; probit 9, 80.2% CL), while an applied dose of 0.5 g^{-3} (300 ppmv) resulted in 0 survivors from 51,210 ± 1167 ($n \pm s$) treated (probit 8.69, 95% CL; probit 9, 80.5% CL).

13.8 Other

The inclusion of preharvest and/or biological factors as components of a systems approach has been directed toward establishing "conditional host" or "non-host" status of the fruit per ISPM 37 guidelines (FAO 2016). The maturity, or ripeness, of fruit impacts the ability to host *D. suzukii*, with infestation not occurring until an "oviposition threshold" is surpassed. For example, plums, nectarines, and peaches will not be infested with *D. suzukii* if they are harvested at an early-enough state of maturity (Bellamy et al. 2013). Transferring this knowledge to regulators requires careful consideration, from both technical and operational perspective, and like any other treatment, requirements must ultimately be approved by the NPPO of the importing country per ISPM 28 (FAO 2007).

References

Agrawal M, Ren Y, Newman J et al (2015) Ethyl formate: a potential disinfestation treatment for eucalyptus weevil (*Gonipterus platensis*) (Coleoptera: Curculionidae) in apples. J Econ Entomol 108:2566–2571

Barkai-Golan R, Follett PA (2017) Irradiation for quality improvement, microbial safety and phytosanitation of fresh produce, Academic, San Diego. 301 pp

Bellamy DE, Sisterson MS, Walse SS (2013) Quantifying host potentials: indexing postharvest fresh fruits for spotted wing Drosophila, *Drosophila suzukii*. PLoS One 8(4):e61227

Bliss CI (1940) The relation between exposure time, concentration, and toxicity in experiments in insecticides. Ann Entomol Soc Am 33:721–766

Bruck DJ, Bolda M, Tanigoshi L et al (2011) Laboratory and field comparisons of insecticides to reduce infestation of *Drosophila suzukii* in berry crops. Pest Manag Sci 67:1375–1385

Chhagan A, Jamieson L, Griffin M et al (2013) Postharvest management of New Zealand flower thrips (*Thrips obscuratus*) on apricots using ethyl formate or pyrethrumbased treatments. NZ Plant Prot 66:63–74

Cini A, Ioriatti C, Anfora G (2012) A review of the invasion of *Drosophila suzukii* in Europe and a draft research agenda for integrated pest management. Bull Insectol 65:149–160

Couey HM, Chew V (1986) Confidence limits and sample size in quarantine research. J Econ Entomol 79:887–890

Diehl J (1995) Safety of irradiated foods, 2nd ed. Marcel-Dekker, New York. 454 pp

Dos Santos LA, Mendes MF, Kruger AP et al (2017) Global potential distribution of *Drosophila suzukii* (Diptera: Drosophilidae). PLoS One 12(3):e0174318. https://doi.org/10.1371/journal.pone.0174318

Follett PA (2009) Generic radiation quarantine treatments: the next steps. J Econ Entomol 102:1399–1406

Follett PA (2014) Phytosanitary irradiation for fresh horticultural commodities: generic treatment, current issues, and next steps. Stewart Postharvest Review 3:1, 7 pp

Follett PA, Neven LG (2020) Phytosanitary irradiation: does modified atmosphere packaging or controlled atmosphere storage creating a low oxygen environment threaten treatment efficacy? Rad Phys Chem 173:108874

Follett PA, Janson A, Price DK (2014) Postharvest irradiation treatment for quarantine control of *Drosophila suzukii* (Diptera: Drosophilidae) in fresh commodities. J Econ Entomol 107:964–969

Follett PA, Swedman A, Mackey B (2018) Effect of low oxygen created by modified atmosphere packaging on radiation tolerance of *Drosophila suzukii* (Diptera: Drosophilidae) in sweet cherries. J Econ Entomol 111(1):141–145

Food and Agricultural Organization (FAO) (2002) International Standards for Phytosanitary Measures (ISPM) No. 14. 2002. The use of integrated measures in a systems approach for pest risk management. https://www.ippc.int/sites/default/files/documents/20140108/ispm_14_2002_en_2014-01-08_2014010810%3A33-272.97%.pdf

Food and Agricultural Organization (FAO) (2007) International Standards for Phytosanitary Measures (ISPM) No. 28. 2007. Phytosanitary treatments for regulated pests. pp 3–11. http://www.furs.si/law/FAO/ZVR/ENG/184208_ISPM28_2007_E.pdf

Food and Agricultural Organization (FAO) (2016) International Standards for Phytosanitary Measures (ISPM) No. 37. 2017. Determination of host status of fruits to fruit flies (Tephritidae)

Food and Drug Administration (FDA) (2019) GRAS Substance (SCOGS) Database. https://www.fda.gov/food/generally-recognized-safe-gras/gras-substances-scogs-database. Accessed 4 May 2020

Haritos VS, Dojchinov G (2003) Cytochrome c oxidase inhibition in the rice weevil Sitophilus oryzae (L.) by formate, the toxic metabolite of volatile alkyl formate. Comp Biochem Physiol C 136:135–143

Horn F, Horn P (2004) Fresh fruit fumigation with phosphine as alternative for methyl bromide. Proc. Methyl Bromide Alternatives and Emissions Res. Conf. Orlando, Florida. http://mbao.org/2004/Proceedings04/058%20HornF%20Abstract%20%20Dr.%20Franziskus%20Horn%20%20PH3%20as%20alternative%20for%20MB%20%2023.6.04.pdf

Jamieson L, Griffin M, Page-Weir N et al (2014) Developing ethyl formate treatment for disinfesting pipfruit. NZ Plant Protect 67:96–102

Kawagoe JC, Abrams AE, Lourie A et al (2020) Ethyl formate fumigation to control brown marmorated stinkbug (Hemiptera: Pentatomidae). Pest Manag Sci, in press

Kruger AP, Schlesener DCH, Martins LN et al (2018) Effects of irradiation dose of sterility induction and quality parameters on *Drosophila suzukii* (Diptera: Drosophilidae). J Econ Entomol 111:741–746

Kwon TH, Park CG, Zarders D et al (2021) Ethyl formate fumigation and ethyl formate plus cold treatment combination as potential phytosanitary quarantine treatments of *Drosophila suzukii* in blueberries. Journal of Asia-Pacific Entomology (in press)

Lee JC, Bruck DJ, Dreves AJ et al (2011a) In focus: spotted wing drosophila, *Drosophila suzukii*, across perspectives. Pest Manag Sci 67(11):1349–1351

Lee JC, Bruck DJ, Curry H et al (2011b) The susceptibility of small fruits and cherries to the spotted-wing drosophila, *Drosophila suzukii*. Pest Manag Sci 67(11):1358–1367

Leesch JL, Tebbets JS, Obenland DM et al (1999) Dose-mortality and large-scale studies for controlling codling moth (Lepidoptera: Tortricidae) eggs on "d'Agen" plums by using methyl bromide. J Econ Entomol 92(4):988–993

Liquido NJ, Griffin RL (2010) Quarantine treatment statistics, United States Department of Agriculture, Center for Plant Health Science and Technology. Raleigh. http://cqtstats.cphst.org/index.cfm. Accessed 5 Mar 2013

Maindonald JH, Waddell BC, Petry RJ (2001) Apple cultivar effects on codling moth (Lepidoptera: Tortricidae) egg mortality following fumigation with methyl bromide. Postharvest Biol Technol 22:99–110

Mastrangelo T, Parker AG, Jessup A et al (2010) A new generation of x-ray irradiators for insect sterilization. J Econ Entomol 103:85–94

Miller FJ, Schlosser PM, Janszen DB (2000) Haber's rule: a special case in a family of curves relating concentration and duration of exposure to a fixed level of response for a given endpoint. Toxicology 149:21–34

Park M, Park C, Yang J et al (2020) Ethyl formate as a methyl bromide alternative for phytosanitary disinfestation of imported banana in Korea with logistical considerations. J Econ Entomol. (in press) 113:1711

Phytosanitary Measures Research Group (PMRG) (2019) Cold treatment research guideline. https://www.ippc.int/static/media/files/partner_publication/2019/03/16/PMRG_Cold_Treatments_Research_Guidelines_2019-03-12.pdf

Pidakala PP, Wilkinson RT, Nangul A et al (2018) Potential use of ethyl formate treatment to control surface pests of "Hass" avocado fruit. NZ Plant Protect 71:121–128

Roberts PB, Follett PA (2018) Food irradiation for phytosanitary and quarantine treatment. In: Ferreira CFR, Antonio AL, Cabo Verde S (eds) Food irradiation technologies: concepts, applications and outcomes. Royal Society of Chemistry, Cambridge, pp 169–182

Simmons P, Fisher CK (1945) Ethyl formate and isopropyl formate as fumigants for packages of dry fruits. J Econ Entomol 38:715–716

Simpson T, Bikoba V, Mitcham EJ (2004) Effects of ethyl formate on fruit quality and target pest mortality for harvested strawberries. Postharvest Biol Technol 34:313–319

Smit R, Jooste MM, Addison MF et al (2000) Ethyl formate fumigation: its effect on stone and pome fruit quality, and grain chinch bug (*Macchiademus diplopterus*) mortality. Sci Hort 261:108845

United Nations Environment Program (UNEP) (2010) Report of the Methyl Bromide Technical Options Committee (MBTOC)—2010 assessment. http://www.mma.gov.br/estruturas/ozonio/_publicacao/130_publicacao19082011113643.pdf

USDA-APHIS (2006) Treatments for fruits and vegetables. Fed Regist 71(18):4451. Rules and Regulations

Van Timmeren S, Isaacs R (2013) Control of spotted wing drosophila, *Drosophila suzukii*, by specific insecticides and by conventional and organic crop protection programs. Crop Prot 54:126–133

Walse SS, Bellamy DE (2012) Combining fumigation and cold-treatment for postharvest control of spotted wing drosophila, *Drosophila suzukii*, in California-grown table grapes. USDA-APHIS report for Biosecurity Australia and Department of Agriculture, Forestry, and Fisheries, April 13, 2012

Walse SS, Bellamy DE (2013) Distribution of spotted wing drosophila life stages. https://www.ars.usda.gov/pacific-west-area/parlier/sjvasc/cpq/docs/spotted-wing-drosophila-development/services/

Walse SS, Jimenez LJ (2020) Postharvest fumigation of fresh citrus with cylinderized phosphine to control bean thrips (Thysanoptera: Thripidae). Postharvest Biol and Technol, in press

Walse SS, Krugner R, Tebbets JS (2012) Postharvest treatment of strawberries with methyl bromide to control spotted wing drosophila, *Drosophila suzukii*. J Asian Pacific Entomol 15:451–456

Walse SS, Myers SW, Liu YB et al (2013) Postharvest treatment of fresh fruit from California with methyl bromide for control of light brown apple moth (Lepidoptera:Tortricidae). J Econ Entomol 106(3):1155–1163

Walse SS, Jimenez LR, Hall WA et al (2016) Optimizing postharvest methyl bromide treatments to control spotted wing drosophila, *Drosophila suzukii*, in sweet cherries from Western USA. J Asian Pac Entomol 19:223–232

Walse SS, Tebbets JS, Leesch JG (2019) Postharvest methyl bromide fumigation of Japanese plums to control codling moth (Lepidoptera: Tortricidae). J Asian Pac Entomol 22(3):807–815

Waterford CJ, Winks RG (1994) Correlation between phosphine resistance and narcotic response in *Tribolium castaneum* (Herbst). In: Highley E, Wright EJ, Banks HJ, Champ BR (eds). Proc. 6th Int. Working Conf. on Stored-Product Protection, Canberra, Australia, April 17–23, 1994, vol. 1. CAB International, Wallingford, pp 221–225

Winks RG (1984) The toxicity of phosphine to adults of Tribolium castaneum (Herbst): time as a dosage factor. J Stored Prod Res 20:45–56

Winks RG (1985) The toxicity of phosphine to adults of Tribolium castaneum (Herbst): phosphine-induced narcosis. J Stored Prod Res 21:25–29

Winks RG, Waterford CJ (1986) The relationship between concentration and time in the toxicity of phosphine to adults of a resistant strain of Tribolium castaneum (Herbst). J Stored Prod Res 22:85–92

Zoffoli JP, Michelow P, Naranjo P (2013) Sensitivity of fruit species to ethyl formate fumigation under quarantine concentrations. Acta Hort 1012:763–767

Chapter 14
Drosophila suzukii Management in the Agriculture 4.0 Context

Flávio Roberto Mello Garcia

Abstract This chapter intends to compile the information presented in previous chapters highlighting the advances in SWD management and to project how research should advance to the management of this pest in a sustainable way in the future. This chapter deals with how should be the *D. suzukii* management in the Agriculture 4 context. Agriculture 4.0 is a group of cutting-edge digital technologies integrated and connected through software, systems, and equipment capable of optimizing agricultural production. The use of drones, GPS, telemetry, and other technological tools should make the management of this pest more efficient and accurate.

Keywords Automated Airborne Pest Monitoring · Artificial Intelligence · Innovation · Digital Agriculture

14.1 Introduction

The concept of Agriculture 4.0 is derived from the concept of Industry 4.0. At the Hannover Conference, in 2011, the term Industry 4.0 was created, which consists of the use of technologies such as Big Data Analytics, Cloud Services, 3D Printing, Cyber Security, stand-alone robots, Internet of things, Wireless sensors, Augmented reality, Simulation, horizontal integration, and Vertical integration. It allows the transformation of how organizations operate together with major changes in business models and manufacturing processes (Ribeiro et al. 2018).

Thus, Agriculture 4.0 consists of a new revolution in agriculture, where the newest technologies are applied to promote the increase in food production, reduce costs and rationalize the use of natural resources (Ribeiro et al. 2018). Digital technology is already a reality in fruit-growing in some countries; for example, in orange and apple orchards in New Zealand, there are already autonomous drones equipped

F. R. M. Garcia (✉)
Departamento de Ecologia, Zoologia e Genética, Instituto de Biologia, Universidade Federal de Pelotas, Pelotas, RS, Brazil
e-mail: flavio.garcia@ufpel.edu.br

F. R. M. Garcia (ed.), *Drosophila suzukii Management*,
https://doi.org/10.1007/978-3-030-62692-1_14

with artificial intelligence technology that identify fruit types, imperfections, and their degree of maturation, harvest fruits at the ideal height, perform thinning and pruning at any time of the day, at any location and at any time of the plant (Agrolink 2020).

In Brazil, an interactive digital software called Uzum web is able to perform the rapid diagnosis of pests, diseases, and physiological disorders of apple, strawberry, peach, and grape crops. Uzum uses artificial intelligence to perform diagnosis from access to the online system. By answering key questions and comparing the symptoms of the orchard with the standard photos displayed by the tool, the user in order to arrive at an identification of possible causes of the problem observed at the end of the questionnaire. The system also provides access to specific recommendations on an information page, with details of symptoms, prevention strategies, control, and management, and a list of publication links that provide detailed additional information about each disorder. Currently, the system is capable of diagnosing 53 grape disorders, 22 apple, 23 strawberry, and 33 peach (Fialho et al. 2020a, b, c, d).

The concept of Agriculture 4.0 or Digital Agriculture can be incorporated into an Area-Wide Integrated Pest Management (AW-IPM) approach. In general, these technologies allow a better result in pest management with lower cost and lower environmental impact. The use of technologies in an AW-IPM approach ranges from monitoring, helping the technician to make a more precise decision about when to control a particular pest, on control itself and after control in the evaluation of control efficiency and its economic and environmental effects.

Thus, this chapter deals with how the management of *Drosophila suzukii* Matsumura (Diptera: Drosophilidae) has been carried out through a synthesis of the previous chapters and what the perspectives for the management of this species in the future within the context of Agriculture 4.0.

14.2 The Management of *Drosophila suzukii* in the Present

The SWD is a polyphagous species, with a short life cycle and that causes high economic losses to farmers and countries, both because of the direct damage caused by larvae to fruits. Indirectly due to the fact that its attack allows microorganisms to penetrate the fruits, by the embargo of importing countries on fruit loads which may be infested by larvae of this species. In addition, its control generates costs for farmers. Knowledge of natural hosts, conditional hosts, and non-hosts of *D. suzukii* is essential to assist in the control of this pest in and around the crops, as well as aid in the export of fruits.

Knowledge of the biology and ecology of *D. suzukii* under different conditions, proper monitoring using the best trap, knowledge of the economic impacts and cost of each control strategy, knowledge of host status (natural hosts, conditional hosts, and non-hosts). And adequate legislation is the bases for *D. suzukii* management with greater efficiency, cheaper and less economic and social impact. It is also important to find out if there are plant species resistant to the SWD attack.

The development of simple and sequential sampling plans and the knowledge of the control and economic levels of this species are essential for decision-making appropriate to *D. suzukii* management. In the future, it is also important to determine a level of non-action, that is, the density of natural enemies of *D. suzukii* that allow control measures not to be taken. These sampling plans must be adapted to the different agricultural crops and regions where the species occurs.

Currently, the control of *D. suzukii* is mainly carried out with insecticides from chemical groups spinosyns, pyrethroids, organophosphates, and diamides. However, natural products including thyme, *Leptospermum ericoides*, *Leptospermum scoparium*, erythritol + sucrose, the chitinase of *Euphorbia characias*, and *Perilla* aldehyde had promising results which will be useful in SWD control strategy.

Biological control through the use of parasitoids, predators, and entomopathogen has been an important method in the management of *D. suzukii* mainly due to its low environmental impact and human health. The parasitoids are more promising in the management of the species. In general, parasitoids present some characteristics that leave them in advantage in relation to predators in the biological control, are as follows: absence of diapause, close specific with the host, long-lived adults with high host search capacity, lower thermal threshold very close to the host, and higher number of generations than host (Berti Filho and Ciociola 2002).

One of the major concerns in the use of predators in pest control is intraguild predation (Garcia et al. 2017). Intraguild predation occurs among natural enemies in biological control systems, where one natural enemy (the intraguild predator) attacks another species of natural enemy (the intraguild prey), whereas they also compete for the same pest (Janssen et al. 2006). There are two types of intraguild predation between predators and parasitoids: (1) the predator can directly predate the parasitoid, feeding from the immature phase when externally to the host and adult phase; (2) the predator can predate the parasitic host, directly consuming the host and indirectly the larva of the parasitoid. The effect of the presence of intraguild predators on the intraguild prey was often negative, but sometimes no significant effect was detected (Janssen et al. 2006). The predators have a very important role in conservation biological control, and it is necessary to intensify studies that evaluate or use agricultural techniques that do not affect an assemblage of predators, such as the use of selective pesticides. Currently, 28 species of larval and pupal parasitoids belonging to *D. suzukii* the families Braconidae, Diapriidae, Figitidae, and Pteromalidae in the world, with emphasis on *Ganaspis brasiliensis* (Hymenoptera: Figitidae), *Trichopria drosophilae* Perkins, and *Trichopria anastrephae* Lima (Hymenoptera: Diapriidae).

In addition to biological control, strategies such as attract-and-kill, cultural control tactics, and field sanitation are important strategies for cultivation of organic berries.

The netting can be a viable alternative or companion to insecticide treatment. In some parts of the world, absolutely no larval infestation in berries is the goal for producers because some markets practice a zero-tolerance policy for larval presence.

The use of Sterile Insect Technique (SIT), Incompatible Insect technique, the Tet-off and Tet-on, Erythromycin-off, Biotin-on, Vanilic-acid regulated,

Phloretin-off, Bile acid-off, and the Quinic-acid systems is an important strategy to be incorporated into an IPM-AW approach. Possibly the advances in research in these areas allowed the creation of SWD biofactories in the patterns currently existing for *Ceratitis capitata* (Wied.) (Diptera: Tephritidae) in some countries of the world (Dias and Garcia 2014) and the use of transgenic mosquitoes, *Aedes aegypti* (L.) (Diptera: Culicidae) in Brazil (Zara et al. 2016).

Even after the harvest of the fruits, the management of SWD should continue. Post-harvest treatment with metal-bromides, phosphine, irradiation, cold treatment, among others, can help very much in the export of fruits to avoid the introduction and dispersion of the species in regions or countries where it still does not occur, and allowing export.

It is important that the control of *D. suzukii* be maintained also in grocery stores since the fruits may be infested and allow the population to increase mainly in the period between harvest.

14.3 The Future of *Drosophila suzukii* Management

In the future of the management of *D. suzukii* in the context of Agriculture 4.0, there will be several technological tools that can be used, such as the use of drones, artificial intelligence, nanotechnology, and even telemetry.

14.3.1 Drones

Advances in miniaturized sensors, microprocessors, telecommunications, engineering, and digital processing techniques allowed the reduction in costs in novel insect automatic detection and monitoring systems. Some of these devices can be connected to wireless sensors networks (Internet of Things) (Lima et al. 2020).

An advance toward the management of *Drosophila suzukii* in a context of Agriculture 4.0 is the Automated Airborne Pest Monitoring (AAPM). The AAPM consisting of traps which are monitored by means of an Unmanned Aerial Vehicle (UAV) and an automatic image processing pipeline for the identification and count of number of *D. suzukii* per trap location (Fahrentrapp 2020). The AAPM has the following advantages over traditional monitoring methods (Fahrentrapp 2020): (1) labor intensity, (2) sampling interval, (3) automatic integration into DSS, (4) monitoring of diverse and even hardly accessible habitats, and (5) population monitoring in vast areas in relation to climatic and other geo-processed parameters. A multi-variable sticky trap evaluation will allow selecting the most suitable one to attract the target insect. Other commercial automatic trap with a high-resolution camera is manufactured by Trap View, EFOS, Slovenia, for the monitoring of SWD in fruits and grapes (Lima et al. 2020).

Another future possibility will be the use of drones for the release of *D. suzukii* parasitoids in the field. In the near future, it should be possible to release *G. brasiliensis*, *T. drosophiliae*, *T. anastrephae*, or another parasitoid of *D. suzukii* in the field through drones, as already done in Brazil and others, with the release of *Trichogramma* spp. (Hymenoptera: Trichogrammatidae) (Cruz 2015) and *Cotesia flavipes* (Cameron) (Hymenoptera: Braconidae) to control other pests. The use of drones allows a better distribution of parasitoids in the area and lower cost to the farmer.

14.3.2 Telemetry

Telemetry is a technology that allows remote measurement and communication of information between systems through wireless communication devices such as radio waves or satellite signals. Telemetry can provide a wide range of advantages to enhance agricultural production, optimize returns, reduce risks for crops, and minimize the environmental impact (Tseng et al. 2006). The telemetry system based on General Packet Radio Service (GPRS) and Global System for Mobile Communication (GSM) provides operative parameters such as layer status, maps of pathways, forward speed, applied volume, and applied rate through a combination of WebGIS platforms and hardware (Sarri et al. 2017). Currently, it is possible to use telemetry in 18 insect species from eight families of the orders Coleoptera, Hymenoptera, Megaloptera and Odonata (Kissling et al. 2014). Telemetry has not yet been used in any Diptera, due to the small size of its body. However, in the future with the improvement of this technology, it may be possible to use this technology in small body insects, such as *D. suzukii*.

14.3.3 Remote Sensing

The use of highly detailed remote sensing can be useful for future dispersion studies of SWD. Metz et al. (2014) developed a method to reconstruct high-resolution land surface temperature (LST) time series at the continental scale gaining 250-m spatial resolution and four daily values per pixel with remote sensing. This method generates a dataset and its derivatives can be used for the assessment of how temperature influenced *D. suzukii*.

14.3.4 Nanotechnology

Krittika et al. (2019) developed a nanoemulsion with attractants and load in the colophony resin trap to catch *Drosophila melonogaster* (Meigen), which will not produce any harsh effects on the environment. This nanoemulsion has great possibilities of being efficient also in the attraction of *D. suzukii* in the future.

14.3.5 Artificial Intelligence

They are reliable for the totally automated identification of orders and counting of insects (Lima et al. 2020). A new form of obtaining data about *D. suzukii* populations in field is via the combination of images obtained with the use of a camera mounted on an unmanned aerial vehicles (UAV), object detection/object recognition software using machine learning (ML) software and deep learning (DL) algorithms (Lamonst 2020). ML is a type of artificial intelligence where the model is provided with input and the requested output, and A DL algorithm consists of multiple layers that are connected, with each layer having a certain number of nodes (Lamonst 2020). The analysis of the images through Convolutional Neural Networks (CNN) allowed the detection of *D. suzukii* in the field (Lamonst 2020). The CNN is a class of deep neural networks, most commonly applied to analyzing visual imagery (Valueva et al. 2020).

Systems that use image recognition techniques and neural networks are the most reliable for fully automated identification of orders and insect count.

In the future, robots may be used to perform insecticide applications, reducing the possibility of contamination to human health and more precise applications than are currently performed for the control of *D. suzukii*.

14.3.6 Near-Infrared Spectroscopy (NIRS)

Near-infrared spectroscopy (NIRS) is a noninvasive, nondestructive, and rapid analytical tool that is growing in popularity in many disciplines worldwide (Johnson and Naiker 2019). The NIRS instrument emits a full spectrum of NIR wavelengths, which enter the sample. Certain wavelengths are absorbed by specific chemical bonds in the sample (e.g., OH, CH, NH), with the amount of absorbance proportional to the number of bonds irradiated with the NIR spectrum. The remaining wavelengths are reflected back to the instrument and measured (Johnson and Naiker 2019).

Using NIRS, it is already possible to identify the species, gender, age, and the presence of *Wolbachia* infection in some species of *Drosophila* (Aw et al. 2012;

Fischnaller et al. 2012; Aw and Ballard 2013; Aw and Ballard 2019; Johnson and Naiker 2019). Due to the metabolic difference between females and males of *Drosophila melanogaster* and *Drosophila simulans*. Sturtevant is possible to determine the gender through NIRS with >90% accuracy (Aw et al. 2012). The age group (9 days old) of his species was determined with 62%–88% and 91%–98% accuracy, respectively (Aw et al. 2012). The species *D. melanogaster*, *Drosophila obscura* Pomini, *D. simulans*, and *Drosophila subobscura* Collin can be identified by the use of NIRS (Fischnaller et al. 2012; Aw and Ballard 2019). It is also possible to identify the presence of *Wolbachia* infection in males and females of *D. simulans* aged 15 days (Aw et al. 2012). In the future, it will probably also be possible to use this technology to determine the species, gender, age, and presence of *Wolbachia* in *D. suzukii* too.

14.4 Conclusions

An important step in *D. suzukii* management is to determine the level of control and the level of economic damage of this species in several agricultural crops and in different regions, allowing reduction of insecticide application, control costs, and reducing environmental impact. Research aimed at the knowledge of selective insecticides, that is, that allows the discovery of molecules that do not affect the non-target populations of the control, such as predators, parasitoids, and pollinators, is another important advance in the management of this pest. In the future, it is also important to determine control and a level of non-action for the various regions. In addition to sophisticated technologies, the use of push–pull technique can be used in *D. suzukii*'s AW-IPM approach; in the future, push–pull is a strategy to repel a pest from a crop, while tracing it toward an external location. It often relies on pheromones or allelochemicals that are relevant in the ecology of the pest insect and can be exploited as lure or repellent (Alkema et al. 2019); the use of traps and repellant substances has been shown to be promising in the management of *D. suzukii* (Wallingford et al. 2016; Alkema et al. 2019). It should be used when this pest presents low densities and along with other compatible techniques, such as biological control. Repellents may, in future, be sprayed into the crops by drones or robots.

The SWD's Area-Wide Integrated Pest Management approach in the context of Agriculture 4.0 will enable the management of populations of this species with greater precision, better cost-benefit and low environmental impact using nanotechnology, drones, remote sensing, robots, artificial intelligence among other technologies.

Acknowledgments I would like to thank the Brazilian National Council for Scientific and Technological Development (CNPq) research productivity grant and the International Atomic Energy Agency (IAEA) for the project funding entitled evaluating the use of Sterile Insects and pupal parasitoids to managing *Drosophila suzukii* in greenhouse, which I am the coordinator.

References

Agrolink (2020) Drones podem auxiliar na colheita de frutas. https://www.agrolink.com.br/noticias/drones-podem-auxiliar-na-colheita-de-frutas_437830.html?utm_source=agrolink-clipping&utm_medium=email&utm_campaign=clipping_edicao_6721&utm_content=noticia&ib=y. Accessed 11 Aug 2020

Alkema JT, Dicke M, Wertheim B (2019) Context-dependence and the development of push-pull approaches for integrated management of *Drosophila suzukii*. Insects 10(12):454. https://doi.org/10.3390/insects10120454

Aw WC, Ballard JWO (2013) The effects of temperature and diet on age grading and population age structure determination in *Drosophila*. J Insect Physiol 59:994–1000

Aw WC, Ballard JWO (2019) Near-infrared spectroscopy for metabolite quantification and species identification. Ecol Evol 9:1336–1343

Aw WC, Dowell FE, Ballard JWO (2012) Using near-infrared spectroscopy to resolve the species, gender, age, and the presence of *Wolbachia* infection in laboratory-reared *Drosophila*. G3 2:1057–1065

Berti Filho E, Ciociola AI (2002) Parasitoides ou predadore? Vantagens e desvantagens. In: Parra JRP, Botelho PSM, Corrêa-Ferreira BS et al (eds) Controle biológico no Brasil. Manole, São Paulo, pp 29–42

Cruz I (2015) Avanços e desafios no controle biológico com predadores e parasitoides na cultura do milho safrinha. XIII Seminário Nacional de Milho Safrinha. https://www.embrapa.br/busca-de-publicacoes/-/publicacao/1030705/avancos-e-desafios-no-controle-biologico-com-predadores-e-parasitoides-na-cultura-do-milho. Accessed 30 Jul 2020

Dias NP, Garcia FRM (2014) Fundamentos da técnica do inseto estéril (TIE) para o controle de moscas-das-frutas (Diptera, Tephritidae). O Biológico 76:57–62

Fahrentrapp J (2020) Automated airborne pest monitoring AAPM of *Drosophila suzukii* in crops and natural habitats. https://www.zhaw.ch/no_cache/en/research/research-database/project-detailview/projektid/1564/. Accessed 27 Jul 2020

Fialho FB, Antunes LEC, Ueno B et al (2020a) UZUM Morango—Sistema especialista para diagnóstico de doenças, pragas e distúrbios fisiológicos em morangueiros. https://www.cnpuv.embrapa.br/uzum/morango/. Accessed 16 Aug 2020

Fialho FB, Botton M, Garrido LR et al (2020b) UZUM Uva—Sistema especialista para diagnóstico de doenças, pragas e distúrbios fisiológicos em videiras. https://www.cnpuv.embrapa.br/uzum/uva/. Accessed 16 Aug 2020

Fialho FB, Gomes CB, Ueno B et al (2020c) UZUM Pêssego—Sistema especialista para diagnóstico de doenças, pragas e distúrbios fisiológicos em pessegueiros. https://www.cnpuv.embrapa.br/uzum/pessego/. Accessed 16 Aug 2020

Fialho FB, Nachtigall GR, Alves SAM LR et al (2020d) UZUM Maçã—Sistema especialista para diagnóstico de doenças, pragas e distúrbios fisiológicos em macieiras. https://www.cnpuv.embrapa.br/uzum/maca/. Accessed 16 Aug 2020

Fischnaller S, Dowell FE, Lusser A et al (2012) Nondestructive species identification of *Drosophila obscura* and *D. subobscura* (Diptera) using near-infrared spectroscopy. Fly 6:284–289

Garcia FRM, Wollmann J, Krüger AP et al (2017) Biological control of *Drosophila suzukii* (Diptera, Drosophilidae): State of the art and prospects. In: Lewis Davenport (ed). Biological Control: Methods, applications and challenges. Nova Science Publishers, New York, pp 1–27

Janssen A, Montserrat M, Hillerislambers R et al (2006) Intraguild predation usually does not disrupt biological control. In: Brodeur J, Boivin G (eds) Trophic and guild interactions in biological control. Springer, Dordrecht, pp 21–44

Johnson JB, Naiker M (2019) Seeing red: a review of the use of near-infrared spectroscopy (NIRS) in entomology. Appl Spectrosc Rev 55:810. https://doi.org/10.1080/05704928.2019.1685532

Kissling WD, David E, Pattemore DE, Hagen M (2014) Challenges and prospects in the telemetry of insects. Biol Rev Camb Philos Soc 89:511–530. https://doi.org/10.1111/brv.12065

Krittika S, Indhumathi P, Vedha Hari BN et al (2019) Evidence of nanoemulsion as an effective control measure for fruit flies *Drosophila melanogaster*. Sci Rep 9:10578. https://doi.org/10.1038/s41598-019-47045-3

Lamonst T (2020) Object detection for automated airborne pest monitoring of *Drosophila suzukii*. Thesis, Wageningen University and Research Centre

Lima CFM, Leandro MDEA, Valero C et al (2020) Automatic detection and monitoring of insect pests—a review. Agriculture 10(5):161. https://doi.org/10.3390/agriculture10050161

Metz M, Rocchini D, Neteler M (2014) Surface temperatures at the continental scale: tracking changes with remote sensing at unprecedented detail. Remote Sens 6:3822–3840. https://doi.org/10.3390/rs6053822

Ribeiro JG, Marinho YD, Espinosa JV (2018) Agricultura 4.0: desafios à produção de alimentos e inovações tecnológicas. In: II Simpósio de Engenharia de Produção

Sarri D, Martelloni L, Vieri M (2017) Development of a prototype of telemetry system for monitoring the spraying operation in vineyards. Comput Electron Agric 142:248–259

Tseng C, Jiang J, Lee R et al (2006) Feasibility study on application of GSM-SMS technology to field data acquisition. Comput Electron Agric 53:45–59

Valueva MV, Nagornov NN, Lyakhov PA et al (2020) Application of the residue number system to reduce hardware costs of the convolutional neural network implementation. Math Comp Simul 177:232–243. https://doi.org/10.1016/j.matcom.2020.04.031

Wallingford AK, Hesler SP, Cha DH et al (2016) Behavioral response of spotted-wing drosophila, *Drosophila suzukii* Matsumura, to aversive odors and a potential oviposition deterrent in the field. Pest Manag Sci 72:701–706. https://doi.org/10.1002/ps.4040

Zara ALSA, Santos SM, Fernandes-Oliveira ES et al (2016) Estratégias de controle do *Aedes aegypti*: uma revisão. Epidemiol Serv Saude 25(2):391–404, abr-jun 2016. https://doi.org/10.5123/S1679-49742016000200017

Index

F. R. M. Garcia (ed.), *Drosophila suzukii Management*,
https://doi.org/10.1007/978-3-030-62692-1

The manufacturer's authorised representative in the EU is Springer Nature Customer Service Centre GmbH, Europaplatz 3, 69115 Heidelberg, Germany. If you have any concerns regarding our products, please contact ProductSafety@springernature.com

Printed and bound by CPI Group (UK) Ltd, Croydon, CR0 4YY
28/09/2026
02224086-0008